URBAN THEORY AND THE URBAN EXPERIENCE

Encountering the city

Simon Parker

Routledge
Taylor & Francis Group

LONDON AND NEW YORK

First published 2004
by Routledge
2 Park Square, Milton Park, Abingdon, OX14 4RN

Simultaneously published in the USA and Canada
by Routledge
270 Madison Avenue, New York, NY 10016

Reprinted 2005

Routledge is an imprint of the Taylor & Francis Group

© 2004 Simon Parker

Typeset in Garamond 3 and Futura by
Florence Production Ltd, Stoodleigh, Devon
Printed and bound in Great Britain by
Bell & Bain Ltd, Glasgow

British Library Cataloguing in Publication Data
A catalogue record for this book is available from the British Library

Library of Congress Cataloging in Publication Data
Parker, Simon, 1964–
Urban theory and the urban experience: encountering the city/Simon Parker.
p. cm.
Includes bibliographical references and index.
1. Cities and towns.   2. Cities and towns – Philosophy.   3. Sociology, Urban.
4. Sociology, Urban – Philosophy.   I. Title: Urban theory and the urban
experience.   II. Title.
HT151.P35 2004
307.76–dc21   2003008627

ISBN 0–415–24591–5 (hbk)
ISBN 0–415–24592–3 (pbk)

URBAN THEORY AND
THE URBAN EXPERIENCE

Urban Theory and the Urban Experience brings together classic and contemporary approaches to urban research in order to reveal the intellectual origins of urban studies and the often unacknowledged debt that empirical and theoretical perspectives on the city owe one another.

From the foundations of modern urban theory in the work of Weber, Simmel, Benjamin and Lefebvre to the writings of contemporary urban theorists such as David Harvey and Manuel Castells and the Los Angeles school of urbanism, *Urban Theory and the Urban Experience* traces the key developments in the idea of the city over more than a century. Individual chapters explore investigative studies of the great metropolis from Charles Booth to the contemporary urban research of William J. Wilson, along with alternative approaches to the industrial city, ranging from the Garden City Movement to 'the new urbanism'.

The volume also considers the impact of new information and communication technologies, and the growing trend towards disaggregated urban networks, all of which raise important questions about the viability and physical and social identity of the conventional townscape. *Urban Theory and the Urban Experience* concludes with a rallying cry for a more holistic and integrated approach to the urban question in theory and in practice if the rich potential of our cities is to be realised.

For the benefit of students and tutors, frequent question points encourage exploration of key themes, and annotated further readings provide follow-up sources for the issues raised in each chapter.

This book will be of interest to students, scholars, practitioners and all those who wish to learn more about why the urban has become the dominant social, economic and cultural form of the twenty-first century.

Simon Parker is Lecturer in Politics at the University of York, UK.

For Esmé, May Beth and Lily

CONTENTS

FIGURES

TABLES

EXHIBITS

ACKNOWLEDGEMENTS

I wish to thank my colleagues in the Department of Politics at the University of York (UK) for allowing me a sabbatical term during which I was able to focus on the writing of this book. Thanks are especially due to my colleagues Roger Pierce and Jon Parkin for supplying valuable sources. I would also like to thank participants in the Political Science Workshop at the University of York and the members of the graduate programme on the Contemporary European City for helping me to sharpen my understanding of the urban experience. The issues and themes in this volume have benefited from discussions with Larry Bennett, Bob Catterall, Pierre Clavel, Roger Keil, Todd Swanstrom and Peter Terhorst. John Foot and Martin Bright have been generous hosts and supportive friends during my periodic sojourns in London. I would especially like to thank my editor at Routledge, Andrew Mould, for his support and encouragement throughout this project, and also the anonymous referees who offered valuable suggestions and constructive criticism on previous drafts of the manuscript. In accepting sole responsibility for the arguments that follow, I wish to distance any of the above from my occasionally impetuous decisions to ignore their sensible advice.

Acknowledgement is hereby given to London's Transport Museum for permission to use the photograph of Henry C. Beck's 1933 diagrammatic map of the London Underground on page 2; to the Museum of the City of New York for the photograph by Jacob Riis on page 32; to Mrs Norman-Butler and the University of London Library Special Collections for permission to reproduce the photograph of Charles Booth on page 33; to Swarthmore College Library's Peace Collection for use of the photograph of Jane Addams on page 38; and to the University of Chicago Press for permission to reproduce the concentric zone diagram by Edward Burgess in Figure 3.6 on page 42. I also wish to thank the director of the British Architectural Library, RIBA, London, for permission to reproduce the images in Figures 4.2, 4.4, 4.5, 4.6, 4.8 and 4.9, Princeton Architectural Press for permission to reproduce Peter Calthorpe's 'Criteria for new towns' in Figure 4.10 on page 67, and also the Architectural Association Photographic Library and the individual photographers for permission to reproduce the images in Figures 4.7, 4.11, 5.1, 8.1, 8.2 and 9.1.

Finally, this book would not have been written without the love and support of my wife Esmé and my daughters May Beth and Lily. Their forbearance and solidarity has been a constant motivation and inspiration to me. I dedicate this book to them with my sincere thanks, love and admiration.

A note to readers: terms highlighted in bold may be referred to in the glossary on pages 177–8.

1

ENCOUNTERING
THE CITY

You have to be out there all day to be sure of getting it. The remission. The pay-off that makes
urban life worth enduring.

Iain Sinclair, *London Orbital*

- At the beginning of the twentieth century some 10 per cent of the world's population dwelt in towns or
  cities.
- In 1975 this figure had risen to 37.8 per cent of the world's population.
- In 1995 the figure had reached 45.3 per cent of the world's population.
- By the year 2006 every second human being is expected to live in urban settlements.
- By the year 2030 over 60 per cent of the world's population will be urbanised.

When the distinguished American urbanist Lewis Mumford asked what is a city? – his focus was on the great renewal of urban society in Europe and the Western World from the tenth century onwards and, especially, the massive expansion in the number and size of cities since the industrial revolution of the nineteenth century. Since Mumford posed that question just before the Second World War only London and New York had populations approaching eight million. At the start of the twenty-first century there were 22 'megalopolises' with eight-figure populations, while the rate of urbanisation has been so considerable in the last fifty years that the majority of the world's population now live in cities. But in all this time have we come any closer to understanding the nature of the urban experience? In particular, have we produced satisfactory answers to

Mumford's supplementary questions on the nature of the relationship between politics and the city, the factors that have led to the development of certain urban forms, and the role that the city plays in relation to its wider region and, indeed, to the wider world (Mumford, 1938: 10–11)?

This book aims to provide a response to these questions through a critical examination of the ways in which different urban commentators, investigators and visionaries have tried to grapple with these issues over the past 150 years or so. The chronology is important because it spans the period that many social scientists and cultural theorists identify as the era of modernity. It is this interface between modernity and how we both think about, and live in, the city that is the master narrative of this volume, but before anticipating the specific themes of the

goodpoint

book, it is important to explain the purpose of theory in the context of urban studies and why it has more than an academic value to the present and future prospects of city life.

WHY DOES URBAN THEORY MATTER?

> Theory ... does not flow above everyday life in a detached way: It comes from some place, and it is the responsibility of analysis to return it there.
>
> (Liggett and Perry, 1995: 2)

Too often 'theory' is presented as an inaccessible language available only to a highly select and self-referential community of scholars and writers. But just as the *bourgeois gentilhomme* in Molière's epony-

mous farce was speaking 'prose' even though he did not realise it, all of us who live in, work in, or visit the city engage in unconscious theoretical activity on a daily basis.

Let us take the example of a shopping expedition to London's West End to illustrate how theory can make sense of an everyday urban experience. If we begin at our local suburban Underground station we are confronted with a map familiar to millions of Londoners and foreign tourists (see Figure 1.1). But what we see is not, of course, an accurate scaled down diagram of hundreds of miles of track, platforms and walkways, but a topographical abstraction aimed at reducing to its most basic linear form a highly complex engineering system. We know that the distance between the suburban stations is a good deal longer than those in the centre of the capital,

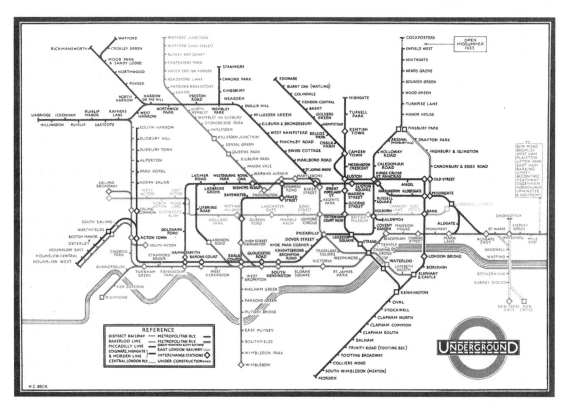

Figure 1.1   Henry C. Beck's design for the 1933 London Underground System.
Copyright Transport for London, courtesy of London's Transport Museum

but the important measure for the London Underground passenger is sequential (how many stops must I travel?) or temporal (how many minutes to Paddington?) rather than distantial (how many miles do I have to go?). Hence, spatial representations are shrunk or stretched to fit the intelligible spaghetti of Harry Beck's 1933 Underground Map.

Beck's Underground Map can be considered from a number of theoretical angles, each of which is discussed in more detail in later chapters:

- As representational space (the imagined city of the Underground provides us with a powerful mental image of London to set alongside other familiar place markers such as Big Ben, Nelson's Column, St Paul's Cathedral, Buckingham Palace, Piccadilly Circus, Tower Bridge and so forth).
- As symbolic space (the London Underground circle and bar motif immediately conjures up an entire system of destinations and rail networks in a single instantly recognisable sign).
- As a narrative space (each station stop and each tube line tells a different story for each passenger – 'where I go to work', 'where I go to shop', 'where I go to watch sport').
- As cultural trace (the classic quality of Beck's 1930's 'map' incorporates with it nostalgic associations of that period – such as families sheltering on Underground platforms during the wartime bombardment – stressing the 'unity' of London and Londoners).
- As commodity fetishism (wear the T-shirt, drink tea from the mug!).
- As an example of state intervention in the urban economy (the London Underground was one of the first examples of a publicly owned wide-area mass transit system).
- As a hidden map of the 'dual city' (for example, seventy years on from the Underground system described in Beck's map, one of London's poorest boroughs, Hackney, is still not served by a single underground station, while wealthy Kensington and Chelsea have six within a short distance of each other).

- As a product of the rational design of high modernity (Le Corbusier would have appreciated the spur and grid-like quality of Beck's design, although he would have deplored the complex engineering required to protect the city's ancient foundations from crumbling).

But what practical benefits can we derive from this theoretically enriched way of seeing the city, and is this any more than an interesting academic undertaking? Paradoxically, the first people to think about the cities in what we might describe as a sociological perspective were not academics but social reformers. Neither Engels, Mayhew, Mearns and Booth in Britain nor Riis or Adams in the US had university jobs, but they changed the way people saw the city by making the 'invisibility' of poverty visible for the affluent classes whose money and power decided the fates of the poor majority. However, none of these early urbanists believed that moral imprecations were enough to achieve their objectives, realising that evidence – especially scientifically defensible evidence – that showed the general cost to society of poverty was worth more than a thousand sermons.

Of course, the theoretical pretensions of the early urban reformers were not as far reaching as those that came after them (with the possible exception of Engels), but that is not to say that their work did not have important theoretical as well as policy implications. Jane Addams' work in the settlement house movement provided an important frame of reference for the 'professional' sociologists of the Chicago school in the 1920s and 1930s, just as Booth's work on London poverty paved the way for the establishment of the social sciences at the London School of Economics. The findings of these men and women helped to shape the worldview of generations of researchers and administrators who incorporated many of their ideas into the welfare policies of governments around the world.

The link between urban theory and urban policy, I would argue, is just as strong today as it was in the early decades of the twentieth century. If we consider a recent policy document such as that produced by the UK government's Urban Task Force under the chairmanship of the architect, Lord Rogers of Riverside, it is evident how much the report has been influenced by arguments in favour of 'dense cities', by debates on what constitutes an archetypal 'sustainable city', by research on how business and city residents can co-exist harmoniously, and how we might tackle threats to the quality of urban life such as crime, poor public services and social exclusion (Urban Task Force, 1999). Each one of these debates has a resonance in and, often, a direct input from, a diverse range of urban theories that have helped planners and policy makers understand the urban complex.

Even in the US, where urban policy has a more chequered history than in northern Europe, professional urbanists working out of policy research bodies such as the Russell Sage Foundation, the Brookings Institute or the Heritage Foundation have long been engaged in 'blue skies' thinking that has influenced the legislative programme of successive federal governments. Politicians on Capitol Hill are familiar with concepts such as 'ghettoisation', 'sprawl', 'smart growth' and 'social capital', and are anxious to see what benefits can be derived for their constituents by making cities work better. Knowledge of how previous investigators and thinkers have engaged with the urban complex, and how cities can best be conceptualised as integrated economies, as sites of social and political identity, as territories of conflict, and as incubators of innovation and creativity is vital if the long-term future, not just of urban communities, but of global society as a whole is to be secured.

THE DEVELOPMENT OF URBAN THEORY

Let us, for the time being, make the audacious presumption that the case for 'urban theory' has been argued and won — why do we talk of theory in the singular and why prefer the adjective 'urban' to 'city', why not for example talk of 'theories of the city'? It is important not to get too hung up on semantics and, indeed, I use the term city and urban, or even metropolitan throughout this volume almost inter-

changeably. But the term 'urban theory' has become accepted in academic circles as shorthand for a range of perspectives and interpretations of the urban world that aim, in their different ways, to provide a general understanding of city life that goes beyond the contingent and the local, while retaining a focus on the essential characteristics of the urban experience. Urban theory can, thus, be considered as a subset of social theory, but for all the shared conceptual vocabulary that social theory and urban theory possess, the latter is distinguished by its conviction that social, cultural, economic and political life is different in the city compared to other types of societies, however precarious and endangered the 'non-urban world' may be. So the first distinction that we ought to note is that urban life is neither universal nor ubiquitous and that, as a consequence, it poses particular theoretical problems for those who wish to understand its complex of activities and functions better.

All urban theory, I would contend, deals with one or more aspect of what we might call 'the Four Cs' of the urban experience — culture, consumption, conflict and community — interpreted in their broadest sense. Hence, 'culture' includes systems of belief, together with the physical built environment (buildings, bridges, streets and parks), the contents and means of communication (newspapers, books, television, radio, the Internet, etc.), as well as traditional cultural production (art, theatre, literature, orchestral music) and popular culture (movies, fashion, comic books, popular music).

Consumption refers not just to the consumption of goods and services, but also to the nature of the exchange and the means by which such goods and services (private and public) are produced. Conflict relates not just to visible, physical violence, such as riots or civil disorder, but to less visible struggles over resources (for example, between urban residents and developers), but also between social classes and different interest and status groups. Community involves all aspects of the social life of cities, from the size of the population to its distribution, demographic make-up, and changing characteristics over time. Community is also a 'value-term' for contiguous association that bears with it a series of assumptions about how we, as humans, should live in close confine with one another. Bearing all four of these Cs in mind I want briefly to explain the ways in which different theorists and theories have worked these themes into an explanation of the urban experience, and how these encounters with the city relate to the general plan of the book.

THE PLAN OF THE BOOK

This volume is both a thematic study of urban theory in its several aspects and an attempt at a 'genealogy of urban knowledge' over the past 150 years or so. Because I believe it important not to privilege one narrative above the other, I have tried as far as I can to make the individual chapters chronologically consistent. However, in order to preserve the conceptual framework of the book I have felt it necessary to introduce certain writers and figures who were important to the early development of urban studies, in later chapters. I have indicated wherever possible the relevant sections that provide a context for more recent analyses.

In Chapter 2 I explore the foundations of urban theory through the writings of four thinkers who, I believe, have had the greatest influence on generations of students of the city. All four could be said to have developed our understanding of what Mumford describes as 'the culture of cities'. For Max Weber it is the culture of the built environment and its historical development from the ancient cities of Greece and Rome, Mesopotamia and Asia to the medieval cities of Europe that offers a benchmark for measuring the expansion of that other crucial 'C' – consumption. Cities are, above all, market settlements and places of commerce and trade, and this is a view with which Georg Simmel would agree – although for Simmel the cultural novelty of cities emerged not so much from their form as from the social mores, habits and self-reflexivity of their inhabitants. Simmel's view of the metropolis is of a society that has achieved the escape velocity from tradition and the timeless world of what a contemporary German sociologist Ferdinand Tönnies termed the '*Gemeinschaft*'

into the pulsating, industrialised mass-society of the '*Gesellschaft*'. This fascination with modernity and the new life-world it offers for the urban citizen is shared by Walter Benjamin who was uniquely able to study consumption and culture through the lens of a **historical materialism** that, in its **Frankfurt School** variant, saw conflict as existing not only in workplace struggles between capitalists and workers but, as Marx wrote, 'in the very surface appearance of things' – or in the reflections of shop windows and the lapidaries of cemeteries.

The final member of the quartet, Henri Lefebvre, shares with Benjamin a **dialectical** understanding of the urban experience drawn from his readings of Marx and Engels, while combining it with the anti-**historicist** philosophy of Heidegger and Nietzsche. Lefebvre almost single-handedly rescues Marxism from its obsession with the temporal by insisting on the spatial dimension of class conflict, not only in a physical sense, but socially and imaginatively. In many ways, Lefebvre's philosophy is a geography of hope because, taking inspiration from revolutionary movements in art and politics such as surrealism and situationism, he believed that a better alternative to the class-divided city is possible if we can only think beyond the realm of **commodified space**.

This belief in the possibility of a better world certainly animated the investigations of the urbanists that we encounter in Chapter 3, but whereas Weber, Simmel, Benjamin and Lefebvre all invoke 'community' but rarely meet it face to face, the empirical researchers and urban sociologists who pioneered the social mapping of the city went out of their way to encounter real communities, or worked directly to improve their conditions. The study of 'communities' came to define urban research in the social sciences in Britain and America for decades afterwards, but critics alleged that this was at the expense of a proper understanding of the nature of conflict, and a failure to grasp the cultural limits to the sorts of urban utopias that many were openly or implicitly propounding.

The search for an alternative urban culture to the industrial 'city of dreadful night' took the new science of town planning in two very different directions.

The anti-metropolitans, led by John Ruskin and Ebenezer Howard in Britain, and Frederick Law Olmsted in the US were convinced that the mass, industrial metropolis celebrated by continental European writers such as Benjamin and Simmel was a social and economic disaster. They urged governments and local authorities to build new 'garden city communities' within reach of the big cities, but surrounded by a belt of green parkland and pasture. Le Corbusier and his disciples in the architectural school that was to become known as the Modern Movement were in favour of such parkland cities, but they rejected the medieval village model in favour of an even more ancient prototype; that of the stilt-supported lakeland dwellings of Palaeolithic Switzerland. Both competing images of the city were to find their realisation in actually existing towns and architectural developments, so, in a very real sense, these 'urban theories' (which is, in essence, what any town plan begins life as) changed the face of cities in a more visible way than applied sociological and ethnographic theories. Here again, though, the meaning of 'community' has owed more to normative assumptions than evidence-based research, and the assumption that 'bad places' and 'bad buildings' make for bad communities is at the core of anti-modernist architectural movements such as **new urbanism**.

The attention of urban sociologists in Britain and the US after the Second World War continued to focus on urban communities but, increasingly, as victims of economic change and disaggregation rather than viable societies in their own right. Three archetypes emerged from this research – the declining 'traditional urban community', the 'emerging sub-urban community', and the proliferation in the US of 'the black ghetto'. Consumption features as a factor in these studies as a shift to an automobile-based service economy accelerates the pace of suburbanisation and sprawl and accentuates the urban divide between the city's haves and have-nots. Welfare state policy in Britain and other European countries helped to ameliorate and slow the pace of economic polarisation among city residents compared to the US, but the trend towards convergence is clear and

unmistakeable. Urban theorists, thus, began to focus on racial segregation and economic deprivation in the US, while in Britain and Europe the attention has been more on the contest for urban living space between different social groups. These locational struggles between the affluent and the socially excluded can be measured in terms of the index of 'gentrification', and in the more hostile law enforcement policies towards vagrants, commercial sex workers and the homeless that frequently follows middle-class incursion into poorer neighbourhoods.

These physical and social characteristics of conflict and consumption need to be understood in the wider context of the political economies of the modern city. Chapter 6 brings together some of the most important writing on the nature of the capitalist city from its origins in the work of Engels and Marx to the more recent work of David Harvey, Manuel Castells and Saskia Sassen. Those with a particular interest in Marxist theories of the city should start here before reading the sections on Benjamin and Lefebvre in Chapter 2. This chapter is principally concerned with a fifth set of 'C words' – commodities and commodification. In Marxist theory, in contrast to neo-classical economics, commodities do not refer simply to tradable goods, but to services, products and artefacts that are the result of an exploitative relationship between labour and capital. A commodity contains a 'surplus' labour value (or that part of the exchange value of a good or service that the employer has not paid the worker), and commodification is the process of transforming an object or service from a use value to an exchange value – in other words by making its value marketable. This conceptualisation of productive relations being manifested through materialism (i.e. in the form of money and fixed or non-liquid capital assets) is fundamentally important to the work of Benjamin and Lefebvre, and also Simmel (though unlike Marx he sees exchange value as being distinct from use value). Materialism has also continued to be the most influential approach among students of contemporary urban political economy.

Theories concerning the dynamics of urbanisation, the spatial division of labour, the 'urban question', the information revolution and the globalised net-work city all have, at their core, an assumption that capitalist markets and relations of production are the chief propellants of new urban configurations. The political economy of cities also provides a crossover point for those who are interested in the work of cities and those whose focus is more on the management of cities and conflicts around private and collective resources. In Chapter 7 we concentrate on the political city – what are the necessary conditions for its administration, why is there such a variety of governance models and why are there so many theories for explaining them? But, because urban politics is not simply about the doings of City Hall, we also need to consider the political processes behind the 're-scaling' of urban government at various levels and Mumford's still relevant question about the relationship between the city and the region. 'Conflict' and 'community' come together in an examination of urban social movements, what motivates urban dwellers to combine around a common cause and what patterns of political mobil-isation we can observe between cities. Finally, how do we relate the 'who gets what, when and how' world of urban political theory to the 'bird's eye' view of the city that the most ambitious urban theory represents?

In the penultimate chapter we return to the theme of urban culture with a focus on the increasingly diverse types of social identities found in the city, and the ways in which representation and meaning – or what we might call 'urban discourse' – are trans-mitted and interpreted. We explore the works of key thinkers on urban culture including Mumford, Zukin, Sennett, Habermas, Bourdieu and Foucault and examine how 'the art of living in cities' has been developed through bourgeois urbanism, why it

is allegedly under threat from middle-class suburban-ism, and the ways in which cosmopolitanism and cultural diversity are re-inscribing urban culture in the new 'urban borderlands'. The growing recogni-tion of cultural practice as a defining feature of the urban experience has been accompanied by the so-called 'culture turn' in philosophy and aesthetic theory. In this chapter, we test its claims to offer a more satisfactory reading of contemporary urban forms and behaviour associated with 'postmodernism' than conventional social theory. Finally, we survey the practice of postmodernity through the study of the symbiotic cultural forms of architecture and cinema.

The concluding chapter has three main purposes – it tries to see whether the urban theories and accounts of the urban experience that we have encountered bring us any closer, if not to a holistic understanding of the city then, at least, to an identifiable common ground from which we can plot future research ques-tions and initiatives. The second section picks up on issues that have been implicit, or even absent, in western urban theory until recent times. Here, we consider questions such as the special theoretical problems posed by urbanism in the developing or majority urban world, the challenge of urban sustain-ability in the face of continuing environmental degradation and urban population growth and loss, together with the long-term impact of new tech-nology on the future viability of towns and cities. In the final section I pose some questions for future research in urban theory and suggest how urbanists might build on the rich heritage of existing urban knowledges while bringing fresh insights into the fascinating and ever changing world of the urban experience.

2

THE FOUNDATIONS
OF URBAN THEORY
Weber, Simmel, Benjamin and Lefebvre

It is often said: stones instead of bread. Now these stones *were* the bread of my imagination, which was suddenly seized by a ravenous hunger to taste what is the same in all places and all countries.

Walter Benjamin, *One-Way Street*

INTRODUCTION

At first sight, the intellectual concerns and development of the four writers under consideration in this chapter appear to have little in common. Max Weber wrote some of his most important contributions to sociology and sociological theory in the late nineteenth and early twentieth centuries, while Henri Lefebvre began writing on the city in the 1930s and continued to produce important contributions right up to his death in 1991. Simmel and Benjamin on the other hand are, perhaps, rather easier to bracket. Both hailed from Berlin, and both were what, today, we would call 'inter-disciplinary' thinkers. They were as familiar with the major debates in philosophy as they were with the hugely exciting and far-reaching developments in every aspect of artistic and creative endeavour in the early decades of the twentieth century. However, although, like Benjamin, it is hard to pigeon-hole Simmel as essentially a sociologist, the latter certainly saw himself as a pioneering social scientist, even though he failed to achieve the same academic recognition as Weber, Tönnies or Durkheim. Walter Benjamin, also remained an outsider to the academic establishment, although he did at one time have aspirations to

become a university professor. The fact that he never succeeded in this endeavour is partly attributable to Benjamin's notorious discomfort at working within prescribed disciplinary boundaries, and his work, in a sense, constituted a reproach to the methodologically conservative world of traditional philosophical and social enquiry.

However, while the critical perspectives of all four thinkers may have widely varied, it is their treatment of the city as an object of critical reflection that makes their work of continuing relevance to contemporary urban studies. This is not to say that the contribution made by each author is equivalent either in quantitative or qualitative terms. Max Weber's essay on the city is the only direct account of urban society produced by a classical sociologist (Durkheim and Marx both failed to give the city any special attention), and for that reason alone Weber must certainly be included among the key classical writers on urban morphology. But it is not as a historian of urban society that Weber is principally known, so much as the author of the monumental studies on the origins of capitalism *The Protestant Ethic* ([1905] 1985) and on the modern state and the social order in *Economy and Society* ([1922] 1968).

Perhaps even more than Weber, the work of Georg

Simmel is profoundly concerned with modern metropolitan life but, like Weber, he only directly addresses the nature of urban life in one, brief, but much quoted essay – 'The Metropolis and Mental Life' (Simmel [1902] 1950). Simmel's *magnum opus*, *The Philosophy of Money* ([1907] 1990) was written with the city as a permanent backdrop for his analysis of the processes and nature of commodities and exchange, but the urban environment remains a largely undeveloped theme in this book. Although Weber and Simmel's social circles intertwined, there is little in the academic record that suggests a correspondence of ideas, but, as David Frisby remarks, this is largely due to the fact that, before 1900, Simmel would not have regarded Weber as a colleague of the new discipline of sociology, although Weber does acknowledge the importance of Simmel's work in *The Protestant Ethic* (Frisby, 1987: 423). Indeed, Simmel remained largely silent on the work of Emile Durkheim (as did Weber), which suggests that the sociological community in these early years was rather more like an archipelago of distant islands than the networked global academy it is today.

Although the links between these writers might seem a slender foundation on which to construct a new theory of urban society, as this chapter seeks to explain, it is the adoption of the city as the microcosm for modern society in general that is important and distinctive in their work. Both Benjamin and Lefebvre gave a special place to the culture, values and rhythms of the city in a series of writings that have inspired subsequent generations of urbanists. If one can identify a common intellectual source it would surely be that of Marxism, but neither was a slavish devotee to official Marxist accounts of the nature of class society promulgated by figures such as Trotsky or Lukacs. Benjamin remained sceptical about the 'deterministic' claims of historical materialism, while Lefebvre's reflections on the city were strongly flavoured by a critical engagement with phenomenology (especially the work of Merleau-Ponty) and with the philosophies of Nietzsche and Heidegger. However different these thinkers' intellectual backgrounds and outlook might have been, it is undoubtedly the case that each writer was attempting to grapple with the transformation of the city wrought by the advent of modernity.

In Weber's account, the explanation offered is evolutionary, the modern city is seen as a special and sophisticated form of the medieval or renaissance city, and he believed that all the components of what we recognise as urban society were in place by the sixteenth century in much of northern Europe. Simmel is more preoccupied with differentiating the city from non-urban society (or the country), and in his reading the city is a rather exclusive society of self-reflexive urbanites that constitute in many ways a new historical community with very different mores and values to traditional society. Benjamin also wants to stress the distinctiveness of urban culture, but it is the possibilities the city offers for re-inventing itself and the lives of its citizens that gives urban life its peculiar quality. Intimacy and anonymity are equally present in urban exchange (whether it be in the shopping arcade, the café or the brothel), and for Benjamin this sets urban society and its constellation of writers, artists and poets apart from both classical and provincial society. Lefebvre's urban imagination tends more towards 'meta-level' analysis than Benjamin's, but their philosophical anchorings in the commercialised world of spectacle and space are remarkably similar. However, Lefebvre's interest in the everyday dynamics of the urban process, aspires towards a more universal formulation of the structures and systems common to metropolitan life in general. As such Lefebvre provides the most important link between 'classical urban theory' and the new urban studies that have developed in recent decades (Harvey, 1989; Soja, 1989; Shields, 1991; Gottdiener, 1992, 1993, 1994, 1996, 2000; Brenner, 1997, 2000; Kipfer, 1998; Borden *et al.*, 2000).

MAX WEBER: THE CITY IN HISTORY

Weber's account of the city owed much to his understanding of the transition from antique society to feudalism and from feudalism to capitalism and it builds on his interest in forms of pre-modern capitalism and the comparative sociology of religion (Nippel, 1995). Weber had been working on the sociology of urban development since 1889, but what Kaesler regards as the 'unfinished' essay *The City* 'was probably written between the years 1911–13' and was posthumously published in the 1922 edition of *Economy and Society* under the title 'Die Stadt'. In the fourth edition of *Economy and Society* edited by Winckelmann, the essay is translated as 'Non-legitimate domination (typology of cities)' (Kaesler, 1988: 42). Although Weber does refer to aspects of urban organisation in other parts of *Economy and Society* such as the division of labour, bureaucracy and religion, the point is usually to territorialise and historicise the sociological phenomenon under consideration.

Weber's geographical compass is extraordinarily wide, taking in not only the whole of Europe and Russia but also the cities of ancient China, southern Asia and the Near East, while his chronology extends back as far as the first recorded urban settlements of Mesopotamia 5,000 years before the birth of Christ. Significantly, however, Weber does not take his narrative past the Medieval period and the reader is left to wonder if, as Jonas suggests, this reluctance to contemplate the city in the industrial era is an intellectual recoil in the face of an unfathomable complexity or an implicit acknowledgement that the characteristics of the urban have, through the development of capitalism, become those of society as a whole (Jonas, 1995: 27).

Weber makes use of 'a multi-dimensional, ideal-typical approach' (Kaesler, 1988: 42) in order to highlight the essential characteristics of an urban form that he believed was quite specific to 'Occidental' civilisation. In this endeavour Weber adopts a rigorously descriptive method, and there is little in his account, other than a quixotic chronology, that would obviously distinguish it from a scholarly work of urban history.[1] By linking together what he considers to be the defining characteristics of the city in different historical epochs, Weber is able to construct a sophisticated historical reconstruction that, despite repeated attempts at generalising its characteristics, nevertheless, retains its contradictory specificities.

In one account, 'the city consists simply of a collection of one or more separate dwellings but is a relatively closed settlement' (Weber, 1958: 65), but he later extends this definition to add that 'the city is a settlement the inhabitants of which live primarily off trade and commerce rather than agriculture' (ibid.: 66). However, he is equally quick to point out that not all localities based primarily on trade and commerce can be called cities. Hence, Weber's technique is to first define the necessary conditions before going on to highlight all the other elements of the urban system that must be in place in order to meet the sufficient requirements of 'cityness'. Implicit in Weber's explanatory framework is the goal of the western capitalist city and, rather like Darwin in *The Origin of Species*, Weber sets himself the task of tracing the evolutionary development of the modern city back to its remote origins in antique society. What distinguishes Weber's narrative from a conventional urban history, however, is the close inter-weaving of evidence from ancient and medieval urban societies that provides a comparative model for analysing subsequent urban formations.

In Weber's account, what gives the city its special character is principally the existence of commerce and trade, together with all the activities associated with it such as the establishment of markets and exchanges. Hence, Weber writes, 'In the meaning of the word here, the city is a market settlement' (Weber, 1958: 67). This does not necessarily mean that early cities were based around 'open markets' in the more contemporary sense. Indeed, Weber is at pains to point out that long before the arrival of the bourgeoisie, princely courts would establish themselves in a particular location where the domestic economy (*Oikos*) of the nobleman created by default an infrastructure of buildings, storage facilities, roads, etc. that operated as a colonial settlement. In some cases these settlements expanded beyond the manorial court as artisans, journeymen, entertainers and other unindentured labourers attached themselves to these burgeoning centres of wealth and employment, forming increasingly autonomous nuclei.

Feudalism, with its relatively decentralised power structure, and its ability to generate money capital through the trade in agriculture and manufactures (such as cloth) and through the extraction of rents, allowed for a much more dynamic form of urban development than was the case in Islamic and Asiatic societies where patrimonial or prebendary[2] social and political relations set tight limits on the growth and organisation of urban centres. As Turner notes, 'In Weber's view, feudalism favoured the development of capitalist relations because, within feudal conditions, free cities, autonomous guilds, an independent legal profession, free labour and commercialisation were able to flourish. By contrast, prebendalism ruled out or limited such developments (Turner, 1996: 247).

Trade and commerce, although important do not complete the picture however, since for Weber, cities must also enjoy a degree of political and administrative autonomy – this would rule out most of the large trading centres set up by Peter the Great in Russia which were little more than income streams for the imperial exchequer. It was also possible, Weber believed, for a locale to exist in a political-economic sense, though it would not qualify as a city economically (Weber, 1958: 74). The attempt to regulate and tax trade, to manage security and fund arbitrational courts meant that a legal and bureaucratic order would quickly develop in the wake of any permanent economic settlement.

As settlements grew in size and affluence it also became imperative to create a means of defence, and history is littered with examples of cities that failed to flourish because their wealth and population were too easily confiscated by marauding armies. Hence, the 'garrison city' or 'fortress city' became a common feature of urban settlements across Europe and far into the Holy Lands in early medieval Europe. As Weber makes clear, 'the city in the past in Antiquity and the Middle Ages, outside as well as within Europe, was also a special fortress or garrison' (Weber, 1958: 75). Only in the latter half of the nineteenth century did the larger European metropolises dismantle their city walls, despite the obstacle to development and circulation that these physical barriers posed.

Here, again, Weber warns against the false syllogism of fortress = city and city = fortress since many fortified settlements existed in the ancient and medieval world that would not qualify as cities in the other senses outlined above. Neither were walls always necessary to defend a city's inhabitants. In the case of Sparta, which Weber describes as 'a permanent open military camp' there was no need to provide fortifications since the population was on a near constant war footing. As with the establishment of urban markets, fortified constructions such as castles or walled towns relied on the resources and initiative of princely or noble elites and the motives for the creation of these defensive structures depended on the threats posed by rival populations as well as the need to protect expeditionary armies from attack – as was the case with the crusader castles built by the Knights of St John and the Knights Templar across the Mediterranean and Holy Land, some of which were to form the basis of important settlements (such as Rhodes) while others became desert ruins.

The juxtaposition of coercive authority in the form of a castle or fortified palace or dwelling and the economic power of a chartered market meant that the city could be, at the same time, a drill-field and a place of exchange. In the case of the Attic *pnyx* that had developed its function from the Greek *agora*, religious and political activities as well as those of trade and commerce could take place in the same space. In Ancient Rome, in Northern Africa, Medieval Europe and Southern Asia it was more usual to separate martial and political activities from purely economic undertakings, although commercial centres were often found close to the sites of legitimate authority (Weber, 1958: 78).

As trade and commerce grew in importance, the merchants who had been attracted to such lordly fortified settlements by the prospect of trade with the noble household and its entourage and by the protection offered by its soldiery, accrued sufficient wealth and power in their own right to establish political authorities of their own. Many, such as the medieval guilds, existed harmoniously within an urban system dominated by princely and Episcopal authority. Other associations of urbanised gentry were more combative and attempted to assert their rights against the dead hand of seigneurial power, as was the case with the city republics of fifteenth- and sixteenth-century Italy. But, as Weber is quick to point out, 'in the Medieval and Ancient Occident, local individual participation in self-administration was out of the question'. Indeed, 'local individual participation in self-administration was often more strongly developed in the country than in the relatively large commercially organized city' (Weber, 1958: 82).

However, the sapling growth of urban community or *civitas* that was to be found in the guild-based mercantile associations of late medieval Europe was not replicated elsewhere. As Weber writes, 'In China, Japan, and India neither urban community nor citizenry can be found and only traces of them appear in the Near East' (Weber, 1958: 83). In Mecca, in the same period, despite the existence of powerful guild-like associations, power remained within a tight network of noble *koreischitic* families and, thus, 'a government by guilds never arose'. For Weber, civic participation, and especially democratic participation is, therefore, a key feature of urban development. But this did not mean that only townsmen could enter democratic life. As Weber argues, 'in Antiquity in Cleisthenes time the peasantry was the foundation of democracy' whereas, '[f]rom the beginning in the Middle Ages commercial strata were the bearers of democracy' (Weber, 1958: 206).

The existence of 'citizens' who were neither rulers nor slaves, whether they be the proletarians of Ancient Athens or the *plebs* of Ancient Rome also had implications for the economic management of cities in that urban elites were obliged to develop schemes for occupying these 'popular classes' in order to reduce the risk of riots and even revolt. Ensuring that the urban population was fed meant that draconian laws forbidding grain exports or imposing tributes on colonised provinces were widely used under the Greek and Roman empires. In Ancient Greece, the urban proletariat were also employed in great public works during the rule of Pericles but, as Weber points out in an insightful historical comparison, existence of an extensive slave economy in Ancient Athens meant that the consumption needs of the nobility and the state were almost entirely met by slave labour, as in the American Deep South, which led to the creation of a disgruntled and politically violent 'poor white trash' with formal political and legal rights but little in the way of economic opportunity (Weber, 1958: 199). Thus, from its first inception, the city was a profoundly political and *politicised* system that could only function effectively when economic, military and administrative life was well integrated and coordinated.

However, the cities of Antiquity were never dominated by economic exigencies alone, as Weber writes, 'ancient economic policy was not primarily concerned with industrial production nor was the polis dominated by the concerns of producers'. Whereas in North Continental Europe, '[t]he establishment of the city was an economic rather than a military affair' (Weber, 1958: 209), and this occurred, Weber believed, because extra-urban power holders lacked the administrative resources to meet the economic needs of the cities (which supplied them with lucrative tax and customs revenues) within their own bureaucratic apparatus. The same could also be said for the Church, and this is why 'all feudal powers beginning with the king viewed the development of cities with extreme distrust' (Weber, 1958: 210).

Thus, the bourgeois society is born in the self-governing cities of medieval Europe and with the growth of mercantile capitalism, the city's pre-eminence and *raison d'être* as a commercial centre is assured. Increasingly, the city emerged as a sphere of 'non-legitimate domination' that operated outside the authority of church and state and, hence, offered a provocative challenge to the state's claim to the monopoly of legitimate authority (Magnusson, 1996: 282–3). The rising affluence of the commercial city stimulates armed struggles for control of its commercial traffic and territorial possessions which, in the case of powerful city-states such as Venice and Genoa, were greater than many kingdoms. In nearly every conflict, the city republics lose their independence and are incorporated within a larger territorial authority, highlighting Weber's major observation that the monopoly of violence is as important to the development of states as commerce and trade.

As a result of improvements in the organisation and extent of armed force by territorial states, itself a consequence of a general economic improvement in the late Middle Ages, the era of the independent city-states proves short-lived. By the fifteenth century the superior military capacity of the Church and Princely Absolutism wins out against the limited democracy of the city republics (with the notable exception of Venice) (Tilly, 1990). Yet the die had been cast, and while the great cities of Europe continued to change hands often with bewildering rapidity, even as late as the nineteenth century, the motive force of economic and political modernisation can be summed up in one word – urbanisation.

Weber's *City* essay helps us to understand how this process of urbanisation leads to the bureaucratic-economic complex of modern capitalism, which forms the subject of *Economy and Society*, but it also serves, as Martindale rightly points out in his introduction to *The City* (Martindale in Weber, 1958: 50), to differentiate Weber's conception of the city from the urban analysis of Georg Simmel. Whereas, Weber implies that all the essential features of the city are present in past societies, only more so, Simmel, as we are about to see, associates the arrival of modernity with the full articulation of the metropolis and with the birth of a new and elusive subjectivity.

GEORG SIMMEL: THE CULTURE OF THE METROPOLIS

Although Simmel did have a significant impact on the new science of sociology in the US in the early decades of the twentieth century (Frisby, 1992b: 155–74), unlike Weber and Durkheim, Simmel's rather quixotic method meant that his ideas were less easily adapted for the purposes of empirical research. The Anglo-American concern with the positivist potential of the social sciences diverged quite radically from what could best be described as an applied moral philosophy in Simmel's case. In this sense, Simmel has more in common with Comte, Marx and Nietzsche than Weber and Durkheim, though his

rejection of the Kantian notion of an a priori moral law made Simmel even more determined to prove that a moral order could be produced synthetically as an outcome of human civilisation (Simmel, 1986: xviii–xix). It is this exultant belief in the transformative powers of modernity (and especially art and aesthetics) that focuses so much of Simmel's attention on the modern metropolis and urban culture in general.

Simmel's celebrated essay 'The Metropolis and Mental Life' was not published in its entirety in English until after the Second World War (Simmel, 1950), and while Simmel's influence on the Chicago School is now well documented, in his 1915 essay 'The City', Robert Park makes no mention of 'the first sociologist of modernity' (Frisby, 1985: 2). On the other hand, Louis Wirth more readily acknowledged Simmel's importance, describing 'The Metropolis and Mental Life' as 'the most important single article on the city from the sociological standpoint' (Wirth, 1925: 219 in Levine *et al.*, 1976: 249) and, as we shall see in the following chapter, he was to develop several of its arguments in his seminal essay 'Urbanism as a Way of Life' (Wirth in Le Gates and Stout, 1996: 189–97) which was to become a classic of American sociology (see Chapter 3). The Chicago School were not the only American sociologists to take Simmel's work seriously and, according to Donald Levine, Talcott Parsons' highly influential *The Structure of Social Action* (1937) was to have contained an article on Tönnies and Simmel. Thus, it would seem that the connections between continental European sociology and 'foundational' American sociology were rather closer than might appear at first sight (Kurtz, 1984: 17; Frisby, 1994: 225).

What is not in question is the growing reputation Simmel enjoyed after the Second World War as a key reference point for students of urban society. Writers such as Fischer have studied Simmel's impact on the Chicago School (and particularly the work of Louis Wirth) while Milgram has used systems analysis to investigate the features of metropolitan mentality that, it was believed, modern urban life was bound to produce (Levine *et al.*, 1976: 250). Simmel's influence extends from Marxist cultural critics such as Fredric

Jameson (Jameson, 1999), to film theorists (Mila, 1998), contemporary urban sociologists (Korllos, 1988; Joas, 1991; Ruggieri, 1993; Haussermann, 1995; Kajaj, 1998), historians of sociology (Jonas, 1991) writers on sexuality and the sex industry (Bech, 1998), social psychologists (Claes, 1994), social geographers (Werlen, 1993), sociologists of philosophy (Bevers, 1982; Gephart, 1992), writers on town planning (Phillips, 1994), space and architectural analysts (Vidler, 1991), and to aesthetic theorists (Baruzi, 1988). This, far from exhaustive, list gives a hint of the range of Simmel's interests, and of his ability to move from micro-level analysis to a metaphysical and metahistorical concern with the nature of human civilisation.

Simmel's metropolis

Born into an affluent and cultured Berlin family in 1856, Georg Simmel remained a resident of what was to become the capital of a united Germany and the administrative and political centre of Kaiser Wilhelm II's expanding empire until the outbreak of war in 1914 (Frisby, 1992b: 99). Thus, Simmel could hardly have been better placed to observe the great advances in technology, communications, commerce and culture that were to signify the birth of the modern era in much of Europe. It was his home city of Berlin, however, that Simmel chose to analyse in the essay 'The Metropolis and Mental Life', originally published in 1903. The context for this rather brief account of the modern city was a series of lectures Simmel gave to accompany an exhibition on the modern metropolis held in Dresden in 1902–3. In this analysis, the city's negative features are not ignored, but Simmel insists that the pressures of metropolitan life are more than compensated by the freedom from parochialism and surveillance that *Gemeinschaft* (small town) existence perpetuates.

Furthermore, the metropolis permits such an intense concentration of capital that the integration of space, time and social actors reaches a hitherto undreamed of complexity. Simmel was one of the first to elucidate the phenomenon that later writers have termed 'time-space compression' (Giddens, 1984;

Harvey, 1989) and to argue that the speed and intensity of social and economic interactions in the city have led to the emergence of a new society (Frisby, 1992b: 113). An example of this necessary synchronisation of activities in the city is provided by the sudden emergence of pocket watches (which became as ubiquitous and indispensable to city dwellers in the early twentieth century as mobile phones have become in the early twenty-first century). Individual time-keeping was vital, Simmel argued, because,

> [t]he relationships and affairs of the typical metropolitan are so varied and complex that without the strictest punctuality in promises and services the whole structure would break down into an inextricable chaos . . . If all clocks and watches in Berlin would suddenly go wrong . . . all economic life and communication of the city would be disrupted for a long time.
> (Simmel, 1950: 412–13)

The key to all exchange in the modern metropolis is, of course, money – the subject of Simmel's major theoretical work – and through this abstraction of power, the integration of even the most complex functions becomes possible. Those who have money wealth are able to secure goods and services without the need for coercion or resort to ideologies of domination as with traditional societies, thus facilitating, 'independence from the will of others'. Simmel goes on to add:

> The inhabitants of a modern metropolis are independent in the positive sense of the word, and even though they require innumerable suppliers, workers and cooperators and would be lost without them, their relationship to them is completely objective and is only embodied in money . . .
> (Simmel, 1978: 300 in Harvey, 1985: 5)

This abstract integrative power also applies in the domain of space where money

> permits agreements over otherwise inaccessible distances, an inclusion of the most diverse persons in the same project, an interaction and therefore a unification of people who, because of their spatial, social, personal and other discrepancies in interests, could not possibly be integrated into any other group formation . . .
> (Simmel, 1978: 347 in ibid.: 14)

Urban character

Like Weber, Simmel views the metropolis as a historical development and his purpose is to identify the general characteristics of urban life *sui generis*, and not only those of relevance to his home city. Unlike Weber, however, Simmel constructs his vision of metropolitan society by observing micro-level behaviour, giving his analysis more of an anthropological or social psychological quality than the 'systemic analysis' preferred by Weber. Thus Simmel tends to focus on 'the consciousness, personality, and "character" of the individual social actor' (Brody, 1982 in Frisby, 1994: 83). Urbanisation is associated with emancipation from traditional forms of social domination that Simmel argues persisted until the eighteenth century. In traditional societies, rigid status hierarchies made it almost impossible for individuals to assert their own autonomy and identity. But with the advent of industrial capitalism in the nineteenth century, the ties that bound the subordinate classes to the land began to loosen and, 'individuals now wish to distinguish themselves from one another'. It was the function of the metropolis, Simmel believed, 'to provide the arena for this struggle, and its reconciliation' (Simmel, 1950: 423).

With this new-found freedom came an indifference to one's fellow city dwellers resulting from the need to escape the tumult of the streets and the inhuman nature of commerce in a cultivated privatism. At the same time the sensory over-stimulation of urban life led to a 'blunting of discrimination' and the development of a 'blasé attitude' in which Simmel suggests urbanites could not draw on the emotional reserves required to empathise or even interest themselves in the lives of other metropolitans (Simmel, 1950: 414). In the city, the individual is able to explore and develop the psychic core of his or her personality. But as Smith reminds us, for Simmel:

> The twin dangers that threaten to frustrate the urban individual's creative search for spiritual refinement are the growing spirit of objectivity and rationalization on the one hand; the growing overexposure of the senses to external stimuli on the other. Simmel traces both of these developments to an emergent metropolitan way of life.
> (Smith, 1980: 91)

Although Berlin was a constant reference point for Simmel's analyses of modern society forming the subject of several self-contained essays on the city, he was a frequent visitor to other European cities such as St Petersburg, Paris, Prague, Vienna, Rome, Florence and Venice. These last three form the subject of essays published between 1989 and 1907 in which Simmel emphasises the haphazard or fortuitous nature of their layout that produces a unique aesthetic quality. The classical Italian city, in Simmel's view, attracts admiration because it can engender some of the wonder felt by encounters with the natural landscape (Frisby, 1992b: 109–10; Simmel, 1996). This is a theme that Benjamin also discusses in his essay (with Asja Lacis) on Naples that we discuss in the following section.

WALTER BENJAMIN: THE EXEGETICAL CITY

Since his tragic suicide in 1940, Walter Benjamin's reputation and importance as a critic and theorist has grown to the extent that it is now possible to talk of a veritable 'Benjamin industry'. Each year several new titles are added to an already substantial secondary literature, while new translations of Benjamin's writings previously unavailable in English continue to appear (although much of Benjamin's *Collected Writings* still remains untranslated). Despite the esteem Benjamin's writings enjoyed from established academic scholars such as Ernst Bloch, Theodor Adorno and Gershom Scholem, only a small percentage of it was published in his own lifetime. Perhaps the most ambitious and important of Benjamin's researches, the *Passagen-Werk* or 'Arcades Project' was published posthumously from the author's notes and, although the themed manuscripts (that Benjamin called *Konvoluts*) are minor works of art in their own right, we get no sense of how each component related to his general argument.

Indeed, the lack of a discernible thesis or rationale that might explain why the Parisian arcades are worthy of such an exhaustive analysis makes it difficult to categorise Benjamin's method or prospectus.

This is because Benjamin's writings are quite consciously seeking to avoid the somewhat dry and sterile investigations of social life associated with the new disciplines of sociology and social psychology. Along with his Frankfurt School colleagues, Benjamin owed his allegiance to the philosophical tradition developed by Hegel and Marx, that of idealism, and although Benjamin's **dialectical** critique differed in important ways from contemporary Marxists, it sets him apart from sociological empiricists whose work had no discernible impact on his city writings. Like Simmel, and even more so Lefebvre, Benjamin was engaged in a philosophical dialogue on the nature of truth, morality and the human condition with the two great 'book ends' of the Enlightenment – Kant and Nietzsche.[3]

Throughout his writing, according to Howard Caygill, it is possible to discern in Benjamin 'the development of a Kantian concept of experience through an extension of a Nietzschean method of active nihilism (Caygill, 1998: xiii). In Benjamin's own researches this ultimately unsuccessful attempt to recast Kantian experience, nevertheless helped him to refine the dialectical method known as **immanent critique**, and by applying this technique to the study of cultural formations, Benjamin hoped fundamentally to change the way in which literary and cultural criticism had hitherto been undertaken. Benjamin found the metropolitan city to be a particularly important stimulus to his reflections on the nature of human experience and, along with his important literary and philosophical work, he bequeathed a corpus of writings ranging from journalistic travel essays and autobiography to the monumental *Arcades* study that continues to inspire the study of the city in all its dimensions.

The 'City Sketches'

Several writers trace the development of Benjamin's interest in the city from the essay on Naples that he wrote with the help of Asja Lacis, and published in the *Frankfurter Zeitung* in 1925 (Buck-Morss, 1989: 8; Gilloch, 1996: 22–3; Leslie, 1999.[4] Though it does have the feel of a travel piece written for an educated middle-class audience, several themes of Benjamin's later writings emerge here. For example, his interest in the visitor's difficult insertion into the closed world of the city's popular districts that Naples so richly evokes. 'The travelling citizen who gropes his way as far as Rome from one work of art to the next, as along a stockade, loses his nerve in Naples' (Benjamin, 1997: 168). In the Naples sketch, Benjamin discusses the porosity of the buildings and the private–public thresholds of courtyards, arcades and stairwells. He is fascinated by the constant change of the urban milieu where '[t]he stamp of the definitive is avoided. No situation appears intended forever, no figure asserts its "thus and not otherwise"' (ibid.: 169). 'Porosity results not only from the indolence of the Southern artisan, but also, above all, from the passion for improvisation, which demands that space and opportunity be at any price preserved'. Against this rather hackneyed view of Neapolitan society must be set the more interesting observation that in Naples, 'Buildings are used as a popular stage. They are all divided into innumerable, simultaneously animated theatres. Balcony, courtyard, window, gateway, staircase, roof are at the same time stage and boxes' (ibid.: 170).[5] 'Porosity is the inexhaustible law of the life of this city, reappearing everywhere. A grain of Sunday is hidden in each weekday, and how much weekday in this Sunday!' (ibid.: 171–2).

In this respect, Benjamin sees Naples as more of a developing city than a European city where the private realm, at least for the popular classes, is hardly in evidence. Like the African kraal where 'each private attitude or act is permeated by streams of communal life . . . the house is far less the refuge into which people retreat than the inexhaustible reservoir from which they flood out' (ibid.: 174). Economic necessity makes a virtue out of this enforced socialisation – 'Poverty has brought about a stretching of frontiers that mirrors the most radiant freedom of thought. There is no hour, often no place, for sleeping and eating' (ibid.: 175). The arrhythmia of the Southern Mediterranean city is a theme that Lefebvre also finds interesting, as we shall later discover but, unlike Lefebvre, Benjamin does not appear to set any special store by climate, topography or geology. Thus, for

Benjamin, Moscow more closely resembles Naples than it does Berlin because its street culture and collective behaviour are closer to peasant societies than the 'princely solitude, princely desolation [that] hang over the streets of Berlin'. Whereas in Moscow, 'goods burst everywhere from the houses, they hang on fences, lean against railings, lie on pavements' just as they do in Naples in contrast to Benjamin's home city where the streets 'are like a freshly swept, empty racecourse' (ibid.: 180). Later in his essay, Benjamin makes the connection between Moscow and Naples explicit when he writes:

> Shoe polish and writing materials, handkerchiefs, dolls' sleighs, swings for children, ladies' underwear, stuffed birds, clothes-hangers – all this sprawls on the open street, as if it were not twenty-five degrees below zero but high Neapolitan summer.
>
> (ibid.: 180–1).

Again, the motifs of 'the inside on the outside' are repeated and put us in mind of the reclaimed urbanism of Aragon's 'Le paysan de Paris' (Paris Peasant). Only in wholly bourgeois cities are the separations between private and public space so neat and distinct. In the proletarian city the personal and the public are intertwined with invention and bravado, so that the streets and the pavements become inscribed with the folk routines of their habitués. The culture of the city and its antagonistic relationship with the rural is perfectly captured in the outskirts of Marseilles – 'the terrain on which incessantly rages the great decisive battle between town and country' (ibid.: 213) – where there is 'the hand-to-hand fight of telegraph poles against Agaves, barbed wire against thorny palms . . . short-winded outside staircases against the mighty hills'. Here, Benjamin is describing the commercial extension of the city into the rural hinterland using language replete with explosions and 'shell splinters', as if the city cannot tolerate the inanimate passivity of the country and must subordinate it using almost military force and violence. In other parts of the Marseilles essay, Benjamin draws our attention to how urban topography is overwritten with languages of class domination and resistance (gaudy advertising hoardings along the walls of the central city, red lettered socialist graffiti in front of dockyards and arsenals). Although these observations are tantalisingly brief, they recur in all the city sketches, combining what would now be called 'urban semiotics' or the analysis of signs with a materialist spatial analysis that anticipates many of the formulations of Marxist urban ecology (see Chapter 6). However, it is in the *Arcades Project* that Benjamin's urban critique reaches its apogee.

The *Arcades Project*

Were we only to have the 'City Sketches' as Benjamin's contribution to critical urban studies, it is doubtful that his readership among strict readers of the genre would have extended beyond its original circulation in the German literary press. However, these journalistic and essayistic forays into the heart of the city prefigured a far more extensive and philosophically ambitious project that was to occupy him for most of his later writing career. In the *Arcades Project*, Benjamin's preparatory notes appear initially to be a random collage (or 'literary montage' as Benjamin himself described it) of themes under which are grouped long quotations culled from a vast library of historical, literary, scientific and journalistic accounts of different urban phenomena. Benjamin assigned upper case and lower case letters to each individual subject (although not all possible letters were used in the collected manuscripts).[6]

It is clear that in the original sketch for the project (which was to have been undertaken with Franz Hessel) every imaginable type of urban phenomena was to be included in the purview of the research. Suggested themes included everything from 'Conveniences and inconveniences (tobacco, mailboxes, tickets, poster pillars, and so forth)' to 'Types of cocottes: streetwalkers, mômes, call girls (deluxe), social relations, tarts, lionesses, girlfriend, sweetheart, artiste' to 'Ridiculous "souvenirs" and bibelots-quite hideous', 'The Sunday of the poorer classes', 'Tea in the Bois', and 'Great and small labyrinths of Paris' (Benjamin, 1999b: 919–21). 'Arcades' was merely listed as one of these themes, but as Benjamin's solitary project continued, these commercial passageways became the *leitmotif* for his critical engagement with the problem of modernity in all of its aspects. The inside-outside world of the dimly lit labyrinth conjured up, for Benjamin, a universe of possibilities and transgressions that only the modern metropolis could offer. As Benjamin writes:

> The city is the realization of that ancient dream of humanity, the labyrinth. It is this reality to which the flâneur, without knowing it devotes himself.
>
> (Benjamin, 1999b: 429–30)

In *Konvolut M*, dedicated to the figure of the *flâneur*, much of the rich seam of Benjamin's contribution to urban critique is to be found. In Benjamin's understanding, the flâneur is not a mere *boulevardier* – a man (the male form of the noun in French is significant as we shall see in Chapter 8) with the means to indulge in the voyeuristic spectacle of café society – but something closer to a secular pilgrim, a seeker after the profane truths of a temporal-spatial universe that has been trampled into the dust by a humanity made dull and inattentive to the hidden wonders of the metropolis.

The *flâneur* goes in search of 'vanished time' like a 'werewolf restlessly roaming a social wilderness' (Benjamin, 1999b: 416, 418). He is in, but not of, the crowd where every face is masked in anonymity, as in 'a masquerade of space'. Like the Neapolitans and the Muscovites, Parisians also have this 'technique of inhabiting their streets', and Benjamin recalls an experience similar to that of Adolf Stahr writing in the 1850s of the spontaneous creation of a small outdoor market beside some roadworks that had formed a sudden human bottleneck. 'Seventy years later, I had the same experience at the corner of the Boulevard Saint-Germain and the Boulevard Raspail. Parisians make the street an interior . . . For if *flânerie* can transform Paris into one great interior – a house whose rooms are the *quartiers*, no less clearly demarcated by thresholds than are real rooms – then on the other hand, the city can appear to someone walking through it to be without thresholds: a landscape in the round' (Benjamin, 1999b: 421–2). The point here is that the city's boundaries are mostly invisible, and it is only the uninitiated who sees the metropolis as a shapeless, undifferentiated totality.

Flânerie is also a function of nostalgia, or rather of the ecstatic remembrance of lost times, sensations, landscapes. Proust represents this notion of 'landscape as epiphany' in *Swann's Way* where he tells of how

> suddenly a roof, a gleam of sunlight reflected from a stone, the smell of a road would make me stop still, to enjoy the special pleasure that each of them gave me, and also because they appeared to be concealing, beneath what my eyes could see, something which they invited me to approach and take from them, but which, despite all my efforts, I never managed to discover . . .[7]

For Benjamin this sudden remembrance of a once forgotten experience can lead to intoxication in much the same way as does the taking of hashish but, as with narcotics, it is a transient impermanent sensation that the *flâneur* or user always seeks to repeat. We come back to the city time and time again in pursuit of that true experience that constantly eludes us, and so, unable to embrace the metropolis in its profound totality, we take refuge in the landscape of memory.

The history of the present

Benjamin's engagement with the city owed much to his childhood and adolescent experiences that he made the subject of an essay written in the 1930s, 'A Berlin Childhood Around 1900' (Lindner, 1986). In this essay and its complement, 'Berlin Chronicle', as Susan Sontag observes, 'Benjamin is not trying to uncover his past, but to understand it: to condense it into its spatial forms, its premonitory structures' (Sontag in Benjamin, 1997: 13). Benjamin shared with Proust a fascination for the redemptive and emancipatory qualities of memory that allow us to reclaim from our past a comprehension of those emotions and perceptions that eluded us at the time. For Benjamin, recollection means precisely the 're collection' of impressions of a fragmented and scattered experience and their reconstitution as a meaningful narrative that seeks in its own imperfect way to assume the dimensions of a social and psychological totality. This work of exegesis, or revelation, often called on unconscious elements (Benjamin's debt to the surrealist artists and poets in this respect is acknowledged throughout his work) and an engagement with the mythical aspects of human development. But, at the same time, Benjamin felt the need to engage with the more rigorous or 'scientific' claims of Critical Theory and historical materialism.[8] In this sense, it could be argued that Benjamin was something of a secular theologian who saw his task as one of divining and explaining the transcendental and contradictory essences of the 'total city'.

Benjamin's historical materialism is often described as 'anti-historicist' – that is to say he rejects the idea that the past can be explained only in terms of the past because history is always seen through the eyes of the present. He writes that 'every image of the past that is not recognized by the present as one of its own concerns threatens to disappear irretrievably'. This is not so much a plea to make the past relevant to the present as a caution against the idea that historians, as Ranke believed, could and should recount the past 'the way it really was' (Benjamin, 1999a: 247). In a much quoted and deeply lyrical passage, Benjamin compares the work of historical materialism to the 'Angelus Novus' in Paul Klee's painting that

> shows an angel looking as though he is about to move away from something he is fixedly contemplating. His eyes are staring, his mouth is open, his wings are spread. This is how one pictures the angel of history. His face is turned towards the past. Where we perceive a chain of events, he sees one single catastrophe which keeps piling wreckage upon wreckage and hurls it in front of his feet. The angel would like to stay, awaken the dead, and make whole what has been smashed. But a storm is blowing from Paradise; it has got caught in his wings with such violence that the angel can no longer close them. This storm irresistibly propels him into the future to which his back is turned, while the pile of debris before him grows skyward. This storm is what we call progress . . .
>
> (Benjamin, 1999a: 249)

Cities as vast, living repositories of the past provide us with the archaeological debris of this 'progress' on a daily basis, but Benjamin urges us to confront history not as a site of 'homogeneous, empty time, but time filled by the presence of the now [*Jetszeit*]'. Ancient civilisations are redeemed and reclaimed by emergent political states in the way that Jacobin Paris 'viewed itself as Rome re incarnate', evoking 'ancient Rome the way fashion evokes costumes of the past' (ibid.: 253). Benjamin's urban imaginary thus offers a strong rebuttal of 'history as progress' accounts of urban development, while at the same time making a strong claim for the place of the subjective imagination and memory in the social construction of urban life. As we shall see in the next section, the critical exploration of the 'imagined city' through the lens of historical materialism is also a central feature of the work of Henri Lefebvre. But more generally, as we shall see in Chapter 8, Benjamin's rich aesthetic and philosophical engagement with the urban experience has inspired a new generation of urban writers who have used Benjamin's insights to attempt new readings of the city as the laboratory of a dynamic and stormy modernity.

HENRI LEFEBVRE: THE PRODUCTION OF THE CITY

> It is at this moment that the *mode of production* dominates the results of history, takes them over and integrates within itself the 'sub-systems' which had been established before capitalism (i.e. exchange networks of commerce and ideas, agriculture, town and countryside, knowledge, science and scientific institutions, law, the fiscal system, justice, etc.), without, however, managing to constitute itself as a coherent system, purged of contradictions. Those who believe in the system are making a mistake, for in fact no complete, achieved totality exists. However, there is certainly a 'whole', which has absorbed its historical conditions, reabsorbed its elements and succeeded in mastering some of the contradictions, though without arriving at the desired cohesion and homogeneity.
>
> (Henri Lefebvre, 1976: 10)

Henri Lefebvre was a genuinely interdisciplinary writer whose intellectual interests included literature, language, history, philosophy, planning and, in later life, even policy-making (Kofman and Lebas,

1996: 6). In all of his reflections on the urban question, Lefebvre also remained a committed Marxist, and his now classic manifesto 'The Right to the City' (Lefebvre, 1968) was first published in the year when the May events heralded a proliferation of social movements centred on the city, giving rise to new forms of creativity and experimentation in urban living. Many of the young activists who read and admired 'The Right to the City' went on to become influential architects, planners, politicians and academics, but until recently Lefebvre's influence has been more or less confined to his native France – partly because few of his books and articles had been translated into English and, indeed, many important aspects of Lefebvre's voluminous output remain untranslated. However, with the growing interest in the work of human geographers such as David Harvey, and in the socio-spatial theories of Manuel Castells, a new readership has developed for a writer who was one of the first to examine in detail 'the politics of space' and the relationship between the physical environment of the city and its many social and economic relations.

For the English-speaking reader trained to distinguish carefully between 'normative' and scientific judgements, Lefebvre can be a frustrating and difficult writer. As with many French philosophers of his generation, Lefebvre prefers an essayistic style that is full of rhetorical flourishes, puzzling metaphors and little of what is somewhat pompously called 'the apparatus of scholarship' (or the comprehensive referencing of sources). A passionate advocate of the need for social justice and the overcoming of bourgeois domination (though his own professional and personal status could only be described as comfortably upper middle-class), Lefebvre often wrote in the manner of the polemical pamphleteer. This was not a characteristic for which he felt any need to apologise, for Marx's most quoted work remains the revolutionary pamphlet he wrote with Friedrich Engels in 1848 (The Manifesto of the Communist Party), and Lefebvre certainly believed that as a philosophy of urban praxis, his ideas should, like those of Marx, be dedicated to changing rather than merely observing society.

The Right to the City

The Right to the City brought together several themes on which Lefebvre had been working for a number of years. Like Marx, Lefebvre saw the city as a location where use value and exchange value meet and are combined in a formal system or as 'relations of production'. Put simply, use value relates to the physical environment and human and raw materials, whereas exchange value relates to the worth of commodities produced for sale by the capitalist mode of production. As the intensity of capitalist development grows, so the transformation of space and the location of activities within urban locations becomes increasingly commodified. The division of labour which was apparent in early capitalism is extended beyond the city or region to the national and international level, while extremes of poverty and wealth, because of the all conquering power of the market, transform the city into a patchwork of central commercial districts, surrounded by underprivileged ghettos and wealthy inaccessible enclaves.

Lefebvre termed the capitalist city 'a bureaucratic society of controlled consumption' (Lefebvre, 1984, chapter two), a term which has become popularised as 'consumer society'. However, the right to the city is not simply a right to consume, since human beings also have a 'need for creative activity, for the *oeuvre* (not only of products and consumable material goods), of the need for information, symbolism, the imaginary and play'. The concept of **oeuvre** is a recurring theme in Lefebvre's work and it refers to the city in all its aspects – the monumental, the aesthetic, the incidental – and is best summarised as the city beyond the realm of commodified space. Conventional social scientific investigations of the city, Lefebvre believed to be limited because

the object, the city, as consummate reality is falling apart. Knowledge holds in front of itself the historic city already modified, to cut it up and put it together again from fragments. As social text, this historic city no longer has a coherent set of prescriptions, of use of time linked to symbols and to a style. . . . The city historically constructed is no longer lived and is no longer understood practically. It is only an object of cultural consumption for tourists, for a [*sic*] estheticism, avid for spectacles and the picturesque.

(Lefebvre in Kofman and Lebas, 1996: 147)

This passage is densely packed with *aperçus* that have now become the stock in trade of cultural and urban studies throughout the world. The notion of the fragmented city that is constantly being reassembled but which has no fixed or historically rooted identity is consistent with many postmodern accounts of contemporary urban life. Similarly, as we shall see in Chapter 8, the concept of the city as a 'social text' or as an urban narrative has been promoted by post-structuralist and post-colonialist commentators on the modern city. While the announcement of the death of the 'historic city' and the rise of what, today, is known as 'the heritage industry' anticipates many contemporary debates on the 'hollowing out' of the city, the commodification of 'historic spaces' through the advent of mass tourism (Urry, 1990), and the irresistible rise of the 'non place urban realm' in the form of the 'edge city' (Webber, 1964; Garreau, 1991).

Moving from analysis to prescription Lefebvre later argues for the adoption of new experimental approaches that will allow 'urban society' to fully emerge. Lefebvre believed that even under late capitalism the rural-agrarian world still limited and stifled the potential for a genuine urban society that is free of rural and 'natural' dominant features (and these were the essentially negative ones of a rigid, hierarchical social order, high levels of surveillance and social control, anti-intellectualism and anti-cosmopolitanism) – an antipathy to which, as we have already noted, Simmel loudly shared. Instead, Lefebvre proposed that the new city builders should adopt the methods of **transduction** and experimental utopia. Transduction is the term Lefebvre uses to describe the process of constructing a theoretical object (such as a new city) on the basis not only of information relating to (say) current urban reality, but by the problematic posed by existing urban realities. As new theoretical imaginings are given concrete form, the feedback mechanism ensures that subsequent projections/projects are informed by this newly altered reality in an endless loop of speculation-investigation-critique-implementation.

Lefebvre's second proposal in *The Right to the City*, experimental utopia, was premised on the idea that any decent urban planner in the late twentieth century should consider herself or himself a utopian. However, utopianism that aimed at the creation of an ideal community required the study of everyday life in actual urban settings. By studying what made for good, happy and playful urban dwellers, Lefebvre believed that the city of tomorrow could be built from the dreams of today (Lefebvre in Kofman and Lebas, 1996: 152). Often though, Lefebvre claimed that architects fell into the dogmatic acceptance of function, form and structure in the guise of their ideological counterparts – functionalism, formalism and structuralism. This meant that architects behaved too often like unsophisticated sociologists – seeing the world in mechanistic terms without appreciating 'the significations perceived and lived by those who inhabit' it (ibid.).

How to remedy this deficiency? Lefebvre's answer was to call for a holistic approach to the study and production of urban society. He wanted to see an end to the sectarian divisions between philosophers of the city and those involved in the more scientific and technical aspects of the urban system, and a proper appreciation of the global and the partial. In other words, attention should be given to the fine detail of individual *milieux* and what general properties and problems have in common with other urban environments throughout the world. To understand the scale of the urban question in all of its historical and spatial dimensions it was also necessary, Lefebvre believed, to develop 'a general theory of *time-spaces*'. This was a major task, and it was one to which Lefebvre dedicated the remainder of his intellectual life. In this context, Lefebvre was able to anticipate several of the key problems addressed by theorists such as Giddens on structuration (Giddens, 1984) and Castells on 'the space of flows' (Castells, 2000a) (see Chapter 6).

On *The Production of Space*

The Production of Space was first published in French in 1974, but it did not achieve publication in English translation until 1991. The declared aim of Lefebvre's book was to bring the philosophy and **epistemology** of space (mental space) into dialogue with real or empirical space. As with so much of Lefebvre's work, his major interlocutors are philosophers such as Descartes, Kant and Hegel, all of whom he felt neglected or mishandled the spatial dimension of human experience. As Katznelson informs us, 'Lefebvre's scheme hinged on a separation of the city and urbanism as theoretical and empirical constructs. The city is an arrangement of objects in space; urbanism a way of life'. Although Lefebvre claimed that 'there is not a strict correspondence between modes of production and the spaces they constitute', he argued nonetheless that 'each epoch produces its own space' (Katznelson, 1992: 96).

If the industrial capitalism of the nineteenth century gave rise to a particular urban form based around the division of labour, Lefebvre believed that in the twentieth century industrial capitalist society had been dialectically transcended by urban society. This was possible because the urbanisation of capital in the form of financial and property speculation had given rise to a second circuit of capital that was becoming increasingly dominant in relation to manufacturing capital (ibid.: 97) (see Chapter 6). Whatever form capital takes it, nonetheless, must result in the exploitation of labour, and the resistance of the working class to their subordination inevitably gives rise to class conflict. As Katznelson once again synthesises, for Lefebvre:

> Space is not just a built-environment but a force of production and an object of consumption. It is also an object of political struggle, because space is an instrument of control by the state.
>
> (ibid.: 98)

However, the control that the state imposes on space is never absolute because, as Lefebvre argues, contradictions between different levels of state authority will always persist, as will conflicts of class.

In this sense, Lefebvre readily acknowledges that political power in the city is 'plural', even though he refuses to embrace 'pluralism' as a democratic model in the American sense (Lefebvre in Kofman and Lebas, 1996: 379) (see Chapter 7). When Lefebvre attaches some hope to the 'things that pluralism lets by', he seems to be moving beyond an orthodox Marxist determinism in identifying what David Harvey has called 'spaces of hope' (Harvey, 2000) where a measure of 'de-commodified' existence might (for a time) be possible. On a practical level the de-commodification of collective space might involve a successful campaign to create a public park or to provide free urban transport, or to recognise the rights of squatters (see Exhibit 7.1, Chapter 7). We could thus interpret Lefebvre's spatial theory, like that of Simmel, as being implicated in the processes of urban consumption as much as in the material production of the commodity form, and in this respect Lefebvre's work relates to the urban theory of Manuel Castells, which we examine in some detail in Chapter 6.

But why Lefebvre remains one of our most enigmatic and challenging urban theorists is his propensity to mix speculative philosophy with anthropology, and to look for the sorts of patterns that would attract an artist or a composer rather than a social scientist as evidenced by his work on the 'rhythmanalysis' of Mediterranean cities in the last years of his life (Lefebvre and Regulier, 1986).[9] For Lefebvre, ritual and rhythm are closely related, and he notes the abundance and importance of religious rituals, festivals, fasts and feasts in Mediterranean society and how these rituals imprint a rhythm on the routines of its citizens. The debt (once again unacknowledged) to Benjamin is evident when Lefebvre suggests how stairs and steps serve the function of providing links between times – 'between the time of architecture (the house and the enclosure) and urban time (the street, the open space, the square and monuments)' (Lefebvre in Kofman and Lebas, 1996: 237). These ancient rhythms can absorb wars and conquests, or the daily tourist invasion in the case of Venice, while still retaining their theatricality and sense of occasion. This strong attachment to private, secret and unyielding rhythms that are quite distinct from the 'metarhythms' of the state has allowed the Mediterranean city to survive substantially unchanged for thousands of years. This explains why in Lefebvre's words, '[e]very form of hegemony and homogeneity are refused in the Mediterranean' (Lefebvre in Kofman and Lebas, 1996: 238).

It is a pity that Lefebvre was not able to develop these rather sketchy and often contradictory ideas into a more coherent and worked out hypothesis on urban typologies because they suggest possibilities for some interesting comparative studies in the cultural geography of cities. However, one can but concur with Katznelson's judgement that Lefebvre's enduring attachment to abstract theorising in his chronology of urban development

> was certainly not based on careful empirical scholarship. He accepted (as did Harvey and Castells) the disappearance of the city as both subject and object from Marxism during the period of industrial capitalism. The realization of urban possibilities seemed almost entirely a matter of voluntary praxis; causal constraints were noted, but forgotten in the interest of exhortation. In all, Lefebvre (re-)created an idolatry of the city by his combination of an asserted historical progression and his assertion that urbanization and its attendant spatial relations together provide the new master processes of modernity.
>
> (Katznelson, 1992: 101)

However, it is not true to say that Lefebvre was inattentive to the ways in which spatial practice played itself out in the urban environment, he simply contested the ways that traditional urbanism viewed the urban question. For Lefebvre:

> Spatial practice regulates life – it does not create it. Space has no power 'in itself', nor does space as such determine spatial contradictions. These are contradictions of society – contradictions between one thing and another within society, as for example between the forces and relations of production – that simply emerge in space, at the level of space, and so engender the contradictions of space . . .
>
> (Lefebvre, 1991: 358–9)

Lefebvre then goes on to provide the example of how, by giving over more and more spaces to cars and roads, we are not easing traffic problems but hastening the removal of public uncommodified space and turning it into profit generating commodified space. As Lefebvre continues:

> There are two ways in which urban space tends to be sliced up, degraded and eventually destroyed by this contradictory process: the proliferation of fast roads and of places to park and garage cars, and their corollary, a reduction of tree-lined streets, green spaces, and parks, and gardens. The contradiction lies, then, in the clash between a consumption of space which produces surplus value and one which produces only enjoyment – and is therefore 'unproductive' . . .
>
> (ibid.)

Here, Lefebvre raises a number of important issues that we return to in later chapters, such as urban sustainability, the struggles to defend public and civic space, and the idea of the city as consumable space – themes that have been taken up by writers as diverse as Jane Jacobs, Richard Sennett, Manuel Castells and Sharon Zukin, and that we will expand on further in later chapters.

CONCLUSION

Despite misgivings about Lefebvre's denial that 'the city was any kind of meaningful entity in modern life' (Harvey in Lefebvre, 1991: 430–1), many prominent contemporary urbanists still regard him as one of the most important urban theorists of the twentieth century. The reason, I would argue, is because works such as *The Production of Space* show the compatibility between Marx's analysis of capitalist power relations and the way these strategies of class domination are embedded, above all, in the fabric of urban society itself. On the other hand, while Max Weber may not receive many plaudits from followers of Lefebvre, his history of urban development set a standard for subsequent urban historians such as Lewis Mumford to follow, as well as for contemporary urbanists such as Peter Hall (1998) and Ed Soja (2000b). At the same time, in terms of contemporary accounts of urban power, it is Weber who continues to provide key reference points in relation to the importance of bureaucratic authority, the existence and operation of markets, and the organisation and reproduction of status and interest groups. It is Weber's insistence that the city is, above all, a 'political entity' that links him directly to more contemporary analyses of urban political institutions, parties and associations (see Chapter 7). As Le Galès writes, '[Weber], alone, proposes an analytical model for cities as social structures, as a site where groups and interests accumulate and are represented' (Le Galès, 1999: 297).

Simmel and Benjamin on the other hand are more philosophical in their treatment of the urban question, and their work also lends itself to anthropological, sociological and cultural analyses of city life, as we shall discover in Chapter 8. But at the heart of the social critique of both writers there is also a careful and critical elucidation of the shaping influence of the money economy on urban form and behaviour. Thus, as we shall see in Chapters 6 and 7, urban researchers writing within a Marxist and urban political economy tradition have also drawn on the concepts and insights of these writers in their attempts to understand the globalising cities of the twenty-first century.

Any patrimony or matrimony in the history of ideas is open to contestation, and it is certainly not the argument of this book to establish Weber, Benjamin, Simmel and Lefebvre as the sole progenitors of what we now understand as 'urban theory'. Instead, the aim of this chapter has been to discuss how these thinkers thought about the city and why their ideas continued to influence successive generations of urbanists and why they are still important to this day. However, in many respects, the most substantial quantitative contribution to our understanding of the urban experience did not come from social theory or critical philosophy but from the direct testimony of urban investigators. In the following chapter we explore how this empirical tradition became the dominant approach to urban studies in the English-speaking world in the twentieth century.

QUESTIONS TO DISCUSS

1 **In what ways do the writers in this chapter draw inspiration from their encounters with the European city?**

2 **Why do all four writers stress the city's potential for realising human freedom?**

3 **Can you identify any other common features of the urban analysis of Weber, Simmel, Benjamin and Lefebvre?**

4 **What links can you make between these 'classical' urban theorists and more contemporary urban observers?**

FURTHER READING

Max Weber

There are some good general introductions to the sociology of Max Weber, such as Frank Parkin's eponymous study (1982) and Anthony Giddens' short study, *Politics and Sociology in the Thought of Max Weber* (1972). *The Cambridge Companion to Weber* edited by Turner (2000) is an authoritative introduction to all aspects of Weber's life and thought. Dirk Kaesler's *Max Weber. An Introduction to His Life and Work* (1988) is an approachable study that includes a useful synthesis of Weber's 'urban sociology' in chapter two. Reinhard Bendix's *Max Weber: An Intellectual Portrait* ([1960] 1998) remains unsurpassed as an intellectual biography, while Gerth and Mills' edited volume *From Max Weber* (1970) is a classic collection of Weber's scholarship. David Frisby's essay 'The Ambiguity of Modernity: Georg Simmel and Max Weber' in Wolfgang Mommsen (ed.) *Max Weber and His Contemporaries* (1987) is a sophisticated comparative study of these two important thinkers by the leading authority on Georg Simmel. One of the few texts to specifically explore the urban sociology of Max Weber is Brian Elliott and David McCrone's *The City: Patterns of Domination and Conflict* (1982).

Georg Simmel

A student-friendly introduction to Georg Simmel, also by David Frisby (1984) provides an annotated bibliography of sources on Simmel and a bibliography of English translations of his major works. For more extensive studies on Simmel, David Frisby's *Simmel and Since. Essays on Georg Simmel's Social Theory* (1992b) contains much of interest for students of urban theory, while his earlier *Fragments of Modernity. Theories of Modernity in the Work of Simmel, Kracauer, and Benjamin* (1988), as the subtitle suggests, situates the work of Simmel alongside contemporaries such as Benjamin. Frisby's *Sociological Impressionism. A Reassessment of Georg Simmel's Social Theory* (1992a), first published in 1981, provides a more general overview of Simmel's contribution to social theory including a particularly revealing chapter on Simmel as the original sociological *flâneur*. Deena Weinstein and Michael A. Weinstein in *Postmodern(ized) Simmel* (1993) offer a counter-interpretation to Frisby's, arguing that Simmel is best seen as a *bricoleur*, a piecer-together of the fragments of the urban experience, and hence a proto-postmodernist. Jaworski (1997) critically evaluates the Simmel reception in American social science including the influence of Simmel's thought on the Chicago sociologists Park and Goffman. Those wishing to study Simmel's collected works in depth will also wish to consult the three-volume edition *Georg Simmel. Critical Assessments* (1994) edited by David Frisby.

Walter Benjamin

Substantial parts of the German edition of Walter Benjamin's Collected Works are now available in English, including the beautifully designed and carefully edited translation of Rolf Tiedemann's edition of *The Arcades Project* (Benjamin, 1999b), by Eiland and McLaughlin. The city sketches are available in monographic form, such as *Moscow Diary* (1986) edited by Gary Smith and in the collections *One-Way Street* (1979) and *Illuminations* (1999a). The secondary literature on Benjamin is too large and wide-ranging to be synthesised in anything other than an arbitrary and partial manner. However, absolute newcomers to the life and work of Walter Benjamin will find Caygill *et al.*'s *Walter Benjamin For Beginners* both instructive and entertaining. Nicely illustrated by Andrzej Klimowski, the text is accessible and comprehensive without being patronising. A more challenging read is Howard Caygill's *Walter Benjamin. The Colour of Experience*, (1998) which repays a familiarity with the philosophy of Kant, Hegel and Benjamin's Frankfurt School contemporaries. Those looking for an authoritative account of the 'first' Frankfurt School should consult Jay (1973). Benjamin's relationships with other important figures in the Institute for Social Research is examined in the chapters in Smith (1991). On Benjamin's urban imaginary, Susan Buck-Morss's *Dialectics of Seeing* (1989) remains an essential reference and source of contemporary criticism, though Graham Gilloch's more recent, *Myth and Metropolis* (1997) proves that the last word will never be written on this enigmatic and multi-faceted thinker.

Henri Lefebvre

An increasing amount of Henri Lefebvre's writing is now available in English. A good place to start is *Henri Lefebvre, Writings on Cities* (1996) edited and translated by Kofman and Lebas who provide a useful introduction to Lefebvre's life and work. Of the major texts, *The Production of Space* (1991) is available in a good English translation by Donald Nicholson-Smith. A further volume of the *Critique of Everyday Life* ([1968] 1992) was published in 2002, and the third and final volume was published in 2003. Rob Shields' *Lefebvre, Love and Struggle. Spatial Dialectics* (1998) is the only book-length treatment of Lefebvre to appear in English, and contains a very comprehensive bibliography of Lefebvre's works. For those looking for a shorter intellectual biography, Alistair Davidson's essay (1992) is sympathetic and informative. Mark Gottdiener's *The Social Production of Urban Space* (2nd edition 1994) first appeared in 1985 and seeks to apply and extend the Marxist spatial analysis of Henri Lefebvre, while providing a critical assessment of more recent approaches to the study of urban society such as globalisation theories, flexible accumulation, postmodernism, the new international division of labour, and the growth-machine perspective (see Chapters 6, 7 and 8 of this volume). Derek Gregory and John Urry's *Social Relations and Spatial Structures* (1985) considers the cross-fertilisation of ideas between geography and sociology, focusing particularly on the work of Lefebvre. *Spatial Practices* edited by Liggett and Perry (1995) offers a series of thought provoking essays on the uses of urban space. More critical responses to Lefebvre that challenge the ascendancy of spatial analyses in urban studies can be found in Saunders (1981, 1995) and Werlen (1993).

Of the many papers and articles on Lefebvre's contributions to urban theory, Kipfer's 'On the Possibilities of the Urban: Rereading Henri Lefebvre's Open and Integral Marxism' (1998) provides a critical survey of Lefebvre's impact on radical geography and critical urban sociology in the Anglo American world, while warning against a postmodern appropriation of Lefebvre's urban Marxism. Crawford and Cenzatti (1998) make interesting use of Lefebvre's 'Right to the City' essay as a starting point for their critique of the 're-appropriation' of public space by the poor of Los Angeles against a background of the 1992 riots (see Chapter 8 of this volume). The kind of postmodern 'misappropriation' of Lefebvre that Kipfer is almost certainly referring to is exemplified by Ed Soja's 'trialectics of space' which appeared as an article in 1996 (Soja, 1996a) and also opens his book *Thirdspace* (1996b). Soja's work is considered in more detail in Chapter 8, but those wishing to trace the influence of Lefebvre's ideas on one of the leading proponents of the 'new geography' should also consult Soja's *Postmodern Geographies. The Reassertion of Space in Critical Social Theory* (1989). De Certeau's *The Practice of Everyday Life* (1984), which has become something of a cult text for radical urbanists openly acknowledges its debt to Lefebvre's urban philosophy of praxis.

3

THE CITY DESCRIBED

Social reform and the empirical tradition in classic urban studies

'You know what the population of London is, I suppose,' said Mr Podsnap.

The meek man supposed he did, but supposed that he had absolutely nothing to do with it, if its laws were well administered.

'And you know; at least I hope you know,' said Mr Podsnap, with severity, 'that Providence has declared that you shall have the poor always with you?'

The meek man also hoped he knew that.

'I am glad to hear it,' said Mr Podsnap with a portentous air. 'I am glad to hear it. It will render you cautious how you fly in the face of Providence.'

Charles Dickens, *Our Mutual Friend*

INTRODUCTION

Although the social theorists we encountered in the previous chapter blazed a trail for a new urban imaginary that could co-exist with the modernist ideas of positive sociology, dialectical materialism and phenomenology, it was a quite different set of preoccupations that drove the true pioneers of urban studies to the investigation of the 'underworld' of the sprawling metropolis. Although the writers that we encounter in this chapter could all be described broadly as 'empirical researchers', that is not to say that a dichotomy between 'theoretical' and 'empirical' urbanists requires practitioners of one technique to avoid use of the other. As we have seen in the previous chapter, Weber's meticulous reconstructions of the ancient and medieval cities of Europe are rich in the sort of detail one would expect from an expert social and economic historian, just as Benjamin's forensic research on the minutiae of Parisian mores and consumption habits would put much

contemporary writing on the sociology of shopping to shame.

However, what distinguishes the work of Weber, Benjamin, Simmel and Lefebvre from what I will call 'the classic empirical tradition' of urban research and writing, is a clear belief that the means of urban enquiry should be subordinate to the task of explaining as far as possible the totality of the urban experience even within the confines of a particular city or group of cities. In these writers the ambition is to identify the 'essence' of the city, and to reduce this great, confusing concentration of humanity into an intelligible schema. Empirical or practical urban studies, on the other hand, tries to make sense of this seemingly chaotic and unstable world in the manner of a naturalist whose task is to count and classify the flora and fauna s/he encounters into distinct classes and categories. The analogy is an appropriate one for the very term 'typology' in the social sciences has its origins in nineteenth-century natural science where notions of 'ecology', 'organism' and 'evolution' provided potent and suggestive metaphors for generations of social researchers who have tried to make sense of 'the human jungle' that many identified in the modern metropolis.

In this chapter we trace the development of empirical urban research from its early manifestations in social reformism, through to the adoption of social research as an academic discipline where the modern city provides recurring examples for the study of 'social pathology'. While most of the research and writing in this chapter is derived from nineteenth- and early twentieth-century British and North American sources, this selection is not meant to imply that the empirical tradition is unique to these nations so much as a particular strength of Anglo-American urban studies in this period.

THE SINFUL CITY: URBAN INVESTIGATION AS SOCIAL REFORM

Ira Katznelson begins his study of urban politics and urban conflict in New York City (*City Trenches*) with a reference to Raymond Williams' work on literary representations of the rapidly expanding populations of the industrial towns and cities in the early nineteenth century (Williams, 1973). Summarising Williams, Katznelson writes, 'English literature maintained a rigid distinction between the processes of rural exploitation – which were dissolved in idyllic portraits of landscapes and green fields – and urban wickedness. The country was innocent and sublime; the city was the dwelling place of iniquity of all kinds' (Katznelson, 1981: 2).

If we think of the greatest writing on the industrial city in Europe and America in the nineteenth century it is to writers of fiction such as Charles Dickens, Edgar Allen Poe, Emile Zola, Fyodor Dostoevsky, Victor Hugo, George Gissing or Herman Melville that we turn for inspiration. What all these writers share is a willingness to confront and explore the dark and desperate side of civilisation that is so palpably a feature of the overcrowded and sensorially overpowering industrial metropolis. Writers in earlier epochs had used their pens to tilt against injustice, and many (such as Defoe, Blake and Wordsworth) expressed humankind's alienation from nature and even humanity in their portrayal of city life, but in the absence of any political mechanism for their mobilisation, these voices were heard only as prophetic echoes from the past.

The mid-nineteenth century provided both a political context (the choice between reform and revolution as Edmund Burke put it) and a literate public hungry for meatier fare than the mannered romanticism of classical literature. The arrival of 'the documentary novel' often published in serial form by popular subscription magazines stimulated a new interest in 'the real city' of the rookeries and flea markets rather than High Society and the Court. Dickens' fictional accounts of the Victorian underclass have been credited with inspiring several calls in Parliament for the reform of the Poor Law and for state action to counter the terrible effects of poverty in Britain's great cities. However, parliamentary commissions and enquiries did little to change the lives of the growing millions who made up the urban poor and about whom little was known in terms of their social and economic characteristics until the London-based *Morning Chronicle* newspaper decided to investigate the problem in an early example of campaigning investigative journalism.

Mayhew's *London Labour and the London Poor*

The *Morning Chronicle* investigation was intended to be a national survey covering many of the regions of England and Wales and was divided into three types: the rural, manufacturing and metropolitan (Razzell in Mayhew, 1980: 3). Henry Mayhew, a jobbing journalist, who had been a founding editor of the satirical magazine *Punch*, became the survey's metropolitan correspondent and London was to provide his case study. With the help of his brother and a small number of journalistic collaborators, in 1849 Mayhew began to write the survey that Thackeray was to describe as 'A picture of human life so wonderful, so awful, so piteous and pathetic, so exciting and

EXHIBIT 3.1   Henry Mayhew, *London Labour and the London Poor*

We then journeyed on to London-street, down which the tidal ditch continues its course. In No. 1 of this street the cholera first appeared seventeen years ago, and spread up it with fearful virulence: but this year it appeared at the opposite end, and ran down it with like severity. As we passed along the reeking banks of the sewer the sun shone upon a narrow slip of the water. In the bright light it appeared the colour of strong green tea, and positively looked as solid as black as marble in the shadow – indeed it was more like watery mud than muddy water; and yet we are assured this was the only water the wretched inhabitants had to drink. As we gazed in horror at it, we saw drains and sewers emptying their filthy contents into it; we saw a whole tier of doorless privies in the open road, common to men and women, built over it, and the limbs of the vagrant boys bathing in it seemed like, by pure force of contrast, white as Parian marble. And yet, as we stood doubting the fearful statement, we saw a little child, from one of the galleries opposite, lower a tin can with a rope to fill a large bucket that stood beside her. . . . As the little thing dangled her tin cup as gently as possible into the stream, a bucket of night-soil was poured down from the next gallery.

(Mayhew, 1980: 37)

terrible, that readers of romances own that they never read anything like to it' (Humpherys, 1977: ix). A passage from Mayhew's first report on the cholera epidemic in the riverside district of Bermondsey (Exhibit 3.1) is enough to confirm Thackeray's verdict.

Mayhew's vivid portraits of the squalor of urban life in Victorian London were all the more compelling because they allowed those who were actually living in these conditions to tell their own stories (see Figure 3.1). In this sense Mayhew was a pioneer of what we now call oral history and his work represented what he described as 'the first attempt to publish the history of the people, from the lips of the people themselves' (Mayhew, 1968: xv; cited in the Introduction to Mayhew, 1980: 2 by Peter Razzell).

The completed study was published some twelve years after Mayhew's first article appeared, and in subsequent accounts he developed a rough schema for his investigations by dividing his subjects into 'those that work, those that can't work, and those that won't work'. Although a journalist by vocation, Razzell argues that it was Mayhew's 'adherence to natural science which led him to such a literal rendering of

THE ORPHAN FLOWER GIRL

Figure 3.1 A character from Henry Mayhew's *London Labour and the London Poor*, 1851.

the evidence given to him by the people he inter-viewed' (Razzell, in Mayhew, 1980: 4). It seems that Mayhew had originally hoped to become an experi-mental chemist and an interest in the scientific method clearly informed his thinking about the London survey as he wrote in a letter to the editor of the *Morning Chronicle* in 1850:

> I made up my mind to deal with human nature as a natural philosopher or a chemist deals with any material object; and as a man who had devoted some little of his time to physical and metaphysical science, I must say I did most heartily rejoice that it should have been left to me to apply the laws of inductive philosophy for the first time, I believe, in the world to the abstract ques-tions of political economy.
>
> (Mayhew, 1850: 6 cited in Mayhew, 1980: 5)

Mayhew clearly intended his work to make a methodological contribution to social investigation and to 'let the facts speak for themselves'. But as Gareth Stedman-Jones points out, although May-hew's *London Labour and the London Poor* always appeared on the reading lists of the Charity Organ-isation Society in the 1870s, 'the passages cited were not those which examined the causes and struc-ture of poverty, but rather those dealing with the elaborate frauds and deceits employed by beggars and vagrants' (Stedman-Jones, 1971: 10). There was, therefore, a danger in over-focusing on the phenomenology of urban poverty without revealing the structural factors that shaped and constrained the lives of the slum dwellers. As we shall see in Chapter 6, although Engels' study of Salford, lacks the narra-tive drive and verve of Mayhew, it does at least offer a structural explanation for how the class-divided city arose and why its effects are so pernicious.

Henry Mayhew died in 1887 a forgotten figure whose work was largely neglected until the 1940s. But today we can appreciate his London studies as 'the fullest and most vivid picture of the experiences of labouring people in the world's greatest city in the nineteenth century' (Neuburg in Mayhew, 1985: xix). Mayhew's reports from the slums also revealed, as Barth reminds us, that 'the biggest press story of the nineteenth century is life in the big city itself' (Barth, 1980: 59 in Lindner, 1996: 9).

The Bitter Cry of Outcast London

However, it was not a journalist but two Congrega-tionalist clergymen by the names of Andrew Mearns and William Carnal Preston who were to finally shake the consciences of the Victorian middle class.[10] Using a similar documentary style to the Anglican preacher Thomas Beames who produced a vivid portrait of the 'Rookery' of St Giles in 1850 (Beames, 1970) and to the more contemporary report of life in a London workhouse by George Sims, *The Bitter Cry of Outcast London* was originally published as a penny pamphlet under the auspices of the London Congregational Union. However, it was only when the pamphlet was excerpted in W.T. Stead's *Pall Mall Gazette* and in the *Daily News* that its sensational accounts of urban deprivation echoed 'from one end of England to the other' (Briggs and Macartney, 1984: 2). After several pages expounding on the poor's worrying aver-sion to church attendance, the reader is invited to follow the narrator's descent into 'the pestilential rookeries' of London where

> tens of thousands are crowded together amidst horrors which call to mind what we have heard of the middle passage of the slave ship. To get to them you have to penetrate courts reeking with poisonous and mal-odorous gases arising from accumulations of sewage and refuse scattered in all directions and often flowing beneath your feet; courts, many of which the sun never penetrates, which are never visited by a breath of fresh air, and which rarely know the virtues of a drop of cleansing water.
>
> (Mearns in Hill, 1970: 6)

Yet, it is the descriptions of the conditions in which the slum-dwellers themselves were forced to live that still shock even after more than a century:

> In another [room] a missionary found a man ill with small-pox, his wife just recovering from her eighth confinement, and the children running about half naked and covered with filth. Here are seven people living in one underground kitchen, and a little child lying dead in the same room. Elsewhere is a poor widow, her three children, and a child who has been dead thirteen days. Her husband, who was a cab driver, had shortly before committed suicide.
>
> (ibid.: 7)

However terrible it might have been, poverty was no barrier to paradise, whereas to the minds of contemporary Christian reformers the contagion of vice was far more dangerous, for it brought eternal damnation. There was plenty in *The Bitter Cry* to trouble the Salvationists, such as the mothers who turned their young children onto the street until the early hours of the morning in order to rent the room for prostitution. While adult siblings were commonly obliged to share the same bed as their parents so that

> Incest is common; and no form of vice and sensuality causes surprise or attracts attention . . . The vilest practices are looked upon with the most matter-of-fact indifference . . . In one street are 35 houses, 32 of which are known to be brothels. In another district are 43 of these houses, and 428 fallen women and girls, many of them not more than 12 years of age.
>
> (ibid.: 9–10)

The authors of *The Bitter Cry* neglect to mention the moral turpitude of the tens of thousands of 'respectable' clients who abused children in this way, but the pamphlet was not intended to reveal the hypocrisy of Victorian society, so much as galvanise it to action, if for no more self-interested reason than to abate the spiralling rise in urban crime, communicable diseases, drunkenness and disorder that threatened to trouble 'the respectable city'. The *Bitter Cry* certainly struck a chord and Mearns' insistence that 'without State interference nothing can be performed upon any large scale' (Mearns in Hall, 1988: 18–19) was even grudgingly accepted by the conservative *Times* whose editor conceded that '*laissez-faire* is practically abandoned and that every piece of state interference will pave the way for another' (cited in Wohl, 1977: 234 by Hall, 1988: 19).

With such an endorsement of the need for change, the way could be paved in 1885 for a Royal Commission on the Poor, chaired by the Liberal MP Charles Dilke, which included such luminaries among its members as the Prince of Wales, the Conservative Party leader Lord Salisbury, and Cardinal Manning. Lord Shaftesbury, the anti-slavery campaigner, was one of the Commission's most convincing expert witnesses. It was not difficult to amass considerable corroboration for Mearns' account –

although the Commission was reassured to learn that the incidence of sexual vice may not have been as high as was first thought (Hall, 1988: 19) – and the 'something must be done' chorus did result in some ad hoc measures to encourage and cajole local authorities into pulling down slums and building working men's dwellings with decent amenities. But only small numbers of affluent working-class families were able to benefit from such reforms, while the lives of the vast mass of the urban poor went on in the same desperate fashion so that the Poor Law Commission of 1909 was unable to report a qualitative improvement in the plight of the underclass in over 70 years of 'reforms'.

How the other half lived in the American city: Jacob Riis and the New York slum

Meanwhile, on the other side of the Atlantic, the photo-journalist Jacob Riis was preparing the manuscript for his illustrated study of the slums of Manhattan. *How the Other Half Lives*, which was published in 1891 had the same shock impact on American society as the *Bitter Cry of Outcast London* had on the British educated public. Like his contemporary, the poet and novelist Stephen Crane, Riis was a crusader for social reform, and believed strongly in the powers of words and images to produce a new moral and ethical susceptibility among the affluent classes that would persuade them to meet their social obligations towards the less fortunate. At the same time, for all his ready stereotyping of different immigrant nationalities, unlike some of his English counterparts Riis was not prepared to moralise or deprecate the poor. He saw value in the street-style of 'the tough' and 'the gang' and even though he deplored slums, he respected the functioning chaos of popular architecture in much the same way that Jane Jacobs was to 70 years later (see Chapter 5) (Gandal, 1997: 9).

Although Riis' narrative style is often sensational, he tempers it by a careful attention to the broader social and economic conditions in which the people of 'the Bend' (the area around Mulberry Street and

Baxter Street in New York City where he conducted his investigations) find themselves. For example, death rates and infant mortality figures are used to demonstrate the stark effects of poor hygiene and overcrowding (see Figure 3.2),[11] and Riis – unlike his British contemporaries is prepared to point the finger of accusation at the wealthy, respectable families who make a comfortable living from the rents these poor migrants are forced to pay for their squalid quarters, and the city authorities that tolerate abuse of building and sanitation regulations (Riis, 1891: 64).

To modern ears, Riis' talk of engulfing 'colored tides', the 'dull gray . . . Jew', or the 'dirty stain' of the Arab tribe, and the 'unhesitating mendacity' of the poor immigrant seems like the worst type of prejudicial and lazy stereotyping.[12] But sociologists have always sought to make generalisations around class, race and gender, it is simply that today, for the most part, social scientists try to represent such differences positively. In his narrative map of the distribution of ethnic groups, and the processes of 'invasion' and 'succession', Riis anticipates the urban surveys of Charles Booth in London and Jane Addams, Edward

Burgess and Robert Park in Chicago, and gives as compelling an account of any immigrant quarter as the later participant-observation studies of the early Chicago School of Sociology (ibid.: 21–7).

Measuring and mapping urban inequalities: the surveys of Charles Booth

Booth was a wealthy Liverpool ship owner who had been moved to action by reading accounts of urban poverty such as *The Bitter Cry*, and he was certainly aware of the survey that H.M. Hyndman, founder of the British Social Democratic Federation, claimed to have conducted that purported to show that 1 in 4 of the population were living in conditions of extreme poverty.

However, although Booth met with a number of leading socialists (among them Hyndman), he did not appear convinced by the remedies advanced by socialism (O'Day and Englander, 1993: 30–1). Instead, Booth described his object simply 'to show the numerical relation which poverty, misery and

Figure 3.2
'Five Cents a Spot'. A lodging house in a New York immigrant quarter. Photographed by Jacob Riis c.1898.
Copyright and courtesy of the Jacob A. Riis Collection, Museum of the City of New York

EXHIBIT 3.2   Profile of Charles Booth

Charles Booth (1840–1916) was born the son of a Liverpool corn merchant. At the age of 22 he joined his brother Alfred's steamship company and Charles went on to become Chairman of the Booth Steamship Company, a post that he retained until 1912. Despite his lack of a university education, Booth maintained a close interest in the newly established discipline of sociology, and particularly in the ideas of its founders Auguste Comte and Henry Spencer (the latter was married to one of Booth's cousins). Although sympathetic to the early manifestations of trade unionism, Booth was wary of socialism largely because he felt that profit and loss was the best measure of a successful enterprise.

Given his parents' involvement in the Unitarian Church it was not surprising that Booth showed an early concern with the conditions of working men, but instead of adopting the moral crusading tone of social reformers such as Mayhew and Mearns, Booth preferred instead to reveal the true extent of poverty in England through the use of official statistics and survey data. He began a preliminary study, *The Tower Hamlets*, on poverty in East London in 1887. The next sixteen years Booth devoted to the publication of the monumental *Life and Labour of the People in London*.

Booth generally avoided prescriptive statements, and this is why he is best described as a social investigator rather than a social reformer. Although he did advocate the adoption of a national system of old age pensions, the formula for which was published in his *Poor Law Reform* and was based on his evidence to the Poor Law Commission (1905–9). Booth can claim a decent share of the credit for the passing of the Old Age Pension Act in 1908, which heralded the birth of the Welfare State in Britain. For his pioneering work on social statistics Booth was elected President of the Royal Statistical Society in 1892, he was fellow of the Royal Society in 1899 and became a privy councillor in 1904.

depravity bear to regular earnings and comparative comfort, and to describe the general conditions under which each class live' (cited in Smith, 1930: 2).

Booth resolved not to confront the problem of social inequality through philanthropy, but to make the case for reform by collating and classifying social 'facts'. The result was the seventeen-volume study, *Life and Labour of the People in London*, the first volume of which was published in 1892 and the last in 1903, which depicted a socially, economically and culturally segregated city in graphic and statistical detail. Though based on the classifications used for the 1891 census, Booth's social categories have been considered a forerunner of modern methods of sociological analysis, and although these methods would be judged deficient by contemporary standards, for

Figure 3.3 Charles Booth in his study, 1902.
Copyright Mrs Norman-Butler, courtesy of the University of London Library, MS 797 II/96/2

their time they represented a great advance in quantitative research techniques (O'Day and Englander, 1993: 4–5).

The volumes were arranged in three series, the first dealing with 'Poverty', the second with 'Industry', and the third 'Religious Influences'. The final volume contained a conclusion and summary of the work as a whole. Every series dealt with the geographical distribution of the theme under consideration, and descriptive sketch maps, as well as detailed survey maps were provided to guide the reader through the maze of statistical detail. Some series, such as the one devoted to poverty also included volumes

on special subjects, such as children or the Jewish community.

Like many Victorian moralists, Booth held to the belief that the history and character of certain localities reproduced patterns of social behaviour through the generations – an ecological presupposition that, as we shall see, was consciously adopted by the Chicago School of urban studies. Booth quotes approvingly from an account of Southwark dating from 1673 that described the borough as 'a settlement for the vilest refuse of humanity', in which, '[d]eeds of darkness were committed with terrible frequency', and '[v]iolence and murder were no

Figure 3.4   Part of Charles Booth's descriptive map of London poverty (1897) showing the North Eastern District (here featuring part of Bethnal Green and Haggerston to the north and east of the City of London). The darker areas represent the 'Lowest Class (Vicious Semi-criminal)', and 'the Very Poor', other lighter shades refer to 'the Poor' (pale blue in the original), 'a Mixture of Classes Some Comfortable Others Poor' (pale pink in the original), 'the Comfortable Working Class (pink in the original)', 'the Middle Class or Well-to-do' (red in the original) and 'the Upper Middle Class or Wealthy' (orange in the original). The main thoroughfares into the City of London tended to be fronted by large townhouses and their owners would have employed several live in servants (shaded dark grey on this map). The smaller streets and terraces were home to some of the poorest families in the capital. Bethnal Green and Whitechapel still include some of the poorest enumeration districts in the country.

uncommon events', and where 'The names of the sur-rounding localities – "Labour-in-vain Alley", "Hang-man's Acre", "Dirty Lane" and "Revels Row" testify to ancient ill-repute'. Booth then notes that, 'Some shadows of this past still rest upon the inner ring of London south of the Thames' (Booth (Series 3, Volume 4), 1902: 4).

Booth's survey produced dozens of neatly colour-coded maps (see Figure 3.4) that remain an invalu-able record of the socio-economic structure of London street-by-street, at a time when it was not just a national capital, but the centre of one of the world's richest and most extensive empires. Booth was there-by able to supplement the reportage of Mayhew and Mearns with the hard fact that in the East End, the poor numbered 314,000 or 35 per cent of the city's population, which meant that over one million Londoners lived on or below the bread line.

Because the poor were so numerous and internally differentiated Booth decided to group them into four categories: Class A who were a very small group (1.25 per cent of the population) of mostly young loafers, semi-criminals, street sellers, street performers, etc. who Booth considered to be at the extreme margins of society and prone to every kind of vice. They were a class that could 'render no useful service . . . and . . . degrade whatever they touch'. Booth's hope was that some way could be found to prevent the gener-ational reproduction of these 'barbarians' echoing a eugenicist view that was to become increasingly fashionable among scientific circles in the ensuing decades. The members of Class B were referred to as 'those in chronic want' and numbered some 100,000 in the East End and perhaps as many as 300,000 in London as a whole. They were described as idle, pleasure-seekers, who immediately spent what little money they came by, and were therefore 'always poor'. This large class was predominantly female (widows, unmarried mothers, young persons and children) and he believed that the best solution was to remove them from the city where they were an infinite burden on the state to colonies in the countryside or abroad where they could undertake work that would provide for their subsistence. Class C were the casual workers who made up 74,000 of the population of the East

End and 250,000 of the London population as a whole, or 8 per cent. They were prone to the most ruthless effects of competition and were often without employment or forced into bare subsistence work. The final category, Class D, included all those who had regular but poorly paid work, amounting to 129,000 East Enders or 14.5 per cent of the general London population. They struggled to make ends meet, but could expect no improvement in the fortunes of their families save by the possibility of their children finding better employment than their parents (Hall, 1988: 30).

Booth was not, however, a naïve supporter of the claim that 'the facts speak for themselves', instead he believed that out of inductive research should emerge a framework that would allow the investigator to reach conclusions based on the relationships between facts (O'Day and Englander, 1993: 35). He was also a firm believer in the utility of the comparative method, which he used extensively as a tool for highlighting the distinct characteristics of each topographical area. It is, therefore, no accident that the authors of the *New Survey of London Life and Labour* who set out to detail forty years' of social and economic change in London since the publication of the first survey should have used as their frontispiece a quotation from the final volume of Booth's original study in which he states that, 'Comparisons with the past are absolutely necessary to the comprehension of all that exists to-day; without them we cannot pene-trate to the heart of things' (Booth [Final Volume] 1902: 31 in Smith, 1930: vi).

In his explorations of the East End and his encoun-ters with its inhabitants, Booth has also been described as a practitioner of 'the arts of participant observation and reporting' (Simey and Simey, 1960: 65 in Pfautz, 1967: 17), and although the narrative account of Booth's survey is mostly subordinated to statistical and cartographic exposition, it was this qualitative method that most impressed the 'pro-fessional' sociologists who came after him.[13] It is also significant that the co-founder of the London School of Economics and leading Fabian socialist, Beatrice Potter (who later took the surname of her husband Sidney Webb), began her apprenticeship as a social investigator and social reformer under Booth's direction.

However, the vital connection between the new developments in social research and experiments in social reform in London in the 1880s and the emerging discipline of sociology in Chicago in the 1890s was established by another self-made social scientist by the name of Jane Addams, and it is her pioneering work and research on the urban migrant and the urban poor in the US that really initiates the systematic, empirical study of the modern American city (Deegan, 1988).

Jane Addams and the settlement movement legacy: civic empowerment from below

Jane Addams was born in 1860 in Cedarville, Illinois, the daughter of a wealthy mill-owner and state senator. As a young woman she attended the prestigious Rockford Female Seminary, where instead of following a typical vocation for school teaching or nursing, she developed a passionate interest in social issues and the practical measures that might be taken to alleviate the condition of the poor. Her father's death in the year of her graduation decided Jane to take stock of her life and widen her horizons by travelling to Europe with her close friend Ellen Gates Starr. In London she followed-up her interest in social work by visiting the Toynbee Hall settlement established by Canon Samuel Barnett in 1884. This Christian foundation, which Barnett resolutely refused to call 'a mission', recruited male students and graduates from the University of Oxford to live and work in the settlement house alongside the resident population.

Addams was impressed by Toynbee Hall's practical intervention in one of London's most deprived neighbourhoods and its emphasis on teaching domestic and practical skills alongside more academic subjects. Although she disagreed with the rather monastic and exclusively male character of the volunteers, she certainly saw the potential benefits for those providing support as much as for those in receipt of help, and was keen to replicate these key elements

in her own settlement project, Hull-House (see Figure 3.5) (Addams, 1967: 89; Meacham, 1987: ix; Deegan, 1988: 5; Dililberto, 1999: 131, Polikoff, 1999: 55).

In a prefatory note to her pioneering survey of the neighbourhoods surrounding the Hull-House settlement, Jane Addams wrote:

> The word 'settlement' is fast becoming familiar to the American public, although the first settlement, Toynbee Hall, in East London, was established so late as 1885. Canon Barnett, the founder urged as the primal ideal that a group of University men should reside in the poorer quarter of London for the sake of influencing the people there toward better local government and a wider social and intellectual life.
>
> Since 1889 more than twenty settlements have been established in America. Some of these are associated with various institutional features, but the original idea of a group of 'residents' must always remain the essential factor.
>
> (Addams, 1895: vii)

The purpose of Hull-House was set out in a pamphlet originally published in 1893 and revised as an appendix for the 1895 Hull-House Maps and Papers volume. In it we read that:

> This centre or 'settlement' to be effective, must contain an element of permanency, so that the neighbourhood may feel that the interest and fortunes of the residents are identical with their own. The settlement must have an enthusiasm for the possibilities of its locality, and an ability to bring into it and develop from it those lines of thought and action which make for the 'higher life'.
>
> (ibid.: 213)

As well as college extension courses and summer schools (chiefly frequented by young women), Hull-House from its earliest years maintained a close association with the labour movement of the city, and Jane Addams had no difficulty in writing that '[i]t is now generally understood that Hull-House is "on the side of unions", (ibid.: 214) despite the settlement's reliance on the patronage of the wives of wealthy business owners and the bitter and often violent labour disputes that had rocked the city in these years. Here, the difference from the Toynbee Hall settlement was most apparent in the contrasting ambition of Canon Barnett's acolytes to wean the working-class away from the depredations of socialism and anarchism by exposure to the civilising culture of liberal Anglican-ism (Briggs and Macartney, 1984).

A second and highly significant antecedent was provided by Charles Booth's social survey of London. Although the contributors from Hull-House had, in Jane Addams' words 'been chiefly directed, not towards sociological investigation, but to construc-tive work', the Hull-House survey was more method-ologically ambitious, if less comprehensive than its London counterpart in geographical scope. The Hull-House survey included large-scale coloured maps to illustrate both household income levels and the distribution of population according to ethnicity and/or geographical origin (Addams, 1895: vii–viii), referring to no fewer than 15 categories ranging from 'English speaking (excluding Irish)', including all the major European nationalities, together with Russian, French Canadian, Chinese and 'Colored'.

In another first for the social survey, the wage maps as well as showing average weekly income per dwell-ing in increments of $5 also plotted the location of neighbourhood brothels. This information would have provided a convincing empirical demonstration of Thrasher and company's observation that deviant social activities tend to be located in marginal or 'interstitial' zones (see below). In the Hull-House map, for example, prostitution appears to be chiefly concentrated in the houses facing the rail freight depots along Clark and Plymouth Streets. Addams limits her remarks to noting that such dwellings 'are almost invariably occupied by American girls . . . the great majority' of whom 'come from the central eastern States'. While, '[t]here are many colored women among them, and in some houses the whites and blacks are mixed' (Addams, 1895: 23). There is none of the moralist sensationalism of Mearns in this account, but the reader is left won-dering what explanation lies behind the ethnic and geographic origin of these women.

Of the ten contributors to the volume all but two were female which, even by today's standards, is an exceptional demonstration of the strength of female scholarship that the Hull-House project was able to generate, let alone for the closing years of the nine-teenth century. The bulk of the investigative research

for the volume was conducted by Florence Kelley who had a privileged entrée into the sweated trades of Chicago by virtue of her position as State Inspector of Factories and Workshops for Illinois. Kelley was also a committed socialist who had corresponded with, and translated, Engels, and it was her associa-tion with Albion Small's mentor, Richard T. Ely who became the editor of the Library on Economics and Politics that offered a publishing opportunity for the Hull-House investigations (Deegan, 1988: 56).

In style and methodological approach, Kelley's account of the tailoring industry in Chicago is similar not only to that of Marx and Engels, as Deegan argues, but also to Henry Mayhew in the detailed descriptions she provides of the manufacturing process, the layout of the 'workshop', and the nature of the workforce employed in the trade. A further source of inspiration for Addams in the Toynbee Hall settlement was Canon Barnett's belief that little could be achieved without the institutional support of local representatives and the civic administra-tion. This was also a subject dear to the hearts of Sidney and Beatrice Webb, who while visiting the Colombian Exhibition of 1893 stayed at Hull-House and were impressed by the seriousness with which Jane Addams set about her task of making a differ-ence to the lives of the working people of Chicago (ibid.: 263–5).

Addams' battles with 'the ward boss' and the Town Hall machine paint a picture of a formidable polit-ical campaigner, but also serve to demonstrate the political nature of social deprivation – the deliberate denial or under-provision of municipal resources to neighbourhoods that lacked either the resources or the votes to make the powerful aldermen of Chicago sit up and listen (Addams, 1898 reprinted in Elshtain, 2002: 118–24; Platt, 2000).

Social and political analyses undertaken at Hull-House included studies of prohibition, immigration, education, juvenile crime and delinquency, the women's movement and the peace movement, all of which demonstrated the breadth and ambition of Addams' now world famous project (Addams, 1930). In her attention to the needs of children and youth, and in the new moral hazards created by the easy

Figure 3.5
Jane Addams with the
children of the Hull-House
settlement.
Copyright and courtesy of
Swarthmore College Peace
Collection

mobility of the city streets and the impersonal materialism of urban capitalism in works such as *The Spirit of Youth and the City Streets*, Addams seems close to Simmel in warning of the state of 'esthetic insensibility' into which urban youth can fall prey as their 'newly awakened' senses are corrupted by the 'gaudy and sensual' world of street music, theatre posters, pulp fiction, and the 'cheap heroics of the revolvers displayed in the pawn-shop windows'. What Benjamin and Baudelaire admired in the sensuality of the licit and illicit commerce of the city street, Addams sees as degenerating working men and women alike, whose over-stimulated sexual imaginations no longer have the capacity to awaken 'the imagination of the heart' (Addams, 1909: 27–8 and in Elshtain, 2002: 126). Nevertheless, Addams' prescription is sympathy and understanding rather than repression and restraint. She knew from the experiences of her female residents at Hull-House that the charms and, hence, the dangers of the city cannot be resisted – but they can be more safely managed – and by understanding how young people interact with their urban environment Addams was convinced that

juvenile protection agencies would have a better chance of preventing further blight to the lives of city youth and their communities. It was this sensitivity to the mutually influential relationship of the urban environment to human behaviour that established Jane Addams as the pioneer of urban sociology in Chicago. In the following section we see how this 'grass roots' tradition of urban investigation linked in with the academic advances in social research that were being made at the city's new university.

QUESTIONS TO DISCUSS

1 **How do the accounts of London poverty provided by Mayhew, Mearns and Booth compare to Jacob Riis' portrait of the slums of New York?**

2 **What did Jane Addams hope to achieve through the establishment of Hull-House in Chicago and in what ways did she contribute to the development of urban sociology?**

URBAN STUDIES AND THE CHICAGO SCHOOL OF SOCIOLOGY

If one had to point to an individual city as the birthplace of modern urban sociology, few would dissent from the nomination of Chicago, Illinois at the turn of the twentieth century.[14] No city offered a better laboratory for the observation of metropolitan life than this mid-western town on the shores of Lake Michigan, which grew from a mere 4,470 inhabitants in 1840 to 1.1 million in 1890 and 3.5 million in 1930 (Wax, 2000: 65). Chicago's economic success depended as much on the commercial calculations of railway companies, ship owners, commodity traders and real estate developers than any natural advantage of its location – and this is why researchers and novelists have found the city such a fertile terrain for investigating the human condition. When the University of Chicago first opened its doors in 1892, its new Department of Social Science and Anthropology[15] soon became the world's leading centre of social science research, while Chicago itself became a study resource for the production of a host of monographs and articles that have changed the course of sociology in general and urban studies in particular.

The use of the term 'Chicago School' to describe the group of researchers who worked in the discipline of Sociology at the University of Chicago from its beginnings up to the 1930s is a convenient shorthand rather than a definition of an 'epistemic community', or a group of individuals who share the same broad understanding and approach to the study of human society. In fact, the celebrated collaborations of Chicago sociologists, such as the edited collection produced by Park and Burgess in 1925 were comparatively rare events (Park et al., 1925), and the more prominent figures took very different approaches to their subject matter both in terms of methodology and research focus (Carey, 1975). As Lee Harvey points out, '[t]he notion of a "Chicago School of Sociology" was not an issue in 1911', while those that worked in the Department during its hey-day in the 1920s do not recall the term being used. Indeed, Howard Becker recalls Louis Wirth's

amazement at his inclusion in the Chicago School by outsiders, since he could not imagine what he had in common with his colleagues (Harvey, 1987: 6).

The Chicago researchers also brought different visions to the study of urban society. The Department's founder, Albion Small and his early collaborator C.R. Henderson, were both ordained Baptist Ministers. As well as serving as the University's chaplain, Henderson was the first to establish the subject of sociology at Chicago within the Divinity School where he taught before taking up a full professorship of Sociology in Small's department. Henderson maintained close links with Jane Addams' Hull-House and also with another leading figure in the Chicago social work movement, Graham Taylor. It was from Henderson, in particular, that Chicago Sociology developed an early interest and expertise in the study of crime and juvenile delinquency (Deegan, 1988: 18–19). Yet, a belief in the benefits of 'amelioration' was not exclusive to Christian reformers nor did this moral outlook preclude a belief in rigorous empirical research into social problems (Harvey, 1987: 25). On the other hand, Small was certainly shrewd enough to emphasise the 'improving' potential of the work his colleagues were engaged in to wealthy benefactors such as the Rockefeller family (Smith, 1988: 1).

All this is not to say that Chicago Sociology was intellectually parochial, since several Faculty members had studied and trained at German universities (including Park, Small, Henderson, Thomas and Zeublin) or had been exposed to continental philosophy and sociology at other American universities. Neither does the common subject matter of the City of Chicago make for a unified approach to its study. But what each of the studies that came out of the Department in the early decades of the twentieth century had in common was a commitment to 'action research' – or the direct collection of empirical evidence and data – allied with a belief in the pragmatic value of the human sciences, where knowledge production is closely tied to its practical applications (Venkatesh, 2001: 277).[16]

Reportage, human ecology and urban ethnography: the work of Park, Burgess and Wirth

America, and perhaps, the rest of the world, can be divided between two classes: those who reached the city and those who have not yet arrived.
(Robert Park in Matthews, 1977: 121)

Robert Park was brought to Chicago by W.I. Thomas whom Park had invited to speak at an 'International Conference on the Negro' in his capacity as secretary to Booker T. Washington, the celebrated African American educator and civil rights leader (Wax, 2000: 69). Like Thomas, Park had studied in Germany where he had been exposed to the new science of sociology, but it was his earlier training as a journalist that proved to be most significant for the approach Park was to develop in his studies of the city (Lindner, 1996). One of his most distinguished students, Nels Anderson, recalls how Park urged him to '[w]rite down only what you see, hear, and know, like a newspaper reporter' (Anderson, 1967: xii). This was why Park's admiration for Charles Booth's study of the London poor rested not so much on

Booth's statistics, but his realistic descriptions of the actual life of the occupational classes – the conditions under which they lived and labored, their passions, pastimes, domestic tragedies, and the life philosophies with which each class met the crises peculiar to it – which made these studies a memorable and permanent contribution to our knowledge of human nature and society . . .
(Park, 1929: 46 in Pfautz, 1967: 5)

In this vein, we can identify another connection between the urban types contained in Walter Benjamin's *Konvoluts* and the *dramatis personae* detailed in Park's 1915 essay on 'The City' where he refers to 'the shop-girl, the policeman, the cabman, the night watchman, the clairvoyant, the vaudeville performer, the quack doctor, the bartender, the ward boss, the strike breaker, the labor agitator, the school teacher, the reporter, the stock-broker, the pawn broker', 'all these' Park declared, 'are characteristic products of city life' (Park, 1915: 586 in Lindner, 1996: 77).

But if these European influences exist they are rarely made explicit in Park's work. For example, we know that Park had certainly read Marx, but he refers to the founder of scientific socialism only four times in his entire published work. Simmel fares better with seven acknowledgements, and Durkheim manages twelve mentions to a mere one for Weber. Significantly, those mainly nineteenth-century sociologists who subscribed to a biological or evolutionary theory of society such as Spencer, Comte and Darwin received more mentions than Marx, Simmel and Weber put together. However, none of these classical figures from the European sociological pantheon could compete with the space devoted to W.I. Thomas and W.G. Sumner – figures who loomed large in the foundation and direction of the Chicago School but whose influence on the other side of the Atlantic was relatively modest (Smith, 1988: 120). Thus, we could say that Park's urban sociology was very much 'home grown' though no less cosmopolitan in its scope and ambition than its European counterparts.

In his observations of the human and anthropological detail of city life, Park was ever conscious of the importance of the urban structure (the planned environment, the layout of buildings and houses, etc.) within which each social transaction was played out. Structure and action are explicit in the divisions of *The City* study (Park, 1915) which deals in turn with: '(i) the City Plan and Local Organization, (ii) Industrial Organization and the Moral Order (iii) Secondary Relations and Social Control and (iv) Temperament and the Urban Environment'. Here, each structural feature is juxtaposed with some aspect of the human character or human relations in an attempt to reinforce Park's view that the city is not just a physical construction but a human community upon which society's complex hierarchies and divisions are mapped (Lindner, 1996: 68). As Park writes in his preface to Nels Anderson's *The Hobo*: '[I]f it is true that man made the city it is quite as true that the city is now making man. That is certainly a part of what we mean we speak of the "urban" as contrasted with the "rural" mind' (Park in Anderson, 1967: xxiii) setting a decidedly Simmelian tone to a study which paradoxically focuses on the one figure, the transient labourer or tramp, who seems capable of successfully negotiating the rural–urban divide.

Like Simmel, Park makes much of the division of labour that results from the concentration of the money economy in the big city, but (contra Durkheim) he did not see the division of labour as the determinant of urban life, preferring instead to emphasise social control as the directive force of human society (Wax, 2000: 70). The emphasis Park places on secondary relationships (i.e. those beyond the immediate family) is a reaction to a social environment in which traditional family units are weakened or attenuated as a result of labour migration and poverty and where survival strategies very often depend on the mediation of third parties (ward bosses, gang leaders, labour officials, etc.). Social control is a recurring problem for city authorities in the face of this unstable and fractured social and economic environment where group loyalty and social integration are much less in evidence than in traditional or 'folk' societies and therefore the sanctions on anti-social behaviour (or anti-authoritarian behaviour) cannot be so easily imposed through group disavowal and, instead, have to be applied using surveillance and coercion.

The wealthy were no less prone to the uncertainties of this brave new world of mass consumption and mass production and, as Stedman-Jones points out, Park was not afraid to link the respectable anarchy of the stock market to the mob or what the Victorian elites referred to as the dangerous classes:

> It is true of exchanges, as it is of crowds, that the situation they represent is always critical, that is to say, the tensions are such that a slight cause may precipitate an enormous effect.
> (Park, 1925: 20 in Stedman-Jones, 1971: 13)

One of the functions of the modern metropolis was to impose and maintain a social order on this fissile world of modernity, and it achieved this objective, argued Park, by the same ecological methods as species achieve equilibrium patterns of distribution in nature. In the ecology of the city, wrote Park:

> There are forces at work within the limits of the urban community – within the limits of any natural area of human habitation, in fact – which tend to bring about an orderly and typical grouping of its population and institutions.
> (Park, 1967: 1 in Dickens, 1990: 34)

The idea of the 'natural area' is related to Ernest Burgess's zone theory in that it is the locale for a broadly similar range of activities and occupations (Guest, 1997: 8). The natural area thus provides a port in what Louis Wirth saw as the stormy sea of the *Gesellschaft* metropolis in offering a sense of community identity where similar confessions, ethnes, and status groups can find mutual recognition and acceptance. The model then of Park's natural area is 'the traditional village community', but because such areas are also prone to the forces of ecological succession (the invasion of the territory of a 'weaker' species by a 'stronger' rival), the moral order of the self-contained urban community, especially on the fringes of expanding business districts, is constantly under threat (ibid.: 9).

Thus, it is easy to see a rather crude Social Darwinism at play that suggests the 'fittest' economic actors seize the prime locations (big businesses and *rentiers* in the central city area, the wealthy middle class in the upwind and upriver, leafy suburbs), with the most socially disadvantaged relegated to the noisiest, most polluted and run down quarters of the city (Dickens, 1990: 34). Thus as competition for space increases, former slums adjacent to high-land value areas can be profitably developed for business or commercial use, while the very presence of large numbers of poor households often induces middle-class 'flight' to the relative safety of the suburbs.

This ecological notion of 'invasion' and 'succession', to which, as we shall see, Burgess can certainly lay joint claim, was much criticised for providing a one-dimensional account of an urban reality that did not correspond to this evolutionary model in many of the world's cities. But it is interesting to note that, despite a change in terminology, contemporary gentrification theorists such as Neil Smith continue to make extensive use of the notion of invasion and succession (Smith, 1996 and see Chapter 5). There are also parallels between Marxist accounts of capitalist urbanisation that aim to show how power and class differentials are mapped onto the city in terms of land-use, tenure and access to resources (Harvey, 1973, 1975 and see Chapter 6), and within the field of economic geography, the locational-spatial analysis known as central place theory (see Chapter 4).

If Park was the narrator of the ecological model of 'urban areas', Burgess was the scholar who literally put the theory on the map. But for all its 'naivety' (Rex, 1973), Burgess's concentric model of urban development represented an enduring contribution to urban sociology and urban geography, if only as a straw man against which potentially more robust theories could be tested. Historians of urban development such as Adna Weber ([1899] 1963) and Lewis Mumford (1924) had produced elaborate surveys of urbanisation and architectural form, but to date no attempt had been made to explain land-use variegation in the modern city. Burgess set out to remedy this deficiency by modelling the pattern of land-use in Chicago with its distinct central business district inside the elevated railway known as 'the Loop', spreading out in a series of concentric circles to include the wealthy residential neighbourhoods of the North Side, while further out to the south and west were industrial and manufacturing zones interspersed with low income housing clustered around ethno-linguistic settlements such as Greek Town, Little Italy, German Town and so forth (see Figure 3.6).

Venkatesh regards Burgess and company's attempts to uncover the human ecology of Chicago as 'a deliberate social construction of the city' (Venkatesh, 2001: 277), while the notion of 'natural areas' (operationalised as the 75 community areas of Chicago) that Burgess promulgated has been long enshrined in the community and neighbourhood politics of the City of Chicago (ibid.: 276). For Burgess's theory to work he needed to show that each 'natural area' was a homogeneous self-aware community that saw itself as quite distinct from its neighbours. Through externally supported research bodies such as the Local Communities Research Committee (LCRC) that Burgess and Park co-founded in 1923 along with colleagues

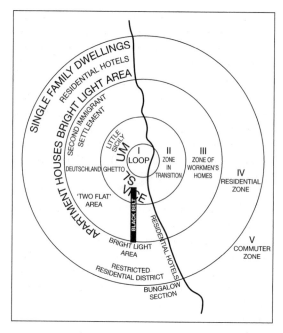

Figure 3.6 'Urban Areas' showing Burgess's concentric zone view of 1920s Chicago.
Source: Burgess in Park et al. (1925)
Copyright and courtesy of the University of Chicago Press

from the Economics and Political Science Departments, Burgess was able to undertake detailed studies of Chicago neighbourhoods on a city-wide scale. Such work had more than an academic purpose, however, because the work of the LCRC demonstrated the practical commitment of Chicago social sciences to foster good community relations and to demonstrate the practical worth of their disciplines (ibid.: 281).

Burgess's descriptors of urban growth constantly evoke biological images and metaphors in much the same way as Lewis Mumford does on a far broader canvas (Mumford, 1938) (and see Chapter 8). For example, Burgess talks of organisation and disorganisation in the city as being 'analogous to the anabolic and katabolic processes of metabolism in the body' (Burgess in Park et al., 1925: 53). He sees 'excesses' of population – i.e. of immigrant black workers from the Southern US, or a higher proportion of one sex to another as 'symptomatic of abnormalities in social metabolism' (ibid.: 54). Thus, it is hard to escape the conclusion that Burgess has an idea of how a normal, well functioning city 'organism' should be, even if he concedes that 'disorganization' is not pathological but the typical pattern of city growth, at least in its early phases. Like Jacob Riis, Burgess finds energy and creativity in the decadence of the slum, but falls into the same stereotypical assessments of national character by arguing that 'racial temperament or circumstance' produces the 'Irish policemen, Greek ice-cream parlors, Chinese laundries, Negro porters, Belgian janitors, etc.' (ibid.: 57).

Burgess acknowledges a debt to Max Weber in recognising the multitude of trades and services the city provides, emphasising above all, the city as a place of economic opportunity and exchange. Burgess takes Weber's notion of the separation and classification of the diverse occupational and status elements of the city and gives it geographical expression. But Burgess is also alive to the dynamic and fluid character of such distribution by emphasising the *mobility* of the urban population – both in and out of the city and within its 'natural areas'. He agrees with Simmel that the psychological impact on the individual of this frequent mobility 'tends inevitably to confuse and demoralize the person'. There is also a loss of contact and affinity with other members of the primary group (family and kinship networks) and a greater reliance on more impersonal secondary group relations (work colleagues, landlords, law enforcement agents). Higher mobility induces a reduction in moral regulation and Burgess believed that it is no surprise that 'areas of mobility are also the regions in which are found juvenile delinquency, boys' gangs, crime, poverty, wife desertion, divorce, abandoned infants, vice' (ibid.: 59).

This thesis is superficially attractive, but there is no real attempt to consider counter-variables. Could it simply be that poverty forces individuals to move at frequent intervals, to lose contact with family and friends, and to engage in crime and vice? Also, what of the mobile wealthy who move from city to city in search of ever more lucrative jobs or investment opportunities, why do they not seem prone to the demoralizsation of the eternal wanderer? And what of the mobile poor who, against all odds, manage to keep their families together? Burgess is not unaware of mitigating factors against urban disorganisation because in his study of the Jewish neighbourhoods of 'the Ghetto' and Lawndale he finds the 'high rate of mobility in social and personal disorganization is counteracted in large measure by the efficient communal organization of the Jewish community' (ibid.: 62). But he fails to draw the obvious conclusion that cultural norms and values can override environmental determinants *sui generis*. This lack of sensitivity to non-spatial determinants of social action and behaviour is now recognised as 'the ecological fallacy', but despite the flaws in Burgess' account of natural areas, the introduction of spatial categories into the conceptual vocabulary of the social sciences has been profound and enduring.

For Wirth, the city could be defined sociologically as: 'a relatively large, dense and permanent settlement of heterogeneous individuals' (Wirth, 1938: 1). But it was the dimension, density and heterogeneity of the urban environment that differentiated the experience of city life from any other. Wirth regarded the immense scale of the urban landscape as the determinant aspect of this environment, arguing that, 'the bigger the city, the wider the spectrum of individual variation and also the greater its social differentiation' (Wirth, ibid.) Such differentiation, he believed, would lead to the loosening of community ties and their systematic replacement by mechanisms of formal and social control. This representation of the social structure of the city conforms closely to the Durkheimian critique of industrial capitalism, which viewed increasing social control and a rise in 'anomie' as inevitable corollaries of rapid urbanisation (Durkheim, 1947). To this, Wirth added other psychological features of the urban personality that included anonymity, superficiality, the transitory character of urban social relations and a lack of participation.

There is clearly a moral prospectus behind such a theorisation that refuses to accept that urban culture can be anything but divisive, fragmentary and corrosive of 'organic identity'. Similarly, cities are considered to be ahistorical entities, lacking a tradition or specificity, they are undifferentiated within and between states, and even internally, variation is only occupational or ethnic while the *pattern* of differentiation is consistent from sector to sector. Taking up Durkheim's theme once again, Wirth claims that such a form of differentiation is in keeping with the diversification of a market economy and a political life based on mass movements. However, unlike European sociologists such as Durkheim and Tönnies, Wirth did believe that cultural heterogeneity could lead to cultural integration within cities but that this required a 'growing self-consciousness of society wide common interests' (Smith, 1988: 5) and also a strongly interventionist role for public authorities in channelling and encouraging civic life.

This classic formulation of urban sociology with its emphasis on the 'pathological' symptoms of urban life (an obsession that gave rise to the famous deviancy studies of anthropological sociologists such as W.F. Whyte (1955)) provided the theoretical resources for a series of 'evolutionist' interpretations of social history that attempted to isolate urban from rural culture and draw out what Redfield termed the 'folk-urban continuum' (Redfield, 1947). Redfield's work was the first systematic attempt to posit the notion of an urban 'sub-culture', even if the urban 'pole' of the continuum was simply defined in terms of a rural anti-type (Miner, 1952).

The importance of ethnography for Chicago sociology related both to the idea that the modern city was as much a puzzle as any 'undiscovered' tribal society, but also that the methods used by anthropologists might usefully be applied to the study of human societies closer to home. It was thanks to the example of Park and Redfield that the field interview and participant-observation research became the most important tools for the generation of sociologists who were to emerge from Chicago in the 1920s and 1930s.

The Chicago field study and urban ethnography

The European influence on Chicago Sociology, derived from encounters with the work of Simmel, Durkheim and Weber, placed stress on the impersonal, anomic dimension of city life. But as the growing sociology faculty at Chicago began to undertake field studies of the surrounding metropolis a rather different picture emerged. Using ethnographic techniques of participant-observation in the 1920s and 1930s, Chicago Sociologists made the study of the discrete urban locale, the urban type (to use Park's definition), or the 'bounded community' its institutional trademark. Chicago was a perfect testing ground for this type of urban investigation, for as the Irish-American writer James T. Farrell wrote:

> the neighbourhoods of Chicago in which I grew up possess something of the character of a small town. They were little worlds of their own. Many of the people living in them knew one another. There was a certain amount of gossip of the character that one finds in small towns. One of the largest nationality and religious groups in these neighbourhoods was Irish-American and Catholic. I attended a parochial school. Through the school and Sunday mass, the life of these neighbourhoods was rendered somewhat more cohesive. My grandmother was always a neighbourhood character, well known. I became known, too, the way a boy would be in a small town.
>
> (Farrell, 1993: xi)

The 'golden age' of the Chicago field study saw the publication of works by Anderson ([1923] 1967), Thrasher (1927), Wirth (1928), Zorbaugh (1929), Cressey (1932), Young (1932), Reckless (1933), and Faris and Dunham (1939). Each of them featured the use of participant-observation in whole or part, and this according to Cavan (1983 cited in Harvey, 1987: 55) was 'the result of Park's interest and influence' and, in particular, his famous injunction to his students to get 'your hands dirty with real research' (Lofland, 1971: 2).

'Participant-observation', as the technique of attaching oneself to the field study for a prolonged period became known, was a very experimental technique for sociologists in the 1920s (although it had been the standard research technique of the anthropologist for many years). The roving working-class labourer, Nels Anderson, was not so much adopting a subject as trying to escape it, but lacking a better alternative he chose to study the twilight world of the itinerants of Chicago's Hobohemia (Anderson,

1967: xiii). His professors soon realised that what Anderson may have lacked in formal sociological training he more than compensated for in terms of the wealth of empirical information he was able to directly access by virtue of his connection with the street.

The resulting account reads rather more like one of George Orwell's documentary essays or a Jack London novel than the 'dispassionate' surveys of Booth, or the detached *flânerie* of Benjamin, and *The Hobo* certainly helped establish the reportage style by which later studies could be identified. Anderson's findings were intended for the Chicago Council of Social Agencies and concentrate on the typologies of homelessness and itinerancy and its concentration in the City of Chicago, and a number of recommendations are made suggesting ways in which care agencies might intervene more successfully in alleviating hardship on a national and local scale. At the same time, Anderson works with the assumptions of Burgess's natural succession model to highlight why 'Hobohemia' is so concentrated in 'an isolated cultural area' around the Madison Street 'main stem', close to the centre of transportation trade – and therefore ideal for casual work and begging – but in an area of low rent and high land value, the 'gray area' in between retreating residential districts and advancing commercial and business development (Anderson, 1967: 14).

Thrasher's study of over 1,300 Chicago street gangs had few antecedents in the social science literature, and with the exception of Nicholas Spykman's 'The Social Theory of Georg Simmel' there was no reference to European sociology in the bibliography. However, Thrasher makes several references to the work of Jane Addams as well as to Charles Booth's 'Life and Labor of the People in London' [*sic*] (from the Macmillan, 1902 edition) confirming the importance of the 'non academic' social researcher for the development of academic sociology in Chicago. Just as Anderson was able to plot in meticulous detail the marginal geography of Hobohemia, Thrasher demonstrated the same 'spatial fix' for Chicago's gangland, which 'represents a geographically and socially interstitial area in the city' (Thrasher, 1927: 22).

Another example of the ecological approach developed by Burgess is John Landesco's contribution to the *Illinois Crime Survey* of 1929, which was subsequently published in its own right as *Organized Crime in Chicago* by the University of Chicago Press in 1968. As Mark Haller writes in his introduction, 'Burgess . . . served as a friend and mentor to Landesco', and using Burgess's concept of 'natural areas' deviant behaviour could be seen 'not as a deviant off-shoot of dominant patterns but rather as the normal pattern resulting from the social life of certain areas of the city' (Haller in Landesco, 1968: x). Or put more straightforwardly, as Landesco writes, 'The gangster is a product of his surroundings in the same way in which the good citizen is a product of his environment' (ibid.: xi). More detailed case studies were produced under the aegis of the Institute for Juvenile Research and the Behavior Research Fund that relied heavily on self-reporting of life-histories such as Clifford Shaw's *The Jack-Roller: A Delinquent Boy's Own Story* (1930) and Shaw *et al.*'s *Brothers in Crime* (1938).

Burgess also contributed a chapter to Shaw *et al.*'s 1938 study in which he attempted to unpick the Gordian knot tying upbringing and surroundings to personality in the make-up of the individual. His conclusions were straightforward and unqualified – 'The entrance and progress of each brother in a delinquent career appears to be almost a direct outcome of the residence of a poverty-stricken immigrant family in a neighbourhood of boys' gangs and criminal traditions' (Burgess in Shaw *et al.*, 1938: 326). Although the author was willing to concede that '[t]his does not mean that in all cases of crime social influences predominate . . . the vast majority of cases of delinquency and crime in American cities is due to social influences' (ibid.).

Yet, for all Burgess's confident advocacy of the ecological-interactionist approach to the study of human behaviour, Louis Wirth felt it necessary to assert that:

> In the rich literature on the city we look in vain for a theory of urbanism presenting in a systematic fashion the available knowledge concerning the city as a social entity.
>
> (Wirth, 1964: 67 in Smith, 1988: 137)

In studying the particular aspects of urban life, did the Chicago sociologists ignore the universal qualities of the urban experience that more abstract theorists such as Simmel and Weber attempted to convey? Not necessarily. In Burgess's introduction to Cressey's *The Taxi-Dance Hall* we read that the study 'raises all the main questions of the problem of recreation under conditions of modern city life, namely, the insistent human *demand for stimulation*, the growth of *commercialised recreation*, the growing *tendency to promiscuity* in the relations of the sexes, and the failure of our ordinary devices of social control to function in a culturally heterogeneous and anonymous society (Burgess in Cressey, 1932: xiii, original emphasis). Burgess's references to the stimulation of urban life and the commercialisation of pleasure have a decidedly Simmelian ring to them. The taxi-dance hall itself is, thus, the perfect expression of the modern metropolis which both produces the problem of estrangement from 'normal' affective relationships and provides a remedy in the form of the 'dime a dance' ballrooms where immediate intimacy can be had – at a price.

Of all the case studies that appeared during the 'golden age' of Chicago sociology, Thomas and Znaniecki's five-volume study, *The Polish Peasant in Europe and America* (Thomas and Znaniecki, [1927] 1984) stands out as a work of landmark importance. Enjoying the generous backing of a local benefactor, William Thomas was able to broaden his study of the Polish community in Chicago to its origins in Europe and, with the help of his collaborator, the Polish philosopher Florian Znaniecki, the authors were able to develop a knowledge of the nature and habits of this community that was innovatory in terms of its comparative approach, and in the use of life histories, personal letters and newspaper reports as documentary sources.

From this major study, Thomas and Znaniecki argued that 'the subject matter of social theory must be acting individuals or groups and not class determinants, codes and structures, statistical quantities, or other abstracted "objective factors"' (Zaretsky in Thomas and Znaniecki, 1984: 3). Instead they proposed a method for observing social processes that

dealt with outside factors, or 'values' such as group rules that provided instructions for how the individual should act in any given situation, together with the 'attitudes', predispositions or inclinations that were specific to that individual. The intention was to put forward a rival theory of structure and agency to that developed by Talcott Parsons and his followers, but as a social theory at least, Thomas and Znaniecki's ideas found little favour in the wider academy – that is at least until the arrival of the French social theorist, Pierre Bourdieu, whose concepts of symbolic capital, field and habitus bear an uncanny resemblance to Thomas and Znaniecki's early work (see Chapter 8).

Although the history of the Chicago School continues to generate misunderstandings about its collaborative and intellectual coherence (Harvey, 1987: 115–17), there is no doubt that the legacy of Chicago sociology continues to attract interest from contemporary urban analysts, a significant proportion of whom are writing in languages other than English (Abbott, 1999: 4, 22). For example, the concentric ring model of urban development has been given new currency by a recent study of the post-war development of Lodz in Poland (Obraniak, 1997), while an increasing number of sociologists (Wax, 2000; Lal, 1990) are rediscovering the Chicago School's importance for the study of ethnic and immigrant culture after its rather disparaging treatment by earlier structuralist Marxism (Castells, 1968, 1977 and see Chapter 6).

Chicago and London

Lacking the abstract traditions of continental social theory, but with its feet firmly planted in the soil of public service and social reform, British urban studies continued its Fabian trajectory of social investigation by drawing increasingly on the methods pioneered by the early Chicago School of the 1920s and 1930s. Just as Chicago became synonymous with urban studies in the US so, writes Bulmer, '[t]he London School of Economics was one of the few centres of concentrated social science research in Europe that could in any way rival Chicago or Columbia.' However:

Social research was not institutionally established in the way it was at Chicago. Various initiatives by Beveridge, such as Lancelot Hogben's work in social biology and the *New Survey of Life and Labour in London* (a replication of Booth), were intended to promote the empirical study of society, but were flawed by a failure to connect empirical data with a significant body of theoretical ideas, as Park was able to do. Only Malinowski's developing school of anthropology was in any way comparable to the achievement of Chicago.

(Bulmer, 1984: 210–11)

This judgement on the British urban research tradition, while accurate, needs to be seen within its historical context. The London School of Economics and Political Science, from its earliest foundation, attempted to bring the world of policy-making or 'public service' and academics into dialogue and, in attracting the support of Edwardian polymaths such as Patrick Geddes, the School established a reputation for 'applied sociology' with a focus on how civic life and the life of cities could be improved for the benefit of all (Dahrendorf, 1995: 101). There was also considerable and early cross-fertilisation between the Fabian founders of the LSE and the Chicago Sociology Department. The Webbs visited the University of Chicago in 1898 (staying once again at its unofficial annexe, Hull-House) and met Zeublin (whom Beatrice Webb found unimpressive) along with several other members of the department. These meetings seemed to confirm Beatrice Webb's view that, as Mary Jo Deegan puts it, the University of Chicago was 'a place of learning without form' (Deegan, 1988: 265).

Hull-House on the other hand seemed more in keeping with the Fabian mission of the London School of Economics, and Jane Addams visited the leading Fabian sociologist Graham Wallas in 1919 which helped prepare the way for the Hull-House sociologist Edith Abbott to study at the LSE for a year and to work directly with the Webbs and their colleagues (ibid.: 266). Although Chicago Sociology was later to become paradigmatic for British urban researchers, it should not be forgotten that Addams, Zeublin and even Henderson were influenced by the Fabian marriage of civics with rigorous empirical social research (ibid.). Typical of this position was the

LSE Professor, T.H. Marshall, who pioneered the entire field of social policy, and saw the social sciences as providing the methodological apparatus for tackling and understanding the great social issues of the day. As we shall see in Chapter 5 the close affinities between American and British urbanists continued to be demonstrated in a series of landmark studies published after the Second World War.

QUESTIONS TO DISCUSS

1 **What do you understand by the term 'urban ecology'?**

2 **What affinities can you identify between the work of Robert Park, Ernest Burgess and Louis Wirth and the European urban theorists encountered in the previous chapter?**

3 **What contribution, if any, do you think Chicago sociology has made to urban theory in general?**

CONCLUSION

We have identified essentially two main threads in this broad loom of transatlantic empirical urban studies. The first is the social reformist agenda that studies the city in order to change it for the better. The second is represented by a humanistic concern with 'social facts' that are, nevertheless, the product of a unique social environment – the city.

Many writers and activists came to their 'urban vocation' as a result of encounters with the city's darker side – the religiously inclined believed that the battle was with vice and sin, others, such as liberals and progressives, chose to emphasise the perils of laissez-faire, the greed of the wealthy, and the indifference of government to its own 'huddled masses'. But be they missionaries or Marxists, all were agreed that publicising and disseminating knowledge about the condition of the urban poor was the chief weapon in any improvement campaign. Newspaper articles, pamphlets and books began to proliferate in the 1880s and 1890s, so that by the new century no educated person, even if she or he read only the fiction of Dickens, Zola or Dreiser, could claim that they were ignorant of the plight of the urban poor.

As we have seen, there were several crossing-points between these two positions. Booth might not have been a socialist, but he knew, as did his Fabian collaborator Beatrice Webb, that reformers would make great capital of the voluminous data he and his team had collected. Jane Addams was a social reformer first and foremost, but she would have been just another Chicago charity patroness if it was not for her commitment to social investigation and to measurable improvements in the lives of the people served by the Hull-House settlement.

This takes us back to the story of academic sociology and Chicago. There can be little dispute that in its early years Chicago sociologists believed that they were producing research that had not just scientific validity, but potentially a social benefit. This may have been a cynical hustle by faculty researchers keen to tap the social conscience of Chicago's affluent philanthropists, but Henderson, Small, Burgess, Park and company genuinely seemed to believe that improved awareness of the urban condition would lead to better social and urban policy. As Abbott neatly summarises, 'the Chicago school thought – and thinks – that one cannot understand social life without understanding the arrangements of particular social actors in particular social times and places' (Abbott, 1999: 196). It is a plea for sensitivity to context, social complexity and history that still has much to teach a social science community for many of whom, to paraphrase Marshall McLuhan, the medium of investigation has become the message.

While Chicago could not possibly recover its earlier pre-eminence in an academy where every major university had a sociology department, as we see in Chapter 5, the post-war generation of Chicago sociologists such as William Julius Wilson have continued to build on the 'golden age' tradition, along with urban researchers in the empirical tradition in Britain and the US in general. However, the Chicago approach has often been criticised as liberal hypocrisy with an agenda of 'devising methods of social control in the urban environment', rather than,

'in the pursuit of justice' (Ellison, [1944] 1972: 303–17 in Lal, 1990: 4). Alvin Gouldner famously denounced Chicago sociology for failing to address issues of class and structural inequality for fear of upsetting powerful financial sponsors. While rival American sociologists derided Chicago for its 'dust-bowl empiricism' (Wax, 2000: 73), both the emerging political economy approach and the outwardly Marxist cohort of post-war urbanists accused Chicago of missing the main story within which patterns of immigrant settlement, levels of vice and criminality, 'natural areas' and the ecology of invasion and succession are mere effects of the great world historical cyclone of capitalist modernity.

This chapter might be accused of making the same prejudicial judgement in bracketing social investigation off from 'theory' as it is traditionally understood. But it does not follow that inductive research informed by a heightened sense of the way time and place shapes the life-world of the urban citizen is inferior to Talcott Parson's notion of a 'social system' or the statistical abstractions favoured by quantitative analysts of human society. What we are really debating are the benefits of using different interpretative lenses for the urban complex – some such as the micro-level case study or participant-observation method give us a vivid and detailed portrait of city life. While macro-level approaches such as Durkheim's concept of 'organic' versus 'mechanical' solidarity, or Simmel's notion of the 'blasé attitude', or Weber's work on status groups (what Abbott calls 'the sociology of variables') aim at a more universal view of modern (urban) society.

What I hope this chapter has shown is that rigorous and careful description of *urban* phenomena is itself an important contribution to the body of knowledge on the nature of the city and urban life. One does not have to be an apologist for liberal capitalism to insist that the city deserves to be studied in its own right. Whatever common features urban life may share with society at large, the metropolis has a distinctive set of social, economic, cultural and topographic features that are not found in smaller, less densely concentrated settlements. The study of the city also has its own techniques, moral preoccupations and prejudices, and it can vary enormously in the quality and worth of its output. But without this archaeology of knowledge, the city would be even more of an impenetrable maze than it seems to us still.

FURTHER READING

Victorian city surveys

One of the best collections of studies on nineteenth-century social investigation and the city is to be found in Englander, D. and O'Day, R. (eds) (1995) *Retrieved Riches: Social Investigation in Britain 1840–1914*. The volume features chapters on Mayhew, Booth and Beatrice Potter (Webb).

Henry Mayhew and **Andrew Mearns** both feature in Gareth Stedman-Jones' excellent social history of Victorian London (Stedman-Jones, 1971, 1976). Anne Humpherys (1977) has produced one of the few comprehensive studies of Mayhew. Useful introductions to *The Bitter Cry* are to be found in Wohl (1970) and also Hill (1970) featuring a note on Octavia Hill and Andrew Mearns by W.H. Chaloner.

Charles Booth: A good selection of material from Booth's *Life and Labour of the People in London* is to be found in Peter Pfauz's *Charles Booth on the City. Physical Pattern and Social Structure* (1967). Fried and Elman's *Charles Booth's London* (1969), is also a good introduction to the subject. The more recent study by Rosemary O'Day and David Englander, *Mr Charles Booth's Inquiry: Life and Labour of the People in London Reconsidered*, (1993) is authoritative and deals with Booth's life and influences as well as the survey itself. More specialist texts that focus on Booth's urban geography include Reeder (1984) and Shepherd (1999), and also Topalov (1993).

The life and work of **Jacob Riis** is well documented by Keith Gandal (1997), who also examines the life and work of Riis' contemporary Stephen Crane. Several anthologies of Riis' photographs exist including Doherty (1981), which offers a complete reproduction of Riis' work and Alland (1975), which includes a preface by Anselm Adams.

Jane Addams: Allen F. Davis, *American Heroine: The Life and Legend of Jane Addams* (1973) is generally regarded as the best account of Addams' life, and an official biography by her nephew, James Weber Linn, *Jane Addams* (1935) was published soon after her death. An edited collection on the history of Hull-House has been compiled by Mary Lynn McCree Bryan and Allen F. Davis (1990). Other publications on Hull-House include Kathryn Kish Sklar, 'Hull-House in the 1890s: A Community of Women Reformers' (1985), and Rivka Spak-Lisak, *Pluralism and Progressives: Hull House and the New Immigrants, 1890–1919* (1989). **Critical studies of Jane Addams**: An important source on the settlement movement in America is Mina Carson's *Settlement Folk. Social Thought and the American Settlement Movement, 1885–1930* (1990) together with Allen F. Davis' earlier study *Spearhead for Reform: The Social Settlements and the Progressive Movement* (1967).

An excellent collection of Jane Addams' essays and writing from her student days to her final years as a Nobel laureate and internationally renowned peace activist is to be found in Jean Bethke Elshtain (ed.), *The Jane Addams Reader* (2002). This volume also contains a useful of chronology of Jane Addams' life and a comprehensive bibliography of her writings together with short introductions to the various phases of her political and literary career.

Chicago and urban studies

A considerable volume of literature now exists on the Chicago School of Sociology. The most comprehensive collection of Chicago School criticism is contained in Ken Plummer's four-volume set, *The Chicago School. Critical Assessments* (1997), which includes essays on almost every aspect of the Chicago Sociology department's origin and development, work, and legacy. For the remainder I signal only those studies that cast light exclusively or partially on the urban studies dimension of 'the first Chicago School'.

A very useful starting point for a general history of Chicago sociology is Andrew Abbott's *Department and Discipline, Chicago Sociology at One Hundred* (1999), which also discusses the equally important social psychological, pragmatist and symbolic interactionalist traditions of the department. Martin Bulmer (1984) offers a highly regarded survey of the Chicago School of Sociology and can usefully be read alongside Dennis Smith's (1988) account of the same tradition. The story of the post-war history of Chicago sociology is well told in the collection edited by Gary Fine, *A Second Chicago School? The Development of a Postwar American Sociology* (1995). A discussion of the impact of Chicago sociology in Europe, particularly in relation to the political economy critique is provided in Lebas (1982). Mary Jo Deegan (1988) convincingly re-sites the origins of the Chicago approach to social investigation within the social reform movement of Jane Addams and Hull-House.

4

VISIONS OF UTOPIA

From the Garden City to
the new urbanism

ita facile confiteor permulta esse in Utopiensium republica quae in nostris civitatibus optarim
verius quam sperarim[17]

Sir Thomas More, *Utopia*

INTRODUCTION

In the previous chapters we saw how both theoretical and empirical writers on the city in what we might broadly call the classic period of urban studies (*c*.1880–1940) had a strong moral vision of what the metropolis stood for and what it might one day become. Linked to the social reform agenda and the emerging concern for the urban condition within the social sciences was another stream of thought and criticism that concerned itself with the form and function of human settlements. Whereas in the past, historians had paid attention to important architecture and the aesthetics of urban design, by the middle of the nineteenth century, the growth of the major European and North American cities had given rise to the new professions of civic engineering and town planning that, of necessity, adopted a holistic approach to urban development. Influential figures such as the biologist Patrick Geddes, believed that advances in the natural sciences could be applied usefully to the study of urban development and the arrangement of functions in the city. Lewis Mumford was to develop Geddes' insights to illustrate how the city could be viewed as a series of organic forms that were dynamically linked to the surrounding region and in which each internal function was inter-related, while Geddes' commitment to community involvement at the earliest stage of planning new developments and in urban regeneration has become a standard feature of local government policy in Britain, and finds an echo in the 'charrette' method of direct user involvement in community design favoured by traditional neighbourhood design architects (see below).

Many attempts at creating the perfect city had been tried in the past, but very few had been achieved in practice. However, the comprehensive re-design of central Paris by Hausmann, together with similar projects for the reconstruction of Berlin and Vienna gave fresh impetus to what became two competing tendencies in utopian urban design, both of which agreed that however successful the industrial metropolis might have been as a conduit for trade and empire, the cost to human civilisation in moral, spiritual, aesthetic and environmental terms could no longer be sustained.

In the first camp we include that chorus of educated opinion that saw in the classic and renaissance eras the pinnacle of human achievement in the form and scale of the modern city. Such artists, writers and intellectuals associated with the Arts and Crafts movement in Britain were critical of the impact of the industrial revolution on the built environment, especially the threat it posed to traditional occupations and local communities. Almost invariably the metropolis was seen in negative terms – unsanitary, overcrowded, dark, noisy, dangerous, criminal, immoral, anarchic, overpowering and cosmopolitan. Both liberal and conservative commentators could find something about the industrial city to dislike, but what unites these traditionalist (or neo-traditionalist) visions is the quest for what Bruno Zevi called the city on 'a human scale' (Zevi, 1950). This is not to say that the anti-modernist voices we encounter in this chapter are 'anti-city' since they mostly all accept that even dense urban communities can be practical and fulfilling places to live. But if the industrial metropolis and the endless sprawl of post-war suburbia are the two heads of the Cerberus of laissez-faire urbanism – the 'anti-moderns' are just as fervently against the urban renewal and redevelopment programmes of post-war planners and architects.

The other camp in the battle for control of the urban landscape appears somewhat later in our narrative, and is associated with the aesthetic and artistic revolution of modernism that had begun to make serious inroads into cultural life by the end of the First World War, though its first base camps might have been established as early as the 1870s. As T.J. Clark artfully remarks, 'modernism' is something you know when you see it, but Max Weber's phrase that he borrowed from Schiller, 'the disenchantment of the world' captures the essence but not the practice of modernism – which is best expressed in Paul Valéry's words as 'Le moderne se contente de peu' (The modern contents itself with little) (Clark, 1999: 7). Re-articulated as Mies van der Rohe's famous epigram 'less is more'; minimalism, with its love of clean lines and abstract space was to become the collective signature of a new generation of urbanists for whom the city appeared as a site of endless experimental possibilities.

While the division between pro-modernist continental European and anti-modernist Anglo-Saxon views on the ideal habitat is inevitably something of a caricature; for example some of the greatest examples and concentrations of modernist architecture are to be found in the US, while celebrated 'neo-traditionalist' architects such as the Luxemburger, Leon Krier, have produced much of their important work in continental Europe. I do want to suggest that a rival camp to modernist urbanism has had a long and influential tradition, especially (though not exclusively) in Britain and in the US as well as the former British dominions of Canada, Australia and New Zealand. Because 'anti-modernism' (and indeed anti-metropolitanism) has been undergoing a revival not just within town planning circles, but at the highest levels of government policy-making in recent years, it is important that this often neglected aspect of urbanism is given proper consideration. As I shall argue in Chapter 8, in order to understand how we arrived at the synthesis of styles and ideologies that is postmodern urbanism it is necessary to have some acquaintance with modernist urbanism and its neo-traditionalist antithesis.

PLANNING UTOPIA: THE CIVIC REVIVALISM OF RUSKIN, HOWARD AND UNWIN

John Ruskin (1819–1900) who was Professor of Art History at Oxford University was perhaps one of the most influential public intellectuals in an age that included such celebrated figures as Carlyle, Macaulay, Cobbett, Darwin and Brunel. Ruskin was not only a profound influence on his chosen disciplines of art and architecture, but also provided intellectuals, writers and planners who deplored the industrial city with their key aesthetic and philosophical arguments. However, Ruskin's influence extended beyond the educated elites, as his disciple William Morris (1834–1896) wrote, to the working class audiences who could 'see the prophet in him rather than the fantastic rhetorician, as more superfine audiences do' (Cole, 1934: 633 in Lang, 1999: xi). According to Lang, Ruskin's prophetic ideas on architecture, urban design and town planning influenced the development of the City Beautiful Movement, the New Towns Movement and concepts in modern urban geography such as central place theory (Lang,

1999: 45). But it is Ruskin's influence on the Garden City Movement, and on its co-founders Ebenezer Howard and Raymond Unwin that really establishes his social thought as the fountainhead of an intelligent but essentially atavistic anti-metropolitanism.

THE GARDEN CITY MOVEMENT

The link between Ruskin's vision and what was to become known as 'the Garden City Movement' is clear from his depiction of what good, modern town planning should incorporate:

> lodging people and providing lodging for them means a great deal of vigorous legislature, and cutting down of vested interests that stand in the way, and after that, or before that, so far as we can get it, through sanitary and remedial action in the houses that we have; and then the building of more, strongly, beautifully, and in groups of limited extent kept in proportion to their streams and wall around so that there be no festering and wretched suburb anywhere, but clean and busy street within and open country without, with a belt of beautiful garden and orchard round the walls so that from any part of the city perfectly fresh air and grass and sight of far horizon might be reachable in a few minutes' walk. That is the final aim.
>
> (Ruskin, 1868 in Lang, 1999: 41)

This passage anticipates by three decades the booklet that Ebenezer Howard was to publish in *To-morrow: A Peaceful Path to Real Reform* (1898). Re-titled four

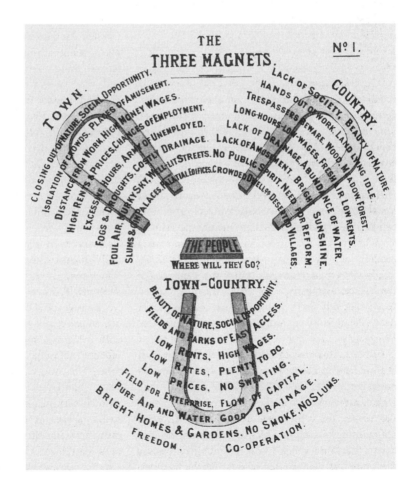

Figure 4.1
'The Three Magnets' from Ebenezer Howard's *Garden Cities of To-morrow*.

years later as *Garden Cities of To-Morrow* (Howard, 1985 [1902]) it became the manifesto for a new town planning movement – the Garden City Association – that Howard had helped to found, and that was to become a key influence on contemporary city planning from Canberra to Columbia, Maryland (Rykwert, 2000: 168–9, 185).

Howard's thesis was simple enough, he contested the view that one had either to reside in the city or the country, believing that there was a third alternative,

> in which all the advantages of the most energetic and active town life, with all the beauty and delight of the country, may be secured in perfect combination; and the certainty of being able to live this life will be the magnet which will produce the effect for which we all striving – the spontaneous movement of the people from our crowded cities to the bosom of our kindly mother earth, at once the source of life, of happiness, of wealth, and of power.
>
> (Howard, [1898] 1985: 9)

Howard was not in favour of a return to low-density village-type communities and would even have tolerated the population densities of inner-London for his new 'Garden Cities', but he did not favour 'garden suburbs' such as those created by his disciple Raymond Unwin in Hampstead. Nor was Howard an authoritarian who wished to 'move people round like pawns on a chessboard', but someone who believed that 'garden cities were merely the vehicles for a progressive reconstruction of capitalist society into an affinity of co-operative commonwealths' (Hall, 1988: 87). Howard's use of the metaphor of the magnet emphasises his view that the 'Garden City' must sell itself, it must be an elective rather than a compulsory community if it is to succeed (see Figure 4.1).

In 1902, Howard put his ideas to the test by purchasing land in Letchworth, a small village some 35 miles north of London. Barry Parker and Raymond Unwin were employed as the architects of the scheme, the money for which was raised from Victorian philanthropists such as Lever and Cadbury, both of whom had built 'model' workers' villages to house the employees who worked in their nearby factories.

The plan for Letchworth incorporated a mix of single villas for middle-class residents, while grouped around a shared grassed quadrangle were numbers of workers' cottages (see Figure 4.2). The design was straight out of the monastic revivalism of William Morris, and there were even two experimental schemes for community living with communal dining rooms and a laundry that appeared to imitate Morris's utopian community in *News from Nowhere*.

There was also to be a public house, but in keeping with Howard's temperant views it served only lemonade and ginger beer. Several of Howard's contemporaries, including Charles Booth, favoured establishing labour colonies for the 'demoralised residuum' as the Toynbee Commission described destitute Britons in 1892. But although he was inspired by Edward Bellamy's socialist utopia *Looking Backward* (1996 [1888]), Howard disliked the idea of the individual's needs being subsumed into that of the group, and looked more to the anarchist ideas of Kropotkin and the cooperative socialism of Ledoux, Owen, Pemberton and Buckingham (Hall, 1988: 90–1). Howard also found Henry George's *Progress and Poverty* (1976 [1880]) inspiring, and while he disagreed with George's claim that all human misery stemmed from the private ownership of land, he did think it important that individuals should have a personal stake in their community by agreeing to become shareholders of the development corporation. This was an idea, as we shall see later, that has been enthusiastically adopted by sections of the American new urbanist movement, although the commercial imperative is not one that Howard would have endorsed (Rykwert, 2000: 241).

The essence of Howard's utopian city is the self-contained community of the feudal village, combined with a limited development of industry and using modern communications to link urban centres to each other. Howard's Garden City would employ only a small number of the population to work in the municipality, while the majority would earn their living in the conventional way by working for the manufacturers, cooperatives and philanthropic societies that Howard hoped to attract – providing both a ready made workforce and a consumer market for

Figure 4.2 The Raymond Unwin designed Garden City of Letchworth.
Copyright and courtesy of the British Architectural Library, RIBA, London

local industry and agriculture (Howard, 1985: 80). But – and this is an important qualification to conventional accounts of the Garden City plan – Howard envisaged these 'satellites' to be connected via railroads to a central city with a population of some 58,000 that would be able to provide all the goods and services that modern industrial society demanded. The ensemble of central city and Garden Cities was what Howard understood by the term 'social city' (Figure 4.3). In order to prevent towns merging into one another, a commonly owned greenbelt of 'field, hedgerow, and woodland' would exist between each settlement to allow the inhabitants to enjoy 'the fresh delights of the country . . . within a very few minutes walk or ride' (Howard, 1985: 105). As with Adelaide in South Australia (the original model for the 'green-belt town extension model'), any new developments would be separated by a similar margin of preferably uncultivated land, and linked via rapid rail transport with a main spur connecting the new settlement to the central city (Hall, 1985: 89).

The idea of a 'green belt' had been put forward by a number of eminent Victorians including, somewhat paradoxically, the father of neo-classical economics, Alfred Marshall. Though as a planning concept the initiative belongs to Raymond Unwin whose vision was enthusiastically endorsed by J.H. Forshaw and Sir Patrick Abercrombie in their 'Plan for London' in 1943, before becoming an orthodoxy of town planning in Britain after the Second World War (Miller, 1992). The second of Howard's projects, Welwyn Garden City, quickly demonstrated the limits of the 'social city' concept. Without the money to complete the purchase of the site, the newly renamed Garden Cities and Town Planning Association was forced to step in and take over the management of the project for which Howard had failed even to acquire the requisite planning permission (Hall, 1988: 107). Welwyn's handsome neo-Georgian villas designed by Louis de Soissons became hugely popular with middle-class commuters who, thanks to the nearby railway station, could reach London in less than an hour. Rather than providing spurs between

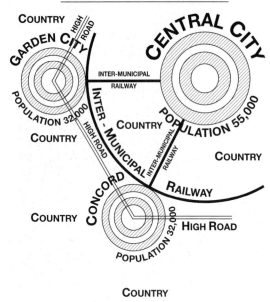

N° 5.
DIAGRAM

ILLUSTRATING CORRRECT PRINCIPLE
OF A CITY'S GROWTH – OPEN COUNTRY
EVERY NEAR AT HAND, AND RAPID
COMMUNICATION BETWEEN OFF-SHOOTS

Figure 4.3 Ebenezer Howard's 'Social City' incorporating a larger central city linked to satellite garden cities.
Source: *Garden Cities of To-morrow*

self-contained, self-managed communities, the railways spread out from London like a spider's web, thus creating a homogeneous middle-class satellite belt around the all-consuming capital. Howard's idea of a local workforce being mostly engaged in local industry, commerce and agriculture was unlikely to succeed given that the price of many of the family sized properties required the professional or managerial salaries that could only be earned in the metropolis.

Howard's architect on the Letchworth scheme, Raymond Unwin, was a more pragmatic visionary and was not afraid to describe himself as a socialist.

He believed that the improvement of the living conditions of the poor was one of the greatest challenges and goals of socialism and his ideas and achievements in town planning were to have a lasting impact on community development, not just in Britain but around the world. If this meant giving up the idea of Howard's '*urbs in rure*' ideal then so be it. For all his affinities with Ruskin and Morris, Unwin believed that getting people out of slums and into decent homes mattered more than roses round the cottage door.

According to Miller, 'Unwin focused on the industrial city as the prime evil to be refuted through management of development, berating the fact that the urbanization process that had assisted the accumulation of national wealth had left a trail of environmental and social problems' (Miller, 1992: 3). In Britain, the introduction of the first Town Planning Act in 1909 owed much to the community planning ideals espoused by Geddes and Unwin, but the results did not conform to the Garden City ideal. Mass housing schemes such as the Becontree development in Essex, which was to become an important source of labour for the new Ford Motor plant at Dagenham was roundly criticised for its vast expanse, its lack of community focus, and its poor transport links – precisely the reverse of what Ebenezer Howard intended.

The reason was quite simply economics. Housing subsidies provided by central government to the London County Council were insufficient to build variegated, well-spaced housing with decent local services and amenities. However, when the first majority Labour government came to power in 1945, the housing minister Aneurin Bevan was determined to give public housing the priority it had lacked in the interwar period (Parker, 1999). New town planning regulations were established allowing for the compulsory purchase of land and property for public purposes, and a planning framework was established that aimed at urban containment through the establishment of a number of green belts around the major towns and cities (Lang, 1999: 158). The actual planning process was left to professional planning officers who could vary the regional structure within certain

EXHIBIT 4.1   Profile of Ebenezer Howard

Ebenezer Howard (1850–1928) was born in the City of London, the third child of a confectioner. He was educated at private boarding schools until the age of 15 when he left to become a clerk for firms of City stockbrokers, merchants and lawyers. He also worked for a brief period as clerk to the Congregationalist Minister Joseph Parker, whose charismatic personality exercised a strong influence on the young Howard.

In 1872, Howard went to find his fortune in the US, working first on a farm in Nebraska, and then later moving to Chicago where his self-taught shorthand was put to good use as a stenographer in the Law Courts. From Chicago it is likely that he borrowed its epithet of 'the Garden City' and he would no doubt have been inspired by the new garden suburb of Riverside designed by Frederick Law Olmsted. During his stay in America, Howard was much affected by the spiritual writings of Emerson, Lowell and Whitman. Another influence on Howard was Edward Bellamy, whose utopian novel *Looking Backward* portrays a futuristic Boston where society is organised according to the principles of collective ownership. Bellamy's book inspired many of the ideas in *To-morrow. A Peaceful Path to Real Reform*, published in 1902 as *Garden Cities of To-morrow*.

Like the socialist, poet and designer, William Morris, Howard believed that true human well-being could only be achieved so long as man and

Figure 4.4   Ebenezer Howard.
Copyright and courtesy of the British Architectural Library, RIBA, London

nature were in harmony. This meant that the overcrowding of the cities and the depopulation of the countryside had to be reversed, if necessary by the intervention of government, in order to contain urban sprawl by the creation of a 'rural belt' or green belt around the larger urban centres. Ebenezer Howard can justly be considered one of the founders of modern town planning, and a good deal of legislation that was introduced after the First World War dealing with urban development bears his hallmark. True to his egalitarian principles, Howard never became wealthy from his new town ventures, and he continued to work as a shorthand writer to meet his family's modest needs right up until his death. He was knighted in 1927 for his services to the public.

limits before submitting their decisions to locally elected councillors for final approval.

However, Lang's judgement that the 'planning system has been remarkably effective at maintaining the level of population of the central cities, stopping suburban sprawl, protecting the integrity of historic towns and the natural environment' (ibid.) does not bear close scrutiny. The population of inner London was deliberately reduced after the war by the County of London plan and several new towns were built such as Harlow and Basildon in Essex to accommodate the exodus of working-class families from London's 'slums' – a process of dislocation that both helped to break-up traditional working-class

communities while failing to establish the 'human scale' Garden Cities that Howard had envisaged. As for the banishment of suburbs, the population of London's outer boroughs doubled after the Second World War (Glass *et al.*, 1964) while the population of inner London declined by almost a quarter between 1961 and 1994 (Urban Task Force, 1999). It is now possible to drive from the centre of London in almost any direction for 30 miles without leaving the built-up area, bringing the vision of H.G. Wells' '100 Mile City' ever closer (Sudjic, 1995).

Howard's vision also found a fertile soil in the US where much of the inspiration for the Garden City was gained. Radburn, in New Jersey was to become

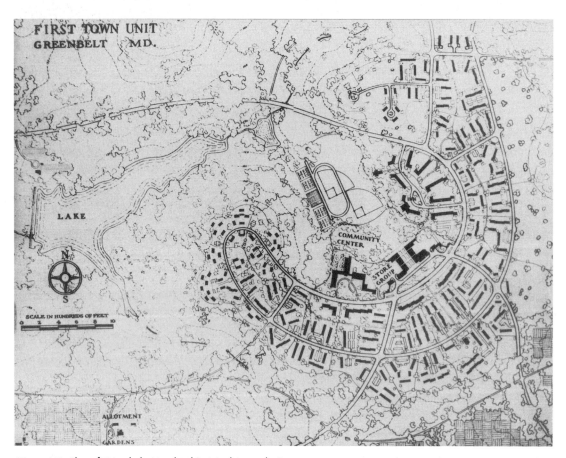

Figure 4.5 Plan of Greenbelt, Maryland (original town plan).
Copyright and courtesy of the British Architectural Library, RIBA, London

Figure 4.6 Greenbelt, Maryland. View of cinema and apartment houses.
Copyright and courtesy of the British Architectural Library, RIBA, London

the first American New Town built on Garden City principles in 1929. Other federally subsidised developments were to follow, all characterised by the signature 'green' in their title – Greenhills, Ohio; Greendale, Wisconsin; and Greenbelt, Maryland (see Figures 4.5 and 4.6). The more arboreal sounding Park Forest, in Illinois was added to the sequence after the Second World War (see Chapter 5). These were decidedly non-metropolitan utopias and they were being built at a time when the office tower began to dominate the skyline of America's great cities to such an extent that it changed forever our view of urban form and function. In order to better understand the spatial imaginaries of America's city builders a brief excursus into the history of American planning will be necessary.

A RELUCTANT MODERNISM: PLANNING THE AMERICAN DREAM

The famous passage from John Donne's 'Meditation XVII' ends with the words 'No man is an island, entire of itself . . . any man's death diminishes me, because I am involved in mankind; and therefore never send to know for whom the bell tolls; it tolls for thee'. In this poem Donne reminds us that we cannot abstain from our common humanity, and that our individual fates are mutually linked. But for many of America's new immigrants fleeing the religious persecutions and state-sanctioned famines of Europe, as Jean-Paul Sartre exclaims in his play 'No Exit' (*Huis-Clos*), '*L'enfer, c'est les autres*' (Hell is other people).

Heaven, then, would be a place where Adam and Eve once stood alone and naked before God. America's Eden turned out to be an inhospitable swamp, and the New World failed to live up to the Utopian dreams of its first European inhabitants. The indigenous tribes, the harshness of the climate, and the difficulty of the terrain encouraged frequent internal movement and migration. If one location failed there were plenty of other trails to try and claims to stake. The shifting frontier threw up and cast down slatboard settlements with such regularity that not until L'Enfant's 1791 design for the national capitol was any city among the thirteen states of the Union purposely planned. The charge that the US's national political elites care little for big cities is not new. Thomas Jefferson can be considered as one of the founding fathers of American anti-urbanism (Tafuri, 1999: 26) in his belief 'that the farm and the small town embodied the true America' (Fishman, 2000: 65). Indeed, it is not until the later nineteenth century and early twentieth century that the search for an urban ideal takes on physical form in the city plans of Frederick Law Olmsted, Daniel Burnham and Thomas Adams.

Olmsted, himself a former farmer, who remained a keen agronomist throughout his life, succeeds in combining elements of American pastoral – the carefully landscaped parks that still bear his name in cities across the US – while providing a flexible and extendable network of streets and subdivisions that allowed for variety and interest in architectural form. As Tafuri notes, 'We might say metaphorically that the ethic of free-trade here encountered the pioneering spirit' (Tafuri, 1999: 40).

Exemplary in this respect were Burnham's 'City Beautiful' inspired *Plan of Chicago* in 1909 and the multi-volume *Regional Plan of New York and Its Environs* funded at considerable expense by the Russell Sage Foundation under the direction of Thomas Adams. Lewis Mumford was particularly scathing of Adams' metropolitan vision, calling it 'obsolete even at the moment of its original formulation', and 'a compendious handbook on how not to approach the future' (Mumford, 1938: 390). The Russell Sage report was planning for a monocentric metropolitan region whereas, and here Mumford's intuition was correct, the default model of American post-war urban development would be that of Los Angeles, not New York or Chicago.

As we shall see in the following chapter, the spatial ideology of the American frontier had changed little since the post-colonial period when Thomas Jefferson's checkerboard or gridiron pattern of rectangular lots imposed itself on the landscape with little or no thought for topography, natural features or visual interest, in order to allow 'a quick parcelling of the land, a quick conversion of farmsteads into real estate, and a quick sale' (Mumford, 1961: 422 in Hommann, 1993: 29). Despite the influential legacy of the Jeffersonian ideal on the collective imaginary of hometown America, by the 1920s the US was further down the road to a ubiquitous modernism in town planning than any other nation.

A number of writers consider the 1893 World Exhibition in Chicago to have been the origin of modern town planning in the US (Benevolo, 1977; Goodman, 1971 in Mattrisch, 2001: 4), and while City Beautiful motifs abounded – from expansive boulevards, to landscaped parks and imposing civic buildings – it was the technology of the second industrial revolution, argues Fishman, that allowed urban forms to express 'the promise of modern urban living' (Fishman, 2000: 11). The combination of new building technologies such as steel frame construction and concrete together with the mass diffusion of the internal combustion engine allowed for the sorts of functional zoning that Burgess and his colleagues were observing in the 1920s and 1930s. Such spatial divisions were not simply the result of urban natural selection so much as the organised preferences of coalitions of industrial magnates, real estate developers, speculators and city bosses keen on increasing their profit margins and expanding their empires – the forerunners of the 'growth coalition' movements that we encounter in Chapter 7. Although it was to become the dominant utopian vision of the city for the remainder of the twentieth century, the functional or Euclidian zoned city had its feet firmly planted in the soil of Mammon.[18]

THE FUNCTIONAL CITY

Men – intelligent, cold and calm – are needed to build the house and to lay out the town.

Le Corbusier, *Towards a New Architecture*, 1986: 127

Functional ideas of the city 'as a large and smoothly operating machine' had already taken hold under the influence of Frederick Winslow Taylor's industrial efficiency, or 'time and motion' studies, while the Euclidean zoning principles that Jane Jacobs so opposed (see Chapter 5) were a 'powerful ideological motivation for urban renewal projects in the 1950s, as well as for suburban sprawl developments' (Wickersham, 2001: 555). On the other side of the Atlantic, Cornelis van Eesteren, the Chief Architect at the Urban Development Section of the Amsterdam Public Works Department had come up with a set of principles that were to form the basis of the International Congress of Modern Architecure's (CIAM) guidelines for the projection of 'the Functional City' which was to form the subject of the group's 'Special Congress' in Berlin in 1931. By now, the research of the team under van Eesteren had become much more oriented towards broader social and economic considerations of how a regionally oriented city functioned. Drawing on the planning methodology of the Regional Plan of New York and its Environs, the first elements of which had appeared in the mid-1920s, the Amsterdam Expansion Plan retained a strong Garden City foundation (complete with rail spurs and green belts), but overlaid with 'rational development methods' that aimed at analysing the way 'functional units' of the metropolis including skyscrapers, parking garages, sports fields, stations and churches, interacted with one another. With the right data and a designated planning area, Van Eesteren believed that the correct urban form could be achieved in five minutes to an hour. He even wrote an article in 1928 based on the same principles entitled 'The Five Minute Building' (Mumford, 2000: 60–1).

However, this was only one feature of the CIAM manifesto, whereas for most observers modernist architecture's defining feature was the shift from the horizontal to the vertical scale, expressed particularly in the 'unité d'habitation' – the geometrically arranged tower blocks which Le Corbusier first planned for central Paris, and which were to become emblematic of the mass housing programmes that began to define the landscape of cities and towns around the world from the 1950s onwards (see Figure 4.7).

Le Corbusier (born Charles-Edouard Jeanneret) epitomised the revolution in urban design that came to dominate European, and then world, architectural practice from the 1920s onwards. His best-selling book *Vers Une Architechture Moderne* (*Towards A New Architecture*) published in 1923 made the famous (or infamous) statement that, just as an airplane is a machine for flying in so 'a house is a machine for living in'. The volume also juxtaposed images of the Parthenon and a motorcar in order to demonstrate the necessity of standards in achieving the perfection of form. Le Corbusier believed that the engineer was the true architect, and that the task of the architect was to understand how the function of a building could be most economically achieved in its construction and design. If the internal combustion engine required the arrangement of pistons, carburettor, fan belt and drive shaft in certain ways, it was no less true of the distribution and arrangement of buildings and the constructed landscape.

The publication of *L'Urbanisme* (Urbanism) in 1925 marked a watershed with the assertion of rationalism above ornamentation, order in place of laissez-faire, and functionality above form. In the same year, Le Corbusier published his 'Plan Voisin' (Neighbourhood Plan) for Paris that envisaged the demolition of almost the entire north bank of the Seine and the erection of 18 cruciform high-rise residential blocks organised in perpendicular clusters divided by broad urban highways, and with the majority of the remainder of the site given over to open space. This was Le Corbusier's application of his 'City of 3 Million' prototype that he first mapped out in his *Ville Contemporaine* (The Contemporary City) of 1922, where commercial, residential and industrial areas are strictly separated and are connected by direct highway links on a grid system. Unlike, Haussmann, Le Corbusier did not succeed in his ambitious scheme, but developers such as Robert Moses in New York

City were able to follow Le Corbusier's masterplan by driving expressways through the heart of long-established urban communities in the cause of 'rational planning' with devastating effects on local neighbourhoods (Berman, 1982).

Ironically, given the animadversion of Garden City enthusiasts to all things to do with Le Corbusier, Jeanneret not only 'acquainted himself with the social arguments of Ebenezer Howard and their practical realization by Barry Parker and Raymond Unwin. He even designed a Garden City in a picturesque curvilinear layout for La Chaux-de-Fonds' (Jencks, 1987: 27). Although Le Corbusier soon rejected the rusticity of the pitched roof and the Hampstead Garden Suburb-style curving street in favour of perpendicular geometries and grids, he persisted with the idea of the parkland city – as we can deduce from the *City of To-morrow* where he refers to Garden Cities

Figure 4.7
Le Corbusier's *Unité d'habitation* apartment block, Marseilles, France. Copyright H.T. Cadbury-Brown, courtesy of the Architectural Association Photo Library

of two million inhabitants. While in the metropolitan centre he insisted on the need to concentrate population densities, thus freeing space for parks and open spaces (Le Corbusier in Le Gates and Stout, 1996: 368–75). This was a principle that was to become orthodoxy in city planning departments across the world in the second half of the twentieth century.

CIAM – vanguard of modernist urbanism

The successful divulgation of Le Corbusier's ideas owed much to the international association most associated with the modernist turn in architectural theory, the International Congress of Modern Architecture (CIAM). CIAM was founded in Switzerland in 1928 by another leading architect of this modernist-producing nation, Sigfried Giedion. As Zykwert writes, CIAM 'aimed to reform not only the design of buildings but the whole fabric of the contemporary city'. Further meetings took place in Frankfurt, Brussels and Barcelona (confirming the movement's continental European bias). But it was at an improvised meeting on a ship bound from Marseilles to Athens (hastily organised after Stalin had provoked the cancelling of the Moscow congress) that the 111 theses of the movement were drawn up. Known as 'The Athens Charter', these principles set the standard for modernist architecture and town planning for decades to come. The theses declared that any city should be analysed according to four basic functions:

1   Dwelling – habitat – well-spaced apartments high-rise apartments to be preferred to other forms
2   Work – to include both offices and factories
3   Recreation and leisure – focusing particularly on sport and, therefore, parks and stadia
4   Circulation – this was to be treated as a separate zone, and given equal weight in planning priorities.
(Rykwert, 2000: 175)

The Garden City, Beaux-Arts, and any kind of explicit 'formalism' were out. 'All forms of traditional urbanism were seen through a regularizing prism that made them appear dark, unsanitary, and chaotic,

unsuited to "modern needs"' (Mumford, 2000: 58). CIAM, thus, made no bones about the fact that it intended to replace 'the existing urban pattern in favour of widely spaced high-rises set in greenery along the lines laid down by Gropius and Le Corbusier and shared by Neutra, Bourgeois, and the other members of CIAM' (ibid.).

Although Le Corbusier has often been the butt of anti-modernist scorn, the architectural school that really launched the 'international style' in architecture is associated with the Bauhaus school of architecture and design established by Walter Gropius after the First World War in Weimar, Germany. Gropius was forced to flee Germany by the Nazis, and he took up a position as professor of architecture at Harvard where he was joined in his American exodus by other eminent architects such as Ludwig Hilberseimer and Mies van der Rohe (Rykwert, 2000: 98). An indigenous American modernism had been well established by the First World War through the work of Irving Gill, Louis H. Sullivan, D.H. Burnham and H.H. Richardson – the latter three founding what was to become known as the Chicago School of Architecture. Sullivan's signature building, the Carson Pirie Scott Store, built in 1899 used revolutionary steel frame building techniques, thus allowing much greater elevation and floor space than conventional construction methods. These design principles were quickly taken up and adopted in new high-rise commercial developments across the US but, unlike in Europe, the rapid arrival of the vertical city was not accompanied by a triumphant assertion of modernist urbanism. Rather, the new cathedrals to mammon such as the Woolworth, Wrigley and Chrysler buildings were cloaked in neo-classical, gothic and beaux-arts detailing in order to restore, if only symbolically, the Jeffersonian ideal of the American pastoral that the Fordist city had done so much to traduce.

Frank Lloyd Wright and Broadacre City

Sullivan's most famous pupil when he was still in partnership with Dankmar Adler was Frank Lloyd

Wright, and it was from his mentor that Wright refined and developed the idea of an 'organic architecture', which rejected conventional styles and techniques of construction in favour of the unison of function, environment and technology. 'Architecture', wrote Lloyd Wright in 1930,

is the triumph of Human Imagination over materials, methods, and men, to put man into possession of his own Earth. It is at least the geometric pattern of things, of life, of the human and social world. It is at best that magic framework of reality that we sometimes touch upon when we use the word 'order'.

(Lloyd Wright, 1930, 1937 in Library of Congress, 1997)

Wright's architectural vision shares many of the naturalist preoccupations of Jefferson (Tafuri, 1999: 27), although his best-known commission, the Guggenheim Museum in New York City, is often held out as the apotheosis of avant-garde modernism. In Wright's 'Broadacres Plan' he attempts to reconcile the democratic individualism of Emerson and Jefferson with the need to achieve an overall harmony and symmetry:

In Broadacres all is symmetrical but it is seldom obviously and never academically so.

Whatever forms issue are capable of normal growth without destruction of such pattern as they may have. Nor is there much repetition in the new city. Where regiment and row serve the general harmony of arrangement both are present, but generally, both are absent except where planting and cultivation are naturally a process or walls afford a desired seclusion. Rhythm is the substitute for such repetitions everywhere. Wherever repetition (standardization) enters, it has been modified by inner rhythms either by art or by nature as it must, to be of any lasting human value.

The three major inventions already at work building Broadacres, whether the powers that over-built the old cities otherwise like it or not, are:

1. The motor car: general mobilization of the human being.
2. Radio, telephone and telegraph: electrical inter-communication becoming complete.
3. Standardized machine-shop production: machine invention plus scientific discovery.

(Wright in Le Gates and Stout (eds), 1996: 377–81)

Figure 4.8 Charles-Edouard Jeanneret (Le Corbusier). Copyright and courtesy of the British Architectural Library, RIBA, London

But, for all Wright's emphasis on 'diversity in unity', critics argue that his vision became corrupted into an instrument of row-house sprawl in the post-war period. The widespread availability of the automobile and the economies of scale that could be derived from uniform pre-fabrication meant that Frank Lloyd Wright's utopian vision of a zoned but successful 'the new urbanism' has been in promoting an alternative vision of town life and what prospects it affords for offering solutions to the widely perceived dystopia of the modern metropolis.

QUESTIONS TO DISCUSS

1 Compare Wright's ideas concerning the perfect city with those of Le Corbusier. On what essentials of the town does Wright agree and on what others does he disagree with Le Corbusier?

2 Why do you think Broadacres has been seen as the forerunner of the post-war American suburb?

THE NEW URBANISM

New urbanism has been described as 'a planning movement that is gaining increasing popularity' (Talen, 1999), it has featured on the covers of *Time, Newsweek, The New York Times*, and *Atlantic Monthly*, and has captured the imagination of the American public 'like no urban planning movement in decades' (Fulton, 1996: 1 in ibid.: 1375). On the other side of the Atlantic, new urbanism has also attracted the support of influential public figures such as the Prince of Wales, who sees it as an urgently needed antidote to the contagions of modernism and sprawl. But new urbanism, known also as 'traditional neighbourhood design' or 'neo-traditional development' is much more than a planning movement, it seeks to recreate the social bonds that its advocates believe have been lost in the maul of unfettered urbanisation by building in 'social capital' (see Chapter 5) to building design, street layout, and community facilities and resources.

Although the aesthetic and philosophical principles of new urbanism date back to the time of Ruskin and Victorian classical revivalism, the re-packaging of neo-traditionalism as a political manifesto can be traced to a more recent gathering at Ahwahnee Lodge in Yosemite in 1991. The so-called 'Ahwahnee Principles' were set out in fifteen short points, but

Figure 4.9 Frank Lloyd Wright.
Copyright and courtesy of the British Architectural Library, RIBA, London

integrated city was not realised in practice. In contrast to Britain and most of Northern Europe, the relative affordability of green field land plots and the general absence of building development controls meant that there was no disincentive to expand low-density suburbia to the infinite horizon of the American frontier. One or two prominent voices such as those of Jane Jacobs and Holly Whyte were raised against the mass homesteading of America in the 1950s and 1960s and saw it as a threat to the vitality of the traditional city and an attack on community life (see the following chapter). But it is only in recent years that architects and planners have attempted to come up with practical alternatives to the ubiquity of sprawl. In the following section we consider how other than an added emphasis on respect for the natural environment, they repeat the propositions advanced in *Thirteen Points of Traditional Neighbourhood Development*[19] by Andres Duany and Elizabeth Plater-Zyberg (or DPZ as their firm is known), the names most associated with 'new urbanism' as architectural practice (MacCannell, 1999: 108).

What the *Principles* set out to challenge is the formless suburban sprawl that has characterised American residential developments since the 1950s. The authors prescribe, instead, a return to the traditional street with mixed land-uses incorporating retail and residential units varying from single bedroom apartments to family town houses. Elementary schools and day care should be within walking distance, children's play areas should also be a short distance from any dwelling, while the streets themselves should be narrow in order to reduce vehicle speed and to encourage pedestrians and cyclists. Porches and verandas are the favoured architectural style by which the authors hope to stimulate good neighbourliness and openness to one's surroundings. Where possible, natural features and drainage are to be preserved in order to have the minimal environmental impact, and infrastructure and services are designed so as to minimise waste and maximise energy efficiency. Each community should be self-governing (ibid.: 108), where possible even at the planning stage. For example, DPZ pioneered the use of five-day brainstorming and opinion sharing known as 'charrettes'[20] that directly involve potential users in all aspects of the master planning process. Although this could be criticised as the equivalent of hiring an 'exterior design' consultant (you get to choose the species of trees outside your house and the colour shade of the street furniture), the same techniques have been used successfully in neighbourhood renewal schemes involving lower income groups in inner-city locations.

Although new urbanists dislike sprawl, they do not suggest that there should be a return to dense cities. Instead, new urbanist planners want to improve on suburbia by introducing more variety to residential blocks, by allowing mixed land-use, and by making towns pedestrian rather than automobile friendly.

The new urbanist advocates are clear that their models for good urban design derive from Garden City pioneers such as Raymond Unwin in Britain and Frederick Law Olmsted in the US (Dutton, 2000: 71–2). Peter Calthorpe quotes Ebenezer Howard approvingly in his *The Next American Metropolis* and refers to the desirability of 'urban growth boundaries' (for which read 'green belts') as interstices between 'satellite' towns and the city centre, and linked (naturally enough) by commuter rail (see Figure 4.10 and compare with Figure 4.3). The principle of not permitting the building of new towns until infill plots have been developed and zoning specific new growth areas within the existing planning boundaries is consistent with '**smart growth**' criteria (American Planning Association, 2002). But when do these constraints become too tight and when must they be loosened in order to prevent rising housing costs and the flight of working families (Calthorpe, 1993: 71)? Unfortunately, we are not offered a satisfactory answer. New towns are good at absorbing sprawl, but does it really matter if the city's workforce moves to a 'sprawl' suburb or a more scenic new town since the negative effect on the city's economy is exactly the same? Indeed, as we shall see in the following chapter the evidence from writers such as Garreau (1992) is that the new suburbia, or the 'exurbs' (Davis *et al.*, 1994) is increasingly economically self-reliant and no longer depends on the traditional city for jobs and services as it once did. We are not seeing satellites around a metropolitan sun, so much as – in the case of cities such as Detroit – a metropolitan red dwarf surrounded by nebulae of self-contained wealthy residential/business clusters.

For all its 'new urbanist' claims, the traditional neighbourhood design (**TND**) movement is anti-metropolitan in practice if not intent, because the majority of schemes incorporating new urbanist planning and design principles have been new, small town developments rather than conversions of existing housing stock or infill development in the larger cities. The effect, therefore, is to discourage the return of high-income middle-class family households to the core cities rather than, as in the case of the Netherlands, upgrading the urban infrastructure and

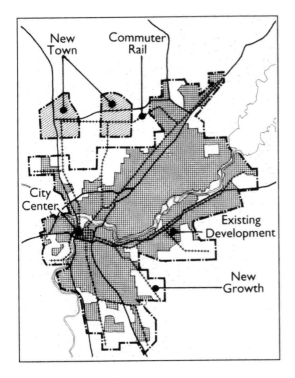

Figure 4.10 Criteria for new towns.
Source: Peter Calthorpe (1993) *The Next American Metropolis*, Princeton Architectural Press, p. 71
Copyright and courtesy of Princeton Architectural Press

social amenities to such a high quality that suburban 'exit' actually appears a less favourable option than remaining loyal to the traditional, dense city (Russell, 2001).

Another Florida development that has stimulated much interest in new urbanist town planning is that of Celebration. Originally designed by the Disney Corporation as a flagship community that represented the best in traditional design and community living, it is now a self-governing municipality with a population that is growing toward its 12,000 target. Celebration is aptly named, for it celebrates rather than plays down its TND credentials. 'Shortly after the first residents settled in to the community in 1996,' we read, 'Celebration soon became a poster-child for "traditional neighborhood design," as it rejuvenated the spirit of community with its neigh-

borhood parks, front porches and pedestrian-friendly design'. The poster worked, for the house journal of TND, *New Urban News* reported that in 2001, '380 such communities are currently planned or constructed', providing homes for people searching 'for places with a strong sense of community, parks for children, neighborhood schools and walkable destinations' (Celebration Company, 2001). Assuming each community to have an average population of 8,000 inhabitants, this means that around three million Americans will be living in a TND community in the next decade. This figure sounds impressive enough, but it represents only one per cent of the total population prediction for the US by 2010 (US Bureau of Census, 2002). New urban living is, therefore, a residential choice or opportunity available to a very small minority of the population, but its ideological prospectus overhangs its private-commercial niche into the architects' graveyard of mass public housing. For example, the Housing and Urban Development department in the US has 'instituted a multi-billion dollar housing programme which funds public housing projects strongly influenced by new urbanist principles' (Talen, 1999: 1375). Examples include Crawford Square in Pittsburgh, Pleasant View Gardens in Baltimore, Park DuValle in Louisville, and Port Royal, South Carolina.[21]

Andrew Ross sees developments such as Celebration and the nearby DPZ flagship community of Seaside as having their roots in the Garden City principle, but with the developer retaining control of unsold plots and further development opportunities rather than the community as a whole, as Ebenezer Howard had envisaged. Restrictive covenants serve to protect property values and to ensure that standards of maintenance and appearance are kept, while strictly prohibiting unilateral change of use. Decisions affecting the community as a whole are often taken via homeowners' associations who can elect their own representatives or a city manager to deal with the day-to-day issues of community government (Ross, 2000: 227–8).

This type of 'privatopia' (Mackenzie, 1994) or common interest development (**CID**) (Lave Johnston and Johnston Dodds, 2002) is consistent with the restricted ratepayer franchise that existed before the reform of local government in England and Wales after the First World War. Its consequence is that the only stakeholders in the decision-making process of the community are homeowners and the developer. Renters are excluded, as are those who live in the household of an employer as home helps or nannies. Votes are restricted to one per property unit, while the developer often retains control of the community administration until two-thirds of the lots are sold, and even then may enjoy plural votes by retaining ownership of small parcels of land. The fact that over 40 million Americans in some 200,000 plus communities choose to live under this form of community confirms Buchanan's (1965) observation that 'consumption ownership-membership arrangements' – such as CIDs, will assent to additional members joining only if the benefit at least balances, and ideally outweighs the costs.[22] In this way, CIDs and new suburban enclaves where a de facto economic exclusion exists through the construction of high-cost dwellings, can insulate themselves from the redistributive demands of the less affluent with serious consequences for the under-funding of collective public goods such as schools, transportation and public policing in long-established urban and suburban neighbourhoods. The obvious economic and security benefits of this form of residential secession for the affluent middle and upper classes have led to a steady increase in common interest developments in other parts of the developed world including Britain and in many of the wealthier enclaves of the majority urban world (for a definition see Chapter 9) (Caldeira, 1996; Landman, 2000; Hook and Vrdoljak, 2002).[23]

Somewhat paradoxically, because New Urbanists preach the importance of social diversity as part of their '12 principles' along with a strong anti-sprawl agenda, they are regarded by conservatives as born again New Dealers aiming to re-impose big government on private development by tough planning laws, and the removal of mortgage, highway and utility subsidies for new 'green field' developments (Shaw and Utt, 2001). However, although the proponents of TND argue in favour of affordable housing in the neo-traditionalist style, precious little of it has been built. Certainly no low-income family could afford to purchase a house in the designer communities of Seaside or Celebration where prices have multiplied since the first lots were sold. In reality, new urbanist villages are little different from the affluent middle-class enclaves that are found all over the US except for a generally higher quality of architectural design and a greater emphasis on Garden City ideals in the physical layout of each development.

A vision not just for Britain: new urbanism's royal warrant

Undoubtedly the most famous advocate of traditional neighbourhood design and the most listened to critic of modern architecture and modernist urbanism is His Royal Highness, Charles Prince of Wales. Prince Charles' interventions in public debates over the future of cities and human habitats are legion, and culminated in 1989 in the publication of his lavishly illustrated book, *A Vision of Britain*. Charles has also been instrumental in establishing a neo-traditionalist school of architecture, the Institute of Architecture, which explicitly rejects the values and designs of CIAM in favour of classical revivalism and a celebration of indigenous building styles and materials.

Charles is not a strong supporter of tall skyscrapers, as he told an audience of London property developers, because, 'These are giant buildings, with immense public visibility, but serving only a private, indeed, a privatised, purpose' (HRH the Prince of Wales, 2001). The irony of this comment in the context of a royal family that cordons off all but a small part of its several palaces from public view was clearly lost on the heir to the throne. 'Such examples', Charles continued, 'and there are countless others – have given credence to the notion that these structures are alien; that in their very scale and their functionalist aesthetics they simply don't "fit" within the city and are doomed to long-term failure' (ibid.). The Prince's predilections are for 'networks of city streets, squares, parks and plazas, all of which require disciplined and well-articulated buildings to form and frame them'.

This, of course, has been the basic building block of urbanism for all of recorded history, and in all cultures, until the 20th century, when this template for the traditional (or timeless) city, became challenged by new notions of urbanism, the most potent of which was the Modernist vision of a city of towers in a new parkland landscape – a city on stilts and steroids.

(ibid.)

Predictably enough, the villain of the piece is

Le Corbusier, who looked forward to the day when the entire city of Paris would be razed and rebuilt; when the – and I quote – 'wretched pitched roofs are swept away, along with the casual cafes and places for recreation . . . that fungus which eats up the pavements of Paris'! The consequences of making this vision a reality, as most now recognise, have been disastrous, producing the shattered urban wastelands that have desolated entire communities and disembowelled some of our greatest cities.

(ibid.)

In an article for *The Spectator* magazine, in August 1998 the Prince of Wales offers the example of his own traditional neighbourhood development, the village of Poundbury in Dorset, designed by the leading figure in European new urbanism, Leon Krier, who also consulted on the Seaside development. Like its American cousins, Poundbury is 'a completely commercial project' backed by the Duchy of Cornwall itself (the Prince's land holding company). The Prince was determined, he said,

to offer good housing for rent and for sale; jobs and training opportunities for people living locally . . . [while] respect[ing] the fine traditions of local and regional architecture, house construction and town planning . . . serv[ing] to enhance the quality of the historic town, [and] blending it with the Dorset landscape.

(HRH the Prince of Wales, 1998)

Figure 4.11 Middlemarsh Street in Poundbury, Dorset, England. Master Design by Leon Krier. Copyright Mike Jones, courtesy of the Architectural Association Photo Library

Yet, Charles' model Dorset village (see Figure 4.11) achieves a density of only 30 dwellings per hectare, compared to some densities in inner London of 400 or even 500. Given that Britain is a crowded island where the vast majority of greenfield sites have been excluded from development, however attractive it may be, the Prince's Vision of Britain is one that only a tiny fraction of Britons will ever be able to experience.

Poundbury's success is not, therefore, to be measured in quantitative terms, but for the fact that 'it is about as far removed from the soullessness of many housing estates and business 'parks' as one could imagine. In short, it is becoming a place with its own spirit and identity, a proper part of the town of Dorchester, and not just a development' (ibid.).

But what of the impact of traditional neighbourhood design on people who have to stay put in the dense, congested metropolis? The Charter of the New Urbanism 'views disinvestments in central cities, the spread of placeless sprawl, increasing separation by race and income, environmental deterioration, loss of agricultural lands and wilderness, and the erosion of society's built heritage as one interrelated community-building challenge' (Congress of the New Urbanism, [1993] 2003). But is there any evidence that as social capital theorists argue (see Chapter 5), the return to traditional architectural and design practices can help build integrated, successful communities where once there was crime, urban decay and 'soullessness'?

Where TND-inspired refurbishments of 'declining' central city locations have occurred, the arrival of young gentrifiers has been encouraged (see Chapter 5 for a discussion of gentrification). As a consequence, a new and more politically active voting population has been able to elect Republican mayors in cities such as New York, Milwaukee, San Diego and New Jersey City where previously only Democrats ruled. As a result, anti-crime measures, taxation, control of the street homeless and other policy issues dear to the hearts of conservative voters have climbed the political agenda with significant consequences for how America's cities are run (Smith in Beauregard and Body-Gendrot, 1999). In other words, the consumers of traditional neighbourhood design have rarely demonstrated the same enthusiasm for the goals of social diversity and community integration as the movement's proponents.

CONCLUSION

All through this chapter we have seen a joust between two competing utopian images of the city – one is essentially pre-industrial, certainly pre-modern, and is built on a whole set of mental images that are filled with village greens, neighbours whose names you know and problems you share in, and with places for social gathering near at hand – a community centre, a church, a sports hall, a café or a pub. The other image of the city is metropolitan, and certainly modern, but it is a modernity, as Walter Benjamin would have it, that is hewn from the marble of ancient stone. Jencks is right to see Le Corbusier's legacy as the tragic view of architecture because it seeks to achieve what Jean-Pierre Vernant observes about the Athenian city where: 'Not only does the tragedy [of the City] enact itself on the stage ... it enacts its own problematics. It puts in question its own internal contradictions, revealing ... that the true subject matter of tragedy is social thought ... in the very process of elaboration' (Vernant in Copjec, 1999: 232).

This sense of the city as a space of tragic drama where events unfold, narratives are played out, and fates are sealed is, I would argue, the essence of modernist urbanism. Contrast this to atavistic utopias such as the Garden City and traditional neighbourhood design, and we move from the dynamic form of the stage to the static image of the tableau where all contradictions are banished and a timeless order prevails. At the same time there is a sinister edge to the notion of the city as the rehearsal room for some great animation in which citizens are expected to play strictly defined parts. It is not hard to see why totalitarian regimes embraced modernist architecture and urban planning so fervently, and why dystopian cinematic dramas from Fritz Lang's *Metropolis* (1927) to Jean-Luc Godard's *Alphaville* (1965) and Ridley

Scott's *Blade Runner* (1982) are set among the summitless towers of the modern metropolis (see Chapter 8). But then to use an image from the opening scene of David Lynch's movie, *Blue Velvet* (1986) is the communitarian fantasy propagated by the anti-modernists really able to insulate itself from the wider conflicts and discontents of the metropolis? The 'ear in the grass'[24] of David Lynch's chilling suburban thriller is a warning against the complacent assumptions of the metrophobes, in just the same way that Arthur Miller's *The Crucible* warned what life in a commonwealth of saints might be like for those accused of witchcraft. In the following chapter we explore how the themes of community and urban pathology that were first advanced by the Chicago School have been developed and explored in Britain and America in the decades following the Second World War.

FURTHER READING

Howard, Unwin and the Garden Cities Movement

A biography of Howard that also provides a critical analysis of Howard's broader social objectives is to be found in Robert Beevers' study, *The Garden City Utopia* (1988). Peter Hall's magisterial study, *Cities of Tomorrow* (1988), is packed with perceptive and informative accounts of Ruskin, Howard and the Garden Cities Movement. Peter Hall's co-authored volume with Colin Ward, *Sociable Cities. The Legacy of Ebenezer Howard* (1998) is also worth referring to for those interested in Howard's influence on urban design and for its stimulating contemporary interpretations of Howard's ideas for sustainable cities. M.H. Lang's *Designing Utopia* (1999) does much to confirm the important legacy of Ruskin in North America while reminding us that the Garden City was originally an American idea. *Raymond Unwin: Garden Cities and Town Planning*, by Mervyn Miller (1992) is a good place to start for those interested in finding out more in relation to the design principles of Garden Cities. Frank Jackson's *Sir Raymond Unwin. Architect, Planner*

and Visionary (1985) contains chapters on Unwin's contribution to town planning in Britain and the US and numerous photographs and diagrams of the housing schemes with which he was involved. Raymond Unwin's *Town Planning in Practice* (1971) though out of print, is still an important reference for town and country planners and is widely cited by new urbanist designers and proponents.

Planning in the US and Britain

On Frederick Law Olmsted's contribution to planning in America, I.D. Fisher's *Frederick Law Olmsted and the City Planning Movement in the US* (1986) is a useful source. Robert Fishman (ed.) *The American Planning Tradition* (2000) is a readable and informative collection of essays that includes general historical surveys of the town planning movement in the US along with individual city studies. Although it was originally published in the 1970s, Manfredo Tafuri's *Architecture and Utopia* (1999) is a modern classic and should be required reading for every architecture and town planning student as well as being of interest to students of urban society and urban history. Christine Boyer's *Dreaming the Rational City* (1983) is a thought provoking study of American planning history and contains a particularly insightful chapter on the Mumford sponsored utopian model of the regional city as derived from the work of the English Garden City builders. Dolores Hayden (2000) breaks down the post-1820 history of the American suburbs into a series of architectural-functional categories such as 'Streetcar Buildouts', 'Mail Order Suburbs' and 'E-Space Fringes'. For those looking for standard works on city planning in America, Scott's *American City Planning* (1971) and Ciucci *et al. The American City* (1979) are both to be recommended. A swift introduction to urban planning in America can be found in Gans (1991) Part Three, which also considers its social and political goals. Peter Hall's *Urban and Regional Planning* (1985) is one of the best historical accounts of planning in Britain. Other relevant texts include Garside and Hebbert, *British Regionalism 1900–2000* (1989) and Gordon Cherry *Town Planning in Britain since*

1900 (1996). Le Gates and Stout (eds) *The City Reader* (2003) 3rd edition, also provides a wealth of key readings on planning theory, history and practice.

Le Corbusier, CIAM and modernist urbanism

A fascinating read even several decades after its initial publication, Le Corbusier's *Towards A New Architecture* (1986) has the capacity to shock with the boldness of its vision and the quirkiness of the juxtaposition of machines and buildings that are such an authentic statement of the modernist aesthetic. Charles Jencks' *Le Corbusier and the Tragic View of Architecture* (1987) offers what I think is still one of the best intellectual biographies of 'Corbu'. Eric Mumford's *The CIAM Discourse on Urbanism, 1928–1960* (2000) is full of fascinating detail on the personnel and ideology of this hugely influential architectural and planning movement. Fishman's *Urban Utopias in the Twentieth Century* (1977) deals with Le Corbusier, Lloyd Wright and Howard as the twentieth century's greatest utopian visionaries – a verdict with which it is hard to disagree. An extract is to be found in various editions of Le Gates and Stout (1996–2003).

Neo-traditional design/new urbanism/gated communities/ anti-sprawl

Robert Davis tells his own story about the making of Seaside in Tod Bressi's *The Seaside Debates* (2002) in conversation with the architects Duany, Plater-Zyberg, Polyzoides and Solomon along with academic critics including the architect Colin Rowe. Eight US and Canadian cities feature in the debates. Leon Krier sets out his case for traditional architecture in *Architecture: Choice or Fate* (1998). Krier also defends the architectural values of Prince Charles in 'God Save the Prince' (Krier, 1988), an essay that is also reprinted in Charles Jencks (1988). Krier's criticism of contemporary architectural values is especially directed at architectural postmodernists such as Peter Eisenman. For those interested in the counter-case, Peter Eisenman, Manfredo Tafuri and Rosalind

Krauss *House of Cards* (1987), though principally an illustrative volume contains critical essays by all three authors.

The journalists Douglas Frantz and Catherine Collins (2000) took their family to Celebration for a year to find out what it was really like to live in Disney's brave new town. Their account is less critical than that of Ross (2000), although the town's attractions were not sufficient to induce them to extend their lease. *Suburban Nation* (Duany *et al.*, 2001) offers a critique of sprawl from a new urbanist perspective and suggests design and planning alternatives as well as examples of TND style communities. Kunstler argues for the new urbanist alternative to sprawl in his *Geography of Nowhere* (1993) while a summary of his argument can be found in 'Home from Nowhere', *Atlantic Monthly*, September 1996 at http://www.theatlantic.com/issues/96sep/kunstler/kunstler.htm

Alex Marshall's *How Cities Work: Suburbs, Sprawl, and the Roads not Taken* (2000), while highlighting the negative effects of sprawl does not see new urbanism as the solution, especially not in its Disney variant of Celebration, Florida. The *Smart Growth Manual* by Duany *et al.* (2003) offers practical illustrations of smart growth planning. *The Charter of the New Urbanism* edited by Michael Leccese and Kathleen McCormick (1999) expands on the principles of new urbanism, and is full of practical examples of alternative growth strategies, many illustrated by maps and plans. William Fulton, *The New Urbanism: Hope or Hype?* (1996) discusses the implications of TND for the future of the American city. *Solving Sprawl: Models of Smart Growth in Communities Across America* by F. Kaid Benfield *et al.* (2001) tells the story of what the authors consider to be successful examples of smart growth strategy across the US.

Two of the most influential books on gated communities and common interest housing developments are Blakely and Snyder's *Fortress America* (1999) and Mackenzie's *Privatopia* (1994). T. Coraghessan Boyle's novel *The Tortilla Curtain* (1996) is set in a hill-top gated community in Southern California and offers a thought provoking account of the starkness of California's socio-spatial divide told through the contrasting lifeworlds of wealthy liberals Delaney and Kyra Mossbacher and the Mexican illegals Candido and America Rincon. Octavia Butler's *Parable of the Sower* (1993) is an even bleaker near-future vision of Los Angeles' defended neighbourhoods at war with one another over increasingly scarce resources. Mike Davis' controversial account of contemporary Los Angeles, *Ecology of Fear* (2000b) is equally apocalyptic about the California dream's nightmarish prospects.

5

BETWEEN THE SUBURB AND THE GHETTO

Urban studies and the search for community in Britain and the United States after the Second World War

A theory limited to 'the city' is too narrow in scope to explain the changing landscape of the city itself.

Scott Greer, *The Emerging City*

INTRODUCTION

From a relatively dense and geographically specific cluster of research centres in the 1920s and 1930s (New York, Chicago, Cambridge/Boston, New Haven, Philadelphia), the academic study of the city in the US rapidly broadened both geographically and intellectually in the period after the Second World War. While European urbanists engaged with Marxism, and especially its structural variant, as the key expository model for explaining social and spatial differentiation in the city; in the US the Chicago heritage continued to attract new practitioners, notably Gans (whose work we discuss below) Greer (1962) and Suttles (1968). However, the dominant trend in American urban studies was away from the ecological and interactionist approaches favoured by Chicago sociology towards 'a critical political economy that gave greater emphasis to the structure of political power than to the mode of production'. While the scholars in this tradition were generally inclined towards radical liberal social objectives they were more influenced by the work of C. Wright Mills and Floyd Hunter than Karl Marx (Walton, 2000: 300–1).

In Britain, meanwhile, the Booth and Webb legacy of urban social investigation was continued after the Second World War at the Centre for Urban Studies, established at University College London in 1958 under the directorship of Ruth Glass (herself an LSE graduate and wife of the LSE sociology professor, D.V. Glass) with the declared aim of

> contributing to the systematic knowledge of towns, and in particular of British towns; to study urban development, structure and society; and to link academic research with social policy . . . the Centre [also] attempts to bring together the interests of the social sciences and those of allied fields, such as public health and town planning.
>
> (Glass, 1960)

However, lacking the strong methodological traditions of French and German sociology, British urban researchers often fell back on the investigative 'mise-en-scène' style of the Victorian moralists, so that community studies risked being regarded in Ruth Glass's words as '[the] poor sociologist's substitute for the novel' (Glass, 1966: 148 in Bell and Newby, 1971: 13). Glass went further in 'castigating community studies for their innumeracy, a simple lack of figures, even, in some cases, such basic ones as population statistics', while, '[a]ccompanying this has been a penchant for a descriptive narrative style', that has, 'meant that community studies can often be read *like* novels and some have, indeed, reached the best-sellers' lists' (ibid., original emphasis). In this chapter, we begin our investigation by sampling what I believe to be two of the most representative and influential series of community studies in Britain and America during the 'boom years' – Young and Willmott's *Family and Kinship in East London* and *Family in Class in a London Suburb* (Willmott and Young) and Herbert Gans' *The Urban Villagers* and *The Levittowners*.

IN SEARCH OF TRADITIONAL COMMUNITY

Family and kinship in East London

The most celebrated and influential 'best seller' of the British community studies school, *Family and Kinship in East London*, ([1953] 1992) written by the founders of the East London based Institute for Community Studies, Michael Young and Peter Willmott, almost certainly typifies the sociology that Glass was contesting. But as Michael Young's biographer, Asa Briggs, reveals, Young himself saw the novelistic style as a strength rather than a weakness of social research. 'I think', wrote Young, 'we all hope to write like novelists, if only we could get somewhere like it' (Briggs, 2001: 110–11). *Family and Kinship in East London* explicitly advertised the 'primitive anthropology' that Young had encountered at the LSE where he pursued the doctoral research on which the later volume was based under the supervision of Richard Titmuss (heir apparent to T.H. Marshall, and from 1950 the LSE's first Chair in Social Administration) (ibid.: 120).

Along with many young LSE sociologists at the time, the authors identified strongly with democratic socialism, and Michael Young helped to draft the 1945 Labour Party Manifesto 'Let Us Face the Future' which brought a majority socialist government to power for the first time in British politics (ibid.: 73). Understandably, therefore, the book empathised with the plight of the East End working-class communities that were experiencing profound structural changes as the jobs that had previously employed the male population either disappeared or moved down river. The authors also felt that the profound effects of post-war economic change were creating a widening gap between the Labour Government and its traditional working-class supporters (Young and Willmott, 1992: xiv).

Family and Kinship was based on a sample of just under 1,000 subjects, 1 in 36 of those appearing on the electoral register for the Bethnal Green borough, and all of whom were interviewed by the researchers (a considerable achievement by today's standards) (ibid.: 13). If the book's declared aim was to assess the impact of comprehensive re-housing on a traditional working-class inner city community, the unintended consequence was to demonstrate the remarkable coping mechanisms of different generations of East End women in holding the extended family network together. Paradoxically, the authors showed that solidarity was greatest among the densely populated urban core and *anomie* highest in the new, out of city estates, which, though they make no mention of Simmel's 'Metropolis and Mental Life', would tend to contradict the view that cities induce a blasé attitude of social withdrawal, or at least not if the experience of the working-class East End was indicative of other central city populations (ibid.: xix).

The fact that these dense associative networks either did not exist or took on a very different aspect in suburban London could be explained, the authors believed, by the very different class character of the outlying districts. As Willmott and Young wrote in their later study of suburban East London, 'social class is the key to understanding many of the differences in the suburb' (Willmott and Young, 1971: 98) (see Exhibit 5.1), but one could equally make the same point for the city as a whole. The fact that working-class families relied on the state for affordable accommodation meant that, as a class, they were far more subject to the caprices of government ministers and technocrats than middle-class suburbanites.

EXHIBIT 5.1   Peter Willmott and Michael Young, *Family and Class in a London Suburb*, 1960, p. 91 and p. 105

Some informants stressed the importance of maintaining social distance – 'We don't get familiar or over-friendly with the neighbours', Mrs. Brady said. And some, who had come from the East End, felt that Woodford was less welcoming to strangers, less friendly and easy-going, than the borough they had left. 'Woodford is a very cold affair compared with Bethnal Green', Mrs. Clarkson said, 'The people aren't unpleasant but they aren't neighbours or friends in a close sense'. It was our impression, too, from interviewing in both places, that though people in the East End might act as host to fewer personal friends outside the family, social relationships were more closely knit and loyalties stronger there' (91)
[. . .]
In the main, the working-class 'enclaves' – parts of South Woodford, for instance, or a network of roads just north of Wansted High Street, or the Council estates. As a man living near Woodford Green put it, 'The less fortunate people don't live in this part of Woodford. They tend to be combined in particular areas, and I must say they seem to get on with each other very well.'
People were well aware of the class character of different districts. 'The working class live on the other side of Well Road, said Mr. Burgess, 'but not this side'. 'You wouldn't get', Mr. Long told us, 'the Crescent and Albion Park mixing with Franklin Road and Arkwright Road'. The distinction between the two districts on opposite sides of the railway is particularly sharp. Mr. Day explained, 'The railway line is the dividing line – those who live below are not thought of as being as high class as those who live above'. 'You wouldn't expect', said Mr. Scott, 'the people here to go hobnobbing with the people on the Council estate on the other side of the line'. 'I'm no snob', another man asserted, 'but it seems to me it's south of the railway line you've got the working classes in Woodford.' (105)

Controversial though its methodology might have been, Bell and Newby defended community studies because it allowed students to get to grips 'with the social and psychological facts in the raw' (Arensberg and Kimball, 1967: 30 in Bell and Newby, 1971: 13), by which is presumably meant without the excess theoretical baggage that tends to obscure this factual universe. Reading Bell and Newby, one has a sense of the close affinities between American urban studies and British community studies where lines cannot easily be drawn between fictional and 'scientific' representations of the social life of towns and cities and where personal and familiar dramas are set against the transformations of social and economic life engendered by capitalist modernity. However, these structural features of social change are often no more than glimpsed through the keyhole of a backyard ethnography that takes participant-observation very often as its sole and legitimating methodology.

Indeed, as Bell and Newby note, the community studies approach owes much to the classical anthropological techniques developed by Malinowski. This was as true for the US as Britain, and Lloyd Warner proudly admitted to having 'studied American communities with the techniques [he] previously used in the investigation of Australian tribes, and other anthropologists have used in Africa, Polynesia and New Guinea' (Warner, 1963 in Bell and Newby, 1971: 62). The point that Bell and Newby make about anthropological methods having been designed to study non-literate peoples is aimed at highlighting the inattention of urban ethnographers to recorded data that have long been utilised by historians and social statisticians in literate societies. Nevertheless, where a community culture is predominantly oral, such as in the Bethnal Green district of London's East End – as Michael Young and Peter Willmott argued, extensive use of field interviews may be the only way to discover how kinship networks work and survive (or fail) under the pressure of increasing geographical mobility.

Young and Willmott, looking back from the vantage point of the 1980s, put the blame for the loss of community on Le Corbusier inspired flatted estates (although many of the former inhabitants of Bethnal Green actually moved to the Raymond Unwin designed, and Garden City inspired new towns such as Dagenham). They also blamed local and central government planners of the time for using slum clearance as a means of reducing urban population densities without recognising that these 'redevelopment zones' represent complex, intergenerational networks that cannot easily be recreated, a sentiment that they shared with Jane Jacobs and Herbert Gans (see below) (Young and Willmott, 1986: xxi–xxiii).

As the signature work of the Community Studies School, *Family and Kinship* still resonates with the empathetic reform-minded investigative style that characterised the urban surveys of the nineteenth century. In adopting an ethnographic method centred on the household and the family they were unwittingly foreshadowing the research agendas of numerous feminist sociologists, anthropologists and oral historians (Stacey in Young and Willmott, 1992: ix). Also, while Willmott and Young, 1992 make no direct mention of American urban studies, the choice of what was effectively a 'ghetto' community – though not one at the time that was demarcated along colour lines – and the mapping of its geographical dispersal to the suburbs was an even stronger feature of urban sociology on the other side of the Atlantic, and we now turn to consider the development of this tradition in post-war America.

The Urban Villagers

Herbert Gans' study, *The Urban Villagers* ([1962] 1982) which focused on the Italian-American population of what was then the deprived West Side of Boston in the 1950s is directly comparable to the Young and Willmott study. Gans clearly saw his book as an investigation of a working-class (rather than an 'ethnic' neighbourhood) and, as in the case of East

London, the focus was on a primarily white working-class enclave in which the community is basically seen as a 'set of people and institutions to serve and protect the family and peer group' (Gans, [1962] 1982: xi). Just as with Bethnal Green, Gans found a remarkable 'class defensiveness' where families would disavow certain residential locations or occupations as being unsuitable for 'people like us' (see Exhibit 5.2 below).

Both the Bethnal Green study and the Boston study seem to emphasise the *Gemeinschaft* or village-like quality of urban life, and both were accused of an overly romantic portrayal of a working-class experience that appeared to be coming to an end. Like Young and Willmott, Gans thought that this was a pity, because there were valuable aspects of community life that would be lost through the slum-clearance programmes, and he hoped that sociological investigations might persuade policy-makers and government to act differently and more sensitively. But at the same time, along with more conservative

> **EXHIBIT 5.2** Herbert Gans, *The Urban Villagers*, 1962, pp. 220–1
>
> The rejection of middle-class ways is . . . based on the West Ender's fears regarding the middle-class world, and on the recognition that he lacks the skills to participate in this world. [H]is inability to understand bureaucratic behavior or object-oriented groups, and the absence of the kind of empathy needed to interact with middle-class people all create discomfort or fear. So does his lack of education, which he feels to be an obstacle in communicating with them. . . . On the few occasions when I tried to bring West Enders into contact with middle-class acquaintances and friends, they either refused the invitation or participated only minimally. . . . The fear of contact with middle-class people is based on feelings of inferiority, and consciousness of some real – and some imagined – deficiencies. But . . . such feeling should not be mistaken for envy.

social analysts such as Charles Murray, the political anthropologist Edward Banfield, and even the liberal urban sociologist William Julius Wilson, Gans believed that what (after David Riesman (1950)) he took to be the 'inner oriented' culture of certain immigrant communities was detrimental to the well-being and economic and cultural flourishing of these communities and to urban society in general (see Exhibit 5.2).

To contemporary ears the ease with which Gans feels able to argue that 'if people could choose their subculture freely, I believe that the professional upper middle-class subculture would be the most deserving of choice' (Gans, 1962: 264) can seem uncomfortably haughty and elitist. Yet, in recommending the middle-class 'object-oriented' values of education, public service and community association, Gans is merely articulating a moral consensus that runs from Chicago settlement sociology to the contemporary 'social capital' formulations of Robert Putnam and his colleagues, as we shall see later in this chapter. However, in contradistinction to earlier sociologists such as Burgess and Park, the emphasis on urban ecology (place determines character) has been replaced by an explicit cultural determinism (character determines place).

In Gans' subsequent study these social identities, based around class, race and ideology are analysed in what he took to be the 'environmentally neutral' context of the new town communities that had begun to spread rapidly during the 1950s and 1960s. But before we look in detail at Gans' suburban portraits, we turn to an examination of arguably the most influential book to be published on the life of cities – Jane Jacobs' *The Death and Life of Great American Cities*.

QUESTIONS TO DISCUSS

1 **Compare the attitude of Woodford residents and the Levittowners to their new surroundings. How are feelings of social difference expressed in the two communities?**

2 **What qualities of 'the city' do those interviewed by Young and Willmott, and Gans believe are missing from the suburbs in which they now live?**

CITIES FOR PEOPLE: JANE JACOBS AND THE CASE FOR THE DENSE METROPOLIS

When it was first published in 1961, Jane Jacobs' *The Death and Life of Great American Cities* proved to be a milestone in urban analysis that has had a profound and lasting impact on the way professionals, politicians and the public at large view the city. Robert Fishman describes its effect as '[t]he most powerful intellectual stimulus to the revival of the American planning tradition', and, 'perhaps the most powerful attack on the idea of planning ever written' (Fishman, 2000: 19). Her book was intended as a counterblast to the vogue for low-density, car dependent suburban neighbourhoods that had emptied the 'great cities' of population, prosperity and prospects. In particular she spoke out against the impact of federal government imposed 'Euclidian zoning' standards where a municipality is divided into standard districts in order to keep densities as low as possible and to separate land-use functions as far as possible (Wickersham, 2001).[25]

Jacobs felt that the perceived defects of the metropolitan environment had been both exaggerated by a cynical media and political class, but also compounded by poor planning and community design. Cities of a traditional, monocentric type that were still found in many parts of Europe and a dwindling number of American cities worked, she believed, because they were densely rather than sparsely populated and provided social and architectural diversity.

Jane Jacobs described her 'method' in the following way: 'I look for examples of behaviour first. Eventually, when I start to see patterns in them, I begin to generalize' (Gee in Allen, 1997: 159). This inductive approach to urban analysis was developed from her training as a journalist and writer on *Architectural Forum*, but the parallels with the Robert Park style of 'dirty hands' research seem rather more obvious to

EXHIBIT 5.3   Jane Jacobs, *The Death and Life of Great American Cities*, 1992, pp. 50–2

The stretch of Hudson Street where I live is each day the scene of an intricate sidewalk ballet. I make my own first entrance into it a little after eight when I put out the garbage can, surely a prosaic occupation, but I enjoy my part, my little clang, as the droves of junior high school students walk by the center of the stage dropping candy wrappers. (How do they eat so much candy so early in the morning?)

While I sweep up the wrappers I watch the other rituals of morning: Mr. Halpert unlocking the laundry's handcart from its mooring to a cellar door, Joe Cornacchia's son-in-law bringing out his sidewalk folding chair, Mr. Goldstein arranging the coils of wire which proclaim the hardware store is open, the wife of the tenement's superintendent depositing her chunky three-year-old with a toy mandolin on the stoop, the vantage point from which he is learning the English his mother cannot speak. . . .

When I get home after work, the ballet is reaching its crescendo. This is the time of roller skates and stilts and tricycles, and games in the lee of the stoop with bottletops and plastic cowboys; this is the time of bundles and packages, zigzagging from the drug store to the fruit stand and back over to the butcher's; this is the time when teen-agers, all dressed up, are pausing to ask if their slips show or their collars look right; this is the time when beautiful girls get out of MG's; this is the time when anybody you know around Hudson Street will go by.

Jacobs' critics than to the author herself. Although Jacobs often refers to the City of Chicago in 'Great American Cities', no mention is made of the Chicago School of urban sociology. This is surprising, because in several respects her work implicitly advocates an ecological approach to the study of urban land-use. For example, her chapter on 'The curse of border vacuums' (Jacobs, 1992: 257–69) highlights the blight-prone nature of 'interstitial spaces' such as railroad tracks and waterfronts that Thrasher and others identified as prone to vice and delinquency. Also, in Jacobs' championing of the notion of a folk community there is more than an echo of Redfield and Wirth's insistence that cities are, above all, the containers of human communities.

Jane Jacobs' recipe for good cities could be boiled down to four basic rules: 'short blocks, mixed uses, old buildings mingled with new, and residential densities of at least 100 units an acre' (Zotti in Allen, 1997: 61). This type of ideal city neighbourhood is to be found in the 'intricate sidewalk ballet' of Hudson Street in Greenwich Village where she lived for many years and where *The Life and Death of*

American Cities was written (Atkinson in Allen, 1997: 52) (see Exhibit 5.3).

For Jacobs the twin threats to vibrant city life came on the one hand from the abstract logic of modernist architects and planners who believed that they could build utopia in splendid isolation from the rest of humanity, and on the other from the withdrawal to suburban Garden Cities (Jacobs, 1958: 142). There is a certain irony here, for Jane Jacobs has become a champion for supporters of the 'new urbanism' (Kunstler, 2001) which, as we have seen in the previous chapter, retains many of the characteristics of Ruskin's anti-metropolitanism and the Garden City model for all its celebration of 'street architecture' and denunciation of suburbia.

Urban community for Jacobs is comprised of two essentials – density and diversity. But, contrary to Garden City advocates, high population densities need not be associated with overcrowding and slums (Park Avenue in New York and Belgravia in London both have high population densities and could hardly be called overcrowded slums). According to Jacobs, poorly housed and socially stressed households do not

produce viable communities, (although Bethnal
Green in the 1950s by any standard would have been
considered an overcrowded slum), but rather, '[d]iver-
sity and its attractions are combined with tolerable
living conditions in the case of enough dwellings for
enough people, and so more people who develop
choice are apt to stay put' (Jacobs, 1961: 208). Jacobs
is sceptical of Garden City type densities as capable
of generating enough spontaneous interaction to
produce that street ballet she so admired on Hudson
Street, and neither did she think that Lewis
Mumford's prescription for 'in between' densities of
100 to 125 persons an acre would work.[26] Her intuition was that vibrant inner city communities did not
need to be surrounded by parks and gardens, and
Greenwich Village provided a good example of relatively high building densities but of mixed types
housing a diversity of populations and offering a
variety of employments and commercial services.
The prototype is unmistakably European and dates
back to the high medieval chartered cities that Weber
associated with the birth of modern capitalism, and
which the great Austrian planner Count Camillo von
Sitte believed was the highest achievement of our
urban civilisation (Sitte, 1965).

To understand why, not only the US, but most
parts of the developed world, abandoned the dense
mixed-use city in favour of functional decentralisation we have to voyage into the twilight world of the
suburbs, the new towns and the edge cities.

QUESTIONS TO DISCUSS

1 **Highlight the ways Jane Jacobs uses
the metaphor of drama to describe the
street life of her Manhattan
neighbourhood.**

2 **Do you find this a convincing portrayal
of city life?**

3 **If Hudson Street was still the kind of
urban experience Jane Jacobs
describes, do you think it would be an
attractive place to live and work, and
why?**

THE SOCIOLOGY OF SPRAWL: SUBURBS, NEW TOWNS AND EDGE CITIES

The Levittowners

In the post-Second World War period, the archetypal
suburban development in the US was 'Levittown',
named rather ostentatiously for (or after) William
Levitt, the builder who bought, designed and developed a large housing tract originally in Pennsylvania,
and subsequently in New Jersey. The eponymous
study of the New Jersey settlement, *Levittown* published in the late 1960s by Herbert Gans became one
of the classics of urban (or perhaps more accurately
suburban) sociology and provides an interesting
comparison with Willmott and Young's study of a
London suburb (Willmott and Young, 1960) that
we looked at previously. Gans spent over a year of
'participant-observation' as a member of the new
town and his findings though highly personal
were consistent with the style of urban reportage
with which students of Chicago School Sociology
were familiar (although Gans himself was professor
of social science at the University of Pennsylvania). If
we contrast the concerns of the residents of American
suburbia with those of the British suburbanites
studied by Willmott and Young, it is obvious how
much more acutely the racial question is felt among
the majority population in Levittown. There are some
obvious demographic reasons for this – large-scale
immigration of colour into East London did not really
occur until the late 1960s, whereas the increasing
presence of African Americans in the larger northern
urban conurbations of the US had been apparent since
the 1940s. Also anti-segregationist federal legislation
had helped to loosen if not dismantle the colour line
that restricted so many black families to inner-city
ghetto neighbourhoods.

Attitudes to the diffusion of what, at the time,
were called 'negro' residents into predominantly
white neighbourhoods were often a signal of the
political or ideological predisposition of the white
population, and those middle-class whites that were
in favour of community integration Gans tends to

refer to as 'cosmopolitans'. Then, as now, they were not a significant part of the majority population and the typical response of the average Levittowner, as Gans points out, was motivated from a fear of declining house values and an openly expressed moral panic around inter-racial personal relationships. In the suburban London example, however, it was equally clear that class and status distinctions remained important even in the more affluent, tree-lined avenues of Woodford where a 'them and us' division prevailed between the established suburban middle class and the *arriviste* ex-East Londoners made good (Willmott and Young, 1960) (see Exhibit 5.1 above). Both Gans, and Willmott and Young, were uncovering what might be termed a 'suburban mentality' that was to become the authentic folk identity of post-war America, best exemplified by Whyte's study of the Broadacre-style new town of Park Forest, Illinois.

Park Forest, Illinois

William Hollingsworth 'Holly' Whyte was a former Marine Corps captain turned magazine editor. Although he had received no formal sociological training, Whyte turned out to be one of the most acute observers of American urban life in the post-war era, producing a series of best-selling books including *The Organization Man* in 1956 which, along with C. Wright Mills' *The Power Elite* (1956) and David Riesman's *The Lonely Crowd* (1950) helped to explain the changing face of American society during the post-war boom.

In 'How the New Suburbia Socialises', Whyte studied the Garden City-inspired new town of Park Forest, Illinois, but he came to a different conclusion to Gans on the relationship between character and environment, seeming to agree rather more with the 'place determines character' thesis of the Chicago school and human ecology. Thus Whyte comments:

> Once people hated to concede that their behaviour was determined by anything except their free will. Not so with the new suburbanites; they are fully aware of the all pervading power of the environment over them. As a matter of fact, there are few subjects they like so much

to talk about; and with the increasing lay curiosity about psychology, psychiatry, and sociology, they discuss their social life in surprisingly clinical terms. But they have no sense of Plight, this they seem to say, is the way things are, and the trick is not to fight it but to understand it.

> (Whyte in LaFarge, 2000: 34)

Like Willmott and Young, Whyte pays close attention to the layout of houses and streets and, especially, the play routines of children, to explain patterns of social interaction. Park Forest comes across as an essentially white, middle-class enclave designed for thirty-somethings in which the women (few of whom appeared to work) created and maintained the social circle, while their husbands 'just tag along' (ibid.: 34). Whyte describes it as a religious melting pot with Jews, Catholics and the non-religious joining in each other's discussion groups and helping with childcare but, as with Levittown, it was certainly not racially tolerant when it came to 'the unresolved controversy over admitting Negroes to the community'. Here, Whyte's disingenuousness, 'that much of the feeling was not rooted in racial ill will so much as in economic and social fears' contrasts unfavourably with Ruth Glass's more clear-eyed assessment of racial attitudes in 1950s' London (see below). Whyte cites as an example the couple who fled from a 'deteriorating ward' (for which read one with an increasing Black presence) in Chicago and who now viewed any change to the all-white character of Park Forest as 'a threat to everything they had been striving for' (ibid.: 38–9).

However, far from having their prejudices washed away by the tides of time as Whyte so confidently predicts, the evidence pointed in the other direction, precisely because – and this is where new urbanist utopias find their origin – an essentially white, essentially middle-class community where people are not too wealthy or too poor has become the folk archetype of the 'typical' American community for several generations. Simmel believed that it was in order to escape the conformism and monotony of such places that young rural dwellers descended on the intoxicating throng of the metropolis. He could hardly have suspected that for many millions of Americans,

Europeans and Australasians, the comfort and security of conformism and monotony seemed a far more attractive prospect than the metropolis' madding crowd.

The rise of sprawl

If the focus of urban sociology had been the classic, monopolar city described by Burgess of dense streets, crowded tenements, bustling markets and the edgy cosmopolitanism of first generation migrants, the concerns of researchers in the 1940s and 1950s began to shift, along with the mostly white population, to America's rapidly growing suburban fringe. As Kenneth T. Jackson writes:

> suburbia has become the quintessential physical achievement of the United States; it is perhaps more representative of its culture than big cars, tall buildings, or professional football. Suburbia symbolizes the fullest, most unadulterated embodiment of contemporary culture; it is a manifestation of such fundamental characteristics of American society as conspicuous consumption, a reliance upon the private automobile, upward mobility, the separation of the family into nuclear units, the widening division between work and leisure, and a tendency toward racial and economic exclusiveness.
>
> (Jackson, 1985: 4)

Cities, since their most ancient foundations, have always given rise to satellite settlements, but up until the eighteenth century, the periphery of the town, or 'suburbe' was associated with the lower social orders, mean dwellings, crime and delinquency (Fishman in Fainstein and Campbell, 1996: 26). However, by the time of the industrial revolution, and with the advent of better and more reliable forms of transportation, the ability to enjoy the advantages of the facilities, goods and employment opportunities of the central city while avoiding the crowds, noise, and costly and unsanitary housing was extended to the new professional and merchant classes. Aided by a combination of river ferries, suburban railroad developments, and the arrival of the omnibus to cities such as New York City, Chicago, Boston, Philadelphia and Washington DC, the US witnessed a massive growth in suburban development from the 1840s which had the effect of far extending the boundaries of these regional cities into the rural hinterland. Thus, by the dawn of the twentieth century, the suburb had become a 'bourgeois utopia' of low density, tree lined residential settlements with no, or little, commerce or industry and with a high degree of class exclusivity (the more working-class residents mostly being confined to service roles as domestic servants, gardeners, nannies, etc.) (ibid.: 27–9). By the second half of the twentieth century, suburbia had become democratised as the new middle classes began to populate the Levittowns and Park Forests in their tens of millions. Writing in the collection of essays put together by *Fortune* magazine in the late 1950s W.H. Whyte warned that:

> [I]n the next three or four years Americans will have a chance to decide how decent a place this country will be to live in, and for generations to come. Already huge patches of once green countryside have been turned into vast, smog-filled deserts that are neither, city, suburb, nor country, and each day – at a rate of some 3,000 acres a day – more countryside is being bulldozed under. . . . Flying from Los Angeles to San Bernardino – an unnerving lesson in man's infinite capacity to mess up his environment – the traveller can see a legion of bulldozers gnawing into the last remaining tract of green between the two cities, and from San Bernardino another legion of bulldozers gnawing westward.
>
> (Whyte, 1958: 115–16)[27]

The rise of these 'smog filled deserts' was no accident, for Whyte showed how the impact of the Federal Highways Act of 1956 with its stated objective – 'to disperse our factories, our stores, our people in short, to create a revolution in living habits' (ibid.) – was instrumental in making sprawl the American mode of urbanisation, a policy that, if anything, has been reinforced by successions of federal administrations up to the present day. As Dreier, Mollenkopf and Swanstrom note, a survey asking 149 leading urban scholars to identify the most important influences on the American metropolitan areas since 1950 identified, 'the overwhelming impact of the federal government on [the] American metropolis, especially through policies that intentionally or unintentionally promoted suburbanisation and sprawl' (Dreier *et al.*, 2001: 93). Not only did such policies 'favour investment in suburbs and disinvestments from central cities', but as we shall discover later in this chapter, '[g]ovenment policies have also favoured concentrating the poor in central cities' (ibid).

The alarm bells had been ringing about the environmental and social cost of sprawl in the US for several decades, but its march across the American landscape has been relentless. Harper's magazine reported a US Department of Agriculture study that showed an average of 1 million acres of rural land had been lost to urban sprawl every year since 1970 (LaFarge, 2000: 123). The dispersal and fragmentation of social, economic and political life that government sponsored laissez-faire urban expansion has brought in its wake has provoked something of a moral crisis, not dissimilar to that which the Chicago sociologists pointed to in the 1920s and 1930s. The communitarian values that many Americans felt they had rediscovered in the Levittowns, were now coming under attack from a variety of fronts. One was the decentring of communities through the loss of downtown retail and office space to malls and business parks (see Exhibit 5.4). Another was the increasing automobile dependency that resulted from extended home to work journeys resulting in major congestion problems and severe environmental impacts (the smog referred to by Whyte being just one of them – see also Chapter 9). However, none of these costs have borne heavily enough on individual households and businesses to halt the relentless exodus to the non-place urban realm (Webber, 1964) of the 'exurbs' and the edge cities – the new frontier of late modern America.

The rise of Edge City

Think of the small communities of the US as they used to be before the Second World War and you have an image of Tysons Corner Virginia, little more than a junction between country roads, with a beer joint and a feed store surrounded by a dairy farm. Today, it is home to the Tysons Corner Center, with six bespoke 'anchor' stores and no fewer than sixteen restaurants catering to every taste and budget. It is the 'largest agglomeration between Washington and Atlanta' (Garreau, 1991: 350), and provides 100,000 jobs and over 25 million square feet of office space, but it has no civic government. Edge cities have been variously called commercial centres, suburban business districts, suburban cores, urban subcentres, pepperoni pizza cities, superburbia, technoburbs, nucleations, disurbs, peripheral centres, urban villages, middle landscapes, multinucleated metropolitan regions and centreless cities. One thing they all have in common is that they are growing faster than any other existing urban form, they are attracting increasing numbers of businesses, residents and consumers and, above all, they are vast private domains in which 'public space' – from parking lots to restrooms to children's play areas – is nowhere to be seen. In these entirely privatised post-urban realms one is either a customer, an employee or an 'undesirable'.

Unlike many critics of the new suburbia, Joel Garreau is optimistic about the capacity of his fellow Americans to re-build community in these boundary-less zones that incorporate residential developments, business parks and shopping malls, but which have no government, no collective voice (such as a local newspaper or radio station), and little in the way of collective identity markers – no 4th of July Parades, no town map, nor even a road sign saying 'Welcome to Edge City. Population rising!'[28] Anticipating Robert Putnam's argument about the existence of network based 'bridging social capital' (see below), Garreau found a number of self-help group organisations even among the scattered technology executives of Orange County, California where:

> People who worked together started playing sports together. Some who had come together for business reasons stayed together to work on the Orange County Philharmonic, Opera Pacifica, and the Boy Scouts. Members found themselves being invited to one another's weddings.
>
> (Garreau, 1991: 289)

Garreau is making the point that community does not have to be about propinquity (or close physical location) but rather it is about sharing a common bond, and a common network. In the new American Frontier, neighbours may remain forever strangers,

---

EXHIBIT 5.4   Joel Garreau, *Edge City. Life on the New Frontier*, 1992, pp. 4–5

Today, we have moved our means of creating wealth, the essence of urbanism – our jobs – out to where most have lived and shopped for two generations. That has led to the rise of Edge City.

Not since more than a century ago, when we took Benjamin Franklin's picturesque mercantile city of Philadelphia and exploded it into a nineteenth-century industrial behemoth, have we made such profound changes in the ways we live, work and play.

Good examples of our more than two hundred new Edge Cities are:

The area around Route 128 and the Massachussetts Turnpike in the Boston region that was the birth-place of applied high technology;

The Schaumberg area west of O'Hare Airport, near which Sears moved its corporate headquarters from the 110 story Sears Tower in downtown Chicago;

The Perimeter Center area, at the northern tip of Atlanta's Beltway, that is larger than downtown Atlanta

Irvine, in Orange County, south of Los Angeles.

By any functional urban-standard – tall buildings, bright lights, office space that represents white-collar jobs, shopping entertainment, prestigious hotels, corporate headquarters, hospitals with CAT scans, even population – each Edge City is larger than downtown Portland, Oregon, or Portland, Maine, or Tampa, or Tucson. Already, two thirds of all American office facilities are in Edge Cities, and 80 per cent of them have materialized in only the last two decades. By the mid-1980s, there was far more office space in Edge Cities around America's largest metropolis, New York, than there was at its heart – midtown Manhattan.

---

but the affluent middle class will rack up thousands of air miles to stay in touch with old college and work friends. The transit lounges of America's major hub airports are what the local grocery store would have been for their grandparents' generation – a place for 'When Harry Met Sally' chance encounters, or for the modern equivalent of the high school reunion – the business convention – at a suitably well-appointed airport hotel and conference facility.[29] The proliferation of such transit and gathering nodes, or 'non places' (Augé, 1995), emphasises Max Weber's point that the urban is characterised above all else by a necessary density of communication and trade. In the past, natural harbours and navigable rivers provided the most opportune location for the expansion of early settlements, now air hubs and interstate highway intersections fulfil the same function. In a nation that is the size of a continent, and a state that is wealthier and more populated than most nations, it is not hard to see why metropolitan Los Angeles can provide

sufficient custom to sustain five commercial airports in the world's most famous edge city agglomeration.

In the following section we consider how the social implications of these locational trends have been conceptualised through the notion of 'social capital', before going on to consider how residential segregation continues to reflect social and economic inequalities to a far greater extent in the US than in other affluent societies.

THE NEW COMMUNITY STUDIES: SOCIAL CAPITAL AND CIVIC EMPOWERMENT

Of all the post-war studies on the life and times of contemporary American communities, Robert Putnam's *Bowling Alone. The Collapse and Revival of American Community* (2000) stands out as a monumental achievement. Unusually for a work of sociology it has reached a mass audience, and Putnam's ideas have even attracted the personal interest of presidents and prime ministers. Deriving from his earlier co-written study of civic culture in Italy, *Making Democracy Work* (1993a), Putnam has been credited with the development and popularisation of the concept of 'social capital', although he points to at least six separate coinages during the twentieth century including Jane Jacobs (who intended the term as an indicator of neighbourliness), and Pierre Bourdieu (see Chapter 8) who saw it as a resource value that was to be found in social networks (such as the notion of 'the old school tie') (Putnam, 2000: 19). Social capital is easily defined, but less easily measured. Putnam contrasts physical capital, which refers to physical objects, and human capital, which refers to the properties of individuals, to social capital which refers to connections between individuals – social networks and the norms of reciprocity and trustworthiness that arise from them'. Putnam agrees that social capital is related to the notion of 'civic virtue', but adds that it is at its 'most powerful when embedded in a dense network of reciprocal social relations. A society of many virtuous but isolated individuals is not necessarily rich in social capital' (ibid.).

Putnam then makes a distinction between what he calls 'bridging social capital' and 'bonding social capital'. Bridging social capital is that which exists between members of distant or weak networks and is based on inclusivity, while bonding social networks such as country clubs, ethnic fraternal organisations and so on tend to be based around some form of exclusivity (ibid.: 23). Putnam has often been accused of being nostalgic for a lost America of white painted churches, marching bands and bake-sales. Fortunately, he is too good a social scientist and historian to fall into such a trap – his is a narrative of community decline, but the same loss of civic capacity has happened before and civic revival has followed it – he rejects 'gauzy self-deception' in favour of the basic instrument of the social scientist – statistics. Putnam produces and analyses a large number of social capital related statistics to examine how the profile of American civic participation has changed since the

1950s. For our purposes, however, the most interesting chapter is that which relates to 'mobility and sprawl'. Here, Putnam produces some fascinating counter-intuitive evidence to show that Americans are not more mobile than they were in the 1950s, and are 'if anything, slightly *more* rooted residentially than a generation ago' (ibid. 205).

But the stereotypical view of the suburban 'Babbit' as an uncivic, family-centred and self-regarding obstacle to community involvement and concern is contested by Putnam who argues that

> residents of small towns and rural areas are more altruistic, honest, and trusting than other Americans. In fact, even among suburbs, smaller is better from a social capital point of view. Getting involved in community affairs is more inviting – or abstention less attractive – when the scale of everyday life is smaller and more intimate.
>
> (ibid.)

As we noted in the previous chapter, this view of the city as degrading of human potential and civility, is arguably the most robust and persistent theme in American behavioural sciences, but as we shall see in Chapter 8, not every contemporary American urbanist shares this view. Richard Sennett, for example, believes that city life helps us to live with strangers, to be more tolerant, and to accept diversity – 'Sameness stultifies the mind; diversity stimulates and expands it' (Sennett, 2000: 1), Sennett's belief that 'society gains equally with individuals when people's experiences are not limited just to those who resemble them in class, race, or ways of life – is a view that even the country dwelling Prince of Wales endorses (HRH the Prince of Wales, 2001) (see Chapter 4). The other chief virtue of city life, according to Sennett, and it is an issue that we explore at greater length in Chapter 8 is that 'the experience of urban life can teach people how to live multiply within themselves'. Thus, he agrees with Simmel that small towns shrink your horizons and diminish the potentialities of self-realisation, cities expand and increase them (Sennett, 2000: 3).

On the other hand, Putnam does acknowledge that suburbanisation has led to racial segregation, but together with a long line of sociologists dating back to the Chicago school who subscribe to the 'neighbourhood effect', Putnam believes that deprived inner city neighbourhoods need to feel and behave more like small towns. The idea of 'urban villages', which has been strongly backed by the Prince's Foundation (an educational charity dedicated to community regeneration supported by the Prince of Wales) aims to produce the type of 'social mixing' that its proponents believe occurred in more traditional/historical settlements (although the historical evidence for this is far from compelling).

The case of Crown Street in the once notorious Gorbals district of Glasgow in Scotland has been widely trumpeted as a successful community regeneration initiative that combines mixed uses (à la Jane Jacobs) with a more diverse resident population that aims at tackling the perceived disadvantages of homogeneous class and ethnic enclaves.[30] However, despite similar schemes in the United States that combine traditional neighbourhood design concepts with pro-social capital generating institutional features (see Chapter 4), the hard fact is that the central cities of the US have remained segregated and ethnically concentrated in terms of African American and Latino residents. As Portes and Landolt remind us, '[i]t is not the lack of social capital, but the lack of objective economic resources – beginning with decent jobs – that underlies the plight of impoverished urban groups' (Portes and Landolt, 1996).

If the suburb had become the community of choice for white middle-class Americans, it was to the ghetto that the nation's blacks had to get back. In the following section we examine the causes and consequences of ghettoisation and the phenomenon of affluent incursion known as gentrification in order to shed light on an urgent and continuing debate.

THE POLARISED CITY: GENTRIFICATION AND GHETTOISATION

As we shall see in the following chapter, the 'dual city' thesis has a long pedigree, from Engels' accounts of mid-Victorian Salford to the depictions of contemporary New York City and London provided by Mollenkopf and Castells (1991) and Fainstein et al. (1992). What each of these narratives highlights is the persistence of social divisions in the city expressed through spatial segregation. In this section we examine how urban researchers have attempted to study this process by exploring concentrations of wealth (gentrification) and concentrations of poverty (ghettoisation).

The bourgeois urban landscape and the struggle for living space

The term 'gentrification' was first used by the sociologist Ruth Glass to describe the arrival of middle-class incomers among a select number of inner-London districts that had been previously occupied by low-income groups (Glass in Glass et al., 1964: xiii–xlii). But the phenomenon, if not the expression, goes back much further to the great 'slum clearance' schemes of Haussmann's Paris and to the 'improvements' of mid-nineteenth-century Britain in which popular working-class neighbourhoods were torn down and replaced by elegant squares and town houses for the urban gentry (Smith, 1996: 34–5). By the 1970s, urban sociologists and urban geographers had begun to adopt the term to describe middle-class 'invasion' of poor and minority urban neighbourhoods in many other parts of the world. Gentrification takes different forms and involves different population groups in many different urban locales. It does not have to involve the sort of violent displacement of the urban poor that Smith refers to in the context of the eviction of the homeless of Tompkins Square Park (ibid.: 4–12). Indeed, Sumka (1979 in Atkinson, 2000: 310) estimated that of the two million mostly white working-class residents being annually displaced in the US in the 1970s, 86 per cent were the results of 'market' determined gentrification – a figure that Hartman (1979 in ibid.) regarded as an underestimate. In all cases this mostly gradual and invisible process is associated with an increase in the number of incomers who are more likely to be homeowners or to be able to afford expensive private sector rents. Such new arrivals are also typically managers or professionals with college educations whose consumption demands 'may alter the social characteristics and services of an area so that residents' social networks are distended while the cost of living may increase as service provision caters for higher income groups' (Atkinson, 2000: 307).

As Exhibit 5.5 below shows, gentrifiers may have different motives for wishing to reside in a central city location – for some it will be the desirability of quick access to place of work coupled with the aspiration to own a larger and more substantial property than would be open to them in premium locations such as Chelsea, Knightsbridge or Hampstead. This is typical of the gentrification Butler and Robson identified in London's Battersea and is primarily driven by economic capital. In the less affluent middle-class family households of Telegraph Hill in southeast London, the social capital network of the residents makes for a strong place identification built around mutuality and a determination to make the local state (especially in terms of education provision) functional in terms of middle-class aspirations. In areas of high social deprivation such as southeast London, these enclaves become beacons for other middle-class householders who cannot, or do not wish to, educate their children privately but who have no desire to become suburban commuters. Once the collective consumption of a public good (such as a local elementary school) has been **embourgeoised** in this way, social capital networks that have been mobilised to improve this aspect of provision will cooperate to demand improvements to other aspects of public provision or public concern such as parks and green spaces, traffic congestion and crime (Butler and Robson, 2001).

The phenomenon of the embourgeoised urban sphere also plays into discussions of globalisation and the re-structuring of the urban landscape by the movement of capital, leading to uneven development. According to Neil Smith, gentrification is a feature of what Harvey and Lefebvre refer to as 'the second circuit of capital' – the capital invested in land and property – and is associated with the tendency of capital to exploit variations in the worth of land (land rent) during periods of fluctuating prices. As capital spreads out from the urban core and 'suburbanises', a 'rent gap' is produced in the older urban core, which becomes devalued by the flight of capital. As Neil Smith writes, '[t]he devalorization of capital in the center creates the opportunity for the revalorization of this "underdeveloped" section of urban space' (Smith in Smith and Williams, 1986: 24). Thus, rundown industrial districts or slum neighbourhoods can suddenly become goldmines for individual investors as well as corporate speculators and investors once a critical mass of redevelopment and 'improvement' is set in train.

However, the rent gap theory has been challenged by writers such as Hamnett (1991, 1992), Ley (1987), Beauregard (1985) and Zukin (1988) who see gentrification as much more of a socio-cultural process in which the characteristics and the consumption demands of the gentrifiers matter just as much as the production of devalued areas.[31] The call for a more contingent and complex view of gentrification (Beauregard, 1986) is supported by accounts of the emergence of an affluent, highly networked, international professional and executive class, dubbed by Micklethwait and Wooldridge as 'the cosmocrats' (Micklethwait and Wooldridge, 2000). Although they do not constitute a high percentage of the city dwelling affluent middle class, because of their resources and specific consumption requirements they are able to shape the urban environment not just of 'isolated suburbs' but also what might be called 'frontier environments' (often between financial districts and deprived neighbourhoods) such as the loft lands of Hoxton in London and SoHo in Manhattan (see Chapter 8).

> The cosmocrats are increasingly cut off from the rest of society: Its members study in foreign universities, spend a period of time working abroad, and work for organizations that have a global reach. They constitute a world within a world, linked to each other by myriad global networks but insulated from the more hidebound members of their own societies . . . They are more likely to spend their time chatting with their peers around the world – via phone or e-mail – than talking to their neighbours in the projects around the corner. Manuel Castells . . . has summed up the problem neatly: 'Elites are cosmopolitan, people are local'.
>
> (Mickelthwait and Wooldridge, 2000: 241–2)

This is the precise demographic to which upmarket real estate vendors are trying to sell seven-figure warehouse conversions in London's Docklands and in New York's Battery Park. But the sorts of services the cosmocrats require are not crèches, local food stores, bus services, and elderly day care – but broadband multimedia cable facilities, 24 hour door-to-door delivery, upmarket bars and restaurants, and on-site gymnasia. There is little evidence that such enclaves encourage social mixing (see Chapter 4) – while the

---

EXHIBIT 5.5   Gentrification in Brixton

Tim Butler and Garry Robson, 'Social Capital, Gentrification and Neighbourhood Change in London: A Comparison of Three South London Neighbourhoods', *Urban Studies*, Vol. 38, No. 12, 2145–62, 2001.

There are . . . two residential areas involved here in our study. Each abut onto Brockwell Park; Tulse Hill to the west, Herne Hill to the east. The latter, however, is not a part of Herne Hill 'proper', but an area close to central Brixton rechristened by estate agents in recent years as 'Poet's Corner'. This small network of streets (Milton, Spenser, Shakespeare Road, etc.) has now been designated a conservation area and contains highly desirable and architecturally interesting properties. The area as a whole runs parallel to Railton Road, Brixton's notorious 'front line' of the 1970s and 1980s and at the centre of the 1981 'riots'. The process of gentrification here is, given this, perhaps the most dramatic in all our areas. Streets adjacent to one of Britain's best-known symbols or urban disrepair have been settled and largely transformed over the past decade by high-income professionals reclaiming its increasingly sought after 'architectural gems'. Brixton Hill, its counterpart on the other side of Brockwell Park, is a larger area containing mostly terraced streets of housing, less spectacular but solid, desirable and more easily accessed from the centre of Brixton. It is much more socially mixed than Poet's Corner but is, in its interior, equally calm and ordered. Both areas therefore offer residents dense but relatively peaceful environments close by one of Britain's most vibrant and volatile inner-urban areas.

It is significant that a majority of middle-class home-owners in Brixton, when asked about the appeal of the area, stressed the importance and attraction of social and cultural diversity.

(Margaret, 38) The best thing about living here is that it's an open community . . . There is no norm. I find that very reassuring.

(Sara, 32) It's quiet in our street, but there's lots to do nearby, you don't need a car. Practically everything you need is in Brixton, the Ritzy [cinema], restaurants, bars, fabulous things. It's a very diverse population; we don't stick out living here as two women living together . . . I wouldn't want to live anywhere else.

[. . .]

But Brixton's celebration of individualism and the freedom from strong norms appears to have a down side with implications for the nature – or absence of the social cohesion experienced by those living in the area. That is there appears to be something of a gulf between a widely rhetorical preference for multicultural experience and people's actual social networks and connections. The model of social cohesion in Brixton, where physical interaction with an extraordinarily heterogeneous social landscape is an unavoidable feature of everyday life, might be characterised as 'tectonic' (Robson and Butler, 2001). That is to say, relations between different social and ethnic groups in the area tend to be of a parallel rather than integrative nature; people keep, by and large, to themselves. The urban landscape in Brixton appears thus to serve, for most of its middle-class residents, as an ideologically charged and desirable backdrop for lives conducted at a remove from its multicultural institutions.

reaction from working-class neighbourhoods to such 'yuppie' encroachments has often been hostile. This, in turn, has accentuated the 'defensive' and 'secure' architectural features of renovated or new build deluxe accommodation and, in many central cities, has further reinforced the visible division between the haves and the have-nots.

What Robert Reich describes as the 'secession of the successful' has meant that those with the power and the resources to make a difference to the quality of life in the metropolis can live, as Butler and Robson term it 'tectonically' – with the socially and economically excluded, in other words existing side by side, but with absolutely no interaction between the two strata. In the following section we look at what Berry (1985) describes as the 'sea of decay' surrounding the gentrified 'islands of renewal' we have just examined – the urban ghettos that have become a compulsory community for millions of underprivileged Americans.

QUESTIONS TO DISCUSS

1 **What appeals to middle-class residents about living in Brixton?**

2 **In what way can the globalised young professionals Mickelthwait and Wooldridge describe and the Brixton middle class be considered 'tectonic' urban communities?**

LIVING IN THE GHETTO: THE RACIALISED CITY IN THE US

Fortification epitomizes the ghetto in America today, just as back alleys, crowded tenements, and lack of playgrounds defined the slum of the nineteenth century. Buildings grow claws and spikes; their entrances acquire metal plates; their roofs get fenced in; and any additional openings are sealed, cutting down on light and ventilation. Glass windowpanes in first-floor windows are rare. Instead, window openings are bricked up or fitted with glass bricks. In schools and in buses, Plexiglas, frosty with scratches, blurs the view outside.
(Camilo José Vergara, 1995: 102)

The historic ghetto

The emergence of the ghetto in the American city in the latter half of the twentieth century marked a change from the ethnic neighbourhoods of the pre-Second World War period such as those recorded by Chicago sociologists into a much more homogeneously segregated metropolis (Jackson, 1965; Wacquant, 1994). Robert E. Park, whose work we encountered in Chapter 3, was one of the first sociologists to study racial segregation systematically, and his dual notion of 'choice and constraint' in ethnic residential formation has been highly influential on subsequent studies of ethnicity and urban space. Additionally, the attention given to competition for space by Burgess through his concepts of 'invasion and succession' and 'patterns of distribution', focused attention on the city as an ecological unit in which different classes and ethnic groups competed among one another for resources and living space, much in the way that animal species were thought to in the natural world (Brown, 1981: 185, 187). It was really only in the period after the Second World War that this ecological paradigm began to be displaced in studies of urban segregation by a political economy approach in which the key dynamic in spatial distribution of populations and land-use was seen as that existing between capital and the state (see Chapters 6 and 7).

But for all the deficiencies of the ecological paradigm, the 'choice/constraint' determinant of residential location retains an intuitive and empirical appeal as a framework for explaining variations in ethnic segregation in the city. John Logan demonstrates the continuity of the ecological paradigm when he writes (in reference to contemporary New York City): 'People gain security by their collective hold on particular positions in the labor market and in residential communities. Groups compete for space, for employment, and for position, and ethnicity is an important collective tool in this competition. We should expect them to relinquish this tool only slowly, if at all' (Logan, 2000: 168).

The ghetto has been described as 'a residential district that is almost exclusively the preserve of one ethnic or cultural group' (Ward, 1982: 258 in Peach, 1996: 216). But Peach is right in this author's view to regard such a definition as unsatisfactory, because it fails to draw a distinction between an ethnic enclave such as a 'China Town' or a 'Little Italy' and that of a 'ghetto' where inhabitants of the latter are denied the locational choice enjoyed by residents of ethnic enclaves. Thus, for example, it would be a mistake to view the Hassidic neighbourhoods of Crown Heights in Brooklyn or Stamford Hill in London as Jewish ghettos, since no municipal ordinance has been issued requiring members of this faith to live in a designated area. Rather, such a tight concentration of identity-sharing subjects constitutes an 'elective community' that has chosen to congregate in a single district because of the religious, cultural, economic and social benefits that such propinquity is perceived to bring.

On the other hand, the Warsaw and Lodz ghettos which were created by the Nazi occupying authorities in Poland during the Second World War represented an enforced or compulsory community – its inhabitants were ordered to reside within a given district, just as in seventeenth-century Venice, Jews were required to live within the precincts of a particular island that still bears the name from which the modern usage derives. In other words, genuine ghettos contain all, or almost all the members of the ethnic group that form part of the broader national territory (whether defined as 'citizens' or not), whereas ethnic enclaves and other such elective communities will allow the possibility of out-migration to other neighbourhoods and also settlement by other ethnic and 'non-ethnic' populations.

Philpott's study of 1930s' Chicago shows that while it was usual to refer to Irish, German or Polish ghettos, in reality besides African Americans only the Poles lived predominantly in Polish neighbourhoods, while for the black population of Chicago, 90 per cent of its population lived in black areas where it constituted over 80 per cent of the population (Philpott, 1978 in Peach, 1996: 217). However, unlike other ethnic groups with significant spatial concentration, the African American population did not experience a dispersion of its members among the general population, either to other parts of the city or to the burgeoning suburbs. In fact, '[s]ubsequent analyses of the 1960, 1970 and 1980 US censuses have shown the position of African Americans as remaining highly segregated' (Peach, 1996: 217). Indeed, by 1990 in the specific case of New York, the segregation of Whites from Blacks in the same census tracts had actually increased since the previous count (Logan, 2000: 178).[32]

However, unlike its medieval and totalitarian precursors, the modern ghetto is not the product of administrative fiat, but of economic compulsion. Poverty, by and large, is the force field that contains what has variously been described as 'the urban underclass' or 'the new urban poor' from venturing beyond the public housing schemes or run-down low rental apartments in which many millions are forced to live in cities across the US and, on a smaller scale, in many urban centres across Europe and the developed world. But if poverty is in a sense a prerequisite for being an inhabitant of the ghetto, it has also been argued that segregated communities built around racial and economic exclusion are themselves a cause of poverty, crime, low educational achievement and poor health (Massey and Denton, 1993: viii). In particular, Massey and Denton's research has also done much to revive the ecological approach to the analysis of urban problems pioneered by Park and Burgess back in the 1920s. Building on the work of Kenneth B. Jackson who argued in the 1960s that, '[t]he dark ghettos are social, political, educational, and – above all – economic colonies' (Jackson in ibid.: 3), the authors of *American Apartheid* contend that:

> Where one lives – especially, where one grows up – exerts a profound effect on one's life chances. Identical individuals with similar family backgrounds and personal characteristics will lead very different lives and achieve different rates of socio-economic success depending on where they reside. Because racial segregation confines blacks to a circumscribed and disadvantaged niche in the urban spatial order, it has profound consequences for individual and family well-being.
>
> (Massey and Denton, 1993: 149)

Unlike the early Chicago school studies of urban segregation, however, more contemporary urban sociology both acknowledges and emphasises the role of structural and political factors in determining residential and business locations. In the wake of the urban riots that swept across the US in the late 1960s, government-sponsored investigations such as the Kerner Report (US National Advisory Commission on Civil Disorders, 1988) candidly warned the majority population that 'white society is deeply implicated in the ghetto. White institutions created it, white institutions maintain it, and white society condones it' (Kerner in Massey and Denton, 1993: 4). But it was only with the publication of the Moynihan Report on the Negro family in 1965 that longer-term trends in the extent and persistence of urban black poverty could be ascertained (Moynihan, 1965). However, by the late 1960s and 1970s, the desire to avoid stigmatising minority racial groups by focusing on the 'pathological' aspects of their social life persuaded many liberal scholars to concentrate their research efforts in other areas (Wilson, 1987: 20–1). This left the field open to conservative social scientists and policy-makers who have not been shy to proclaim that cultural and even biological characteristics of the African American community are the cause of inter-generational poverty and low social mobility rather than residential location (ibid.: 4–5).

William Julius Wilson, as one of the leading black sociologists in the US, has done much to revive the Chicago tradition of ethnographic research aimed at uncovering the social problems endemic to the inner-city. However, unlike many (predominantly white) liberal sociologists he has not been reluctant to talk of crime, teenage pregnancy, and welfare dependency as social pathologies. But unlike conservative critics, Wilson does not see these problems as examples of moral failure so much as the product of determinate circumstances and policies that have helped to create and maintain a sizeable black underclass. In particular, Wilson focuses on the loss of low-skill and manufacturing jobs on which millions of black male workers relied from the 1960s onwards, and the obstacles to alternative employment opportunities that were put in their way due to low educational opportunities (resulting from a disinvestment in urban public schools in the 1970s and 1980s) and

also a lack of residential mobility. For many black households urban poverty was and has remained a catch-22 problem because in order to move to where the jobs are in the suburban commercial and light-manufacturing districts they have to be able to rent or buy an affordable home (Wilson, 1996). But without the economic means to provide a deposit and proof of regular income, mortgages are more likely to be refused to African Americans and Hispanics (Carr and Megbolugbe, 1993).

White suburban municipalities also make use of restrictive rules on plot developments in order to keep housing densities to a level that would allow for only the construction of larger and more expensive family type dwellings. This practice is exposed to particularly telling effect by Mike Davis in his study of contemporary Los Angeles, *City of Quartz* (Davis, 1992, chapter 4), but it is prevalent in many other affluent suburbs of the US, emphasising the ways in which (most often) white suburban residents and local political leaders collude in sealing 'their' community against social and ethnic diversity, and in the case of 'gated communities' by closing them physically as well as politically and economically against 'outsiders' (Blakely and Snyder, 1999; Boudreau and Keil, 2001 – see also Chapter 4 of this volume).[33] Thus, economic restructuring, urban disinvestment, land-use planning controls, 'revanchist' welfare policy (Smith, 1996) and the continuing demand for cheap migrant labour has contributed to the establishment of a peculiarly condensed space of social and ethnic equality that Loïc Wacquant has termed 'the hyperghetto'.

The hyperghetto

The 'hyperghetto' refers to neighbourhoods where 40 per cent or more of the population lives below the poverty line (Vergara, 1995: xiii, see also Wacquant, 1994). However, it is the intense concentration of poverty that distinguishes the traditional ghetto, such as Harlem in the 1920s and 1930s from the hyperghettos of the 1970s, 1980s, 1990s and 2000s (see Figure 5.1). Paradoxically, the civil rights movement having unlocked the door for a sizeable minority of able and educated African Americans to attend college and move into middle-class occupations, thus exacerbated the plight of poor urban blacks. Like the white middle class in the 1960s and 1970s, middle-class African Americans chose to join the movement away from the dense metropolis to the spacious suburbs, leaving behind a residual underclass that struggled to survive in the absence of the manual and unskilled jobs that had been traditionally filled by working-class black males (Wilson, 1987, 1996).

The flight of the black middle class and white capital was a double disaster for poorer African

Figure 5.1
Burnt-out buildings in Harlem, New York City. The ghetto becomes the hyperghetto. Copyright V. Bennett, courtesy of the Architectural Association Photo Library

American communities because it removed potential community advocates with the social standing and resources to liaise with the city authorities, to organise associations and sports and cultural activities, and generally provide role models for the next generation. In addition, Wilson argues that the weakness of the inner-city labour market resulting from the decline of industrial manufacturing jobs adversely impacts on the incidence of co-habiting fathers (Wilson, 1987), while Charles Murray puts the blame squarely on welfare eligibility rules that provide higher levels of support to unwed mothers than married mothers (Murray, 1984). Although Wilson disputes Murray's findings, the two claims do not have to be mutually exclusive. Testa and others have shown an employed black male father is two and a half times as likely to marry the mother of his child than an un-employed father (Testa et al., 1993), largely due to the fact that economic security encourages commitment. But where welfare receipts would diminish as a result of marriage, only very strong peer expectations about the importance of marriage (such as in some Hispanic and Mexican communities) are likely to make marriage prevail over income optimisation.

The question thus has to be addressed – is there 'a culture of poverty' in the American ghetto that sustains and reproduces itself, and why do other (and more recently arrived) ethnic minority groups seem better at escaping the poverty trap than African Americans? Certainly, the Moynihan Report took the view that there were inherent features of the African American family, many resulting from the brutal separation of family members under slavery that tended against the creation of European-style nuclear families. Even radical social theorists such as Manuel Castells accept that if we take his notion of 'communities of resistance' to be an aspect of the culture of poverty (Castells, 1997), then it is certainly meaningful to refer to a 'ghetto culture' that celebrates through genres such as hip hop and rap the often violent and sexually uninhibited world of young urban blacks. Although such cultural representations are built around the pleasures of shock and exaggeration, urban ethnographers have found the value-system to be anchored in reality (Anderson, 1993).

Yet, the very fact that a black middle class exists and that it displays very similar marital and relationship patterns to the majority, white middle class, suggests that cultural factors are not the key to explaining the persistence of urban black poverty (Landry, 1987). In other words, as William Julius Wilson points out in an earlier study, it is class and not race per se, that determines the life chances of the urban poor (Wilson, 1978). According to Wilson, place is also determinant to the extent that the more extreme and homogeneous the poverty, the worse the life chances are for all citizens. If this observation is true, it somewhat turns on its head the traditional notion of the 'self-sufficient poor', who like the residents of East London in the 1940s may have had little money but were rich in Putnam-style 'social capital' (see above). Rather, Wacquant and Wilson's research suggests that the poorer one is in real capital the less social capital there will be to go around as well (Wacquant and Wilson, 1993: 41–2).

In a sense, both liberal and conservative analysts of the American underclass agree that public policy bears a large degree of responsibility for the persistence of the ghetto and its concentration in the form of the hyperghetto. Conservatives such as Charles Murray believe that since the years of Lyndon Johnson's 'Great Society' reforms, social policy has aided and abetted the formation of the underclass, whereas writers such as Wilson, and Massey and Denton point to segregationist housing policy and under-investment in manufacturing jobs as the chief reasons for the spatial concentration of black poverty.

The levels of violence, criminality and social disadvantage that Wacquant describes in Exhibit 5.6 are disproportionately concentrated in the larger urban centres where the majority of America's predominantly black and Latino populations live. According to US Bureau of Justice Statistics figures, between 1976 and 2000 over half of America's homicides occurred in cities with populations of over 100,000, and almost one quarter of the homicides occurred in cities with populations of over 1 million. In the same period, 58.9 per cent of all gun homicides, 68.3 per cent of drug related murders and 69.2 per cent of gang related murders happened in large cities with

---

EXHIBIT 5.6   Loïc Wacquant, from 'Inside the Zone' in P. Bourdieu *et al. The Weight of the World*, p. 146

In 1990, 849 homicides were committed in the city of Chicago (a rate of 28.3 for 100,000 residents, comparable to those of New York City and Los Angeles but well below those of Washington DC and Detroit). Among the victims, 253 were under 21 years of age (and 27 under 10 years), and 9 out of 10 were killed by gunfire. Over half of those youths killed lived in the six police precincts corresponding to the city's historic 'Black Belt' and 186 (or 73.5 per cent) were Afro-Americans.

[. . .]

It is difficult not to postulate a close causal relation between these astronomical rates of crime and mortality, worthy of a muted civil war – recent epidemiological research has established that young black men from Harlem . . . have a higher chance of dying from violence than did soldiers sent to the front-lines at the height of the Vietnam war – and the crushing poverty that pervades this urban enclave shorn of economic activities and from which the government has virtually withdrawn, save for its repressive arm.

---

populations of over 100,000.[34] African Americans are more than two and a half times as likely to be the victims of such murders than their fellow Americans, and of these victims nearly five times as many are likely to be males as compared to females.[35]

Although homicide figures have certainly improved in the 1990s in America's inner cities, they compare very unfavourably with Europe's larger cities (see Table 5.1) and while, as we have noted, the easy availability of lethal firearms is a major cause of America's high homicide rate, the ecology of violent crime associated with a virtually non-existent real economy and a profitable if dangerous illicit economy in the hyperghetto highlights the US as a global anomaly among the advanced industrialised democracies.

Table 5.1 Comparison of homicide rates in selected cities, 1998–2000

Homicides per 100,000 of the population average per year 1998–2000	
Washington DC	45.79
Pretoria[1]	41.12
Moscow[2]	18.2
New York	8.7
Amsterdam	4.09
Paris	2.85
Berlin	2.52
London	2.38
Sydney	1.49

Notes: [1] 1995–1997; [2] 1997–1999
Source: UK Home Office/Barclay and Tavares, July 2002

THE GLOBALISING GHETTO? THE BRITISH AND EUROPEAN EXPERIENCE

In contrast to the US the British experience of ethnic minority presence in its metropolitan centres has been far less determined by the politics of laissez-faire and government indifference. Since the Second World War, the state provided an increasing proportion of residential accommodation, much of it subsidised and, by the 1970s, this provision was almost exclusively directed and, indeed, occupied by individuals and families on low incomes. The nature of the urban labour market, while no less prone to the chill winds of the downturn in manufacturing industries was, nevertheless, ameliorated to a certain extent by a social democratic interventionist strategy that targeted training and resources as far as possible at the most deprived areas (many of which were inner-city districts) (Rex, 1988: 63). Where social welfare, and especially social housing policies have been aimed at reducing social polarisation, especially in Sweden and the Netherlands, there is evidence of inter-class and inter-ethnic mixing which demonstrates the primary role of government policy in combating or assisting the development of ethnically exclusive neighbourhoods (Dieleman, 1994; and Van Kempen and Priemus, 1999 in Marcuse and van Kempen, 2000: 10).

It is also important to remember that sizeable ethnic minority populations are a relatively recent phenomenon in British cities. When Ruth Glass published the first sociological survey of Caribbean migrants in Britain in 1958, she noted that the 'coloured minority' amounted to only 210,000 (including West Indians, Africans, Indians and Pakistanis) amounting to less than one half of one per cent of the total national population (Glass, 1960: 1). Areas of first settlement inevitably tended to be those parts of the city where rooms could be rented at an affordable price, although discrimination and dependence on private landlords meant that new migrants often paid more for inferior accommodation than white tenants. The high cost of accommodation and the difficulty of finding regular and secure employment contributed to the difficulty of finding shelter. As Glass writes:

> The choice of location for West Indian migrants is . . . very narrow. In fact, they have hardly any choice. They have to stay fairly near to the central London labour market: to the majority, the outlying parts of the London Region – the territory of one family houses, owner-occupiers and new municipal estates – are inaccessible. They cannot go to those central London districts – working class, middle class or upper class – whose patterns of occupation have been stabilised, nor to those which are in the stage of physical reconstruction, planned functional change or social upgrading. Therefore the migrants have to go to patches of inner London which have been neglected, and which have been already for some time in the process of decline and downgrading . . .
>
> (ibid.: 48)

As Burgess and others have shown in the case of the American city, it has always been the fate of poor recent immigrants to occupy the most marginal spaces of the city. But interestingly, with the exception of North Kensington in West London and Brixton in South London, Glass and her associates found little evidence of ethnic clustering. Indeed, she was led to remark that twelve years after the arrival of the first organised migration of Caribbeans on the *Empire Windrush*, '[a]s yet London has no 'coloured quarters' of the kind found in American cities' (ibid.: 42). The sort of accommodation available to these new Londoners is invariably the large town houses once occupied by the well-to-do but not wealthy bourgeoisie who were forced to abandon them as the costs of maintenance and domestic servants made staying in the increasingly noisy and polluted central city less attractive than a purpose built villa in the leafy suburbs. The remaining properties were cheaply acquired by landlords and divided into rooms often accommodating 3 or 4 families in the same property, and all sharing a single bathroom, kitchen and toilet (ibid.: 49).

These are what Glass described as 'zones of transition', but it is clear that what was in progress was a process of de-gentrification (or filtering) where an increasingly dilapidated housing stock is occupied by social classes and ethnic groups who, by their very presence, are likely to depress the prices of any remaining freehold properties yet further. In this sense 'ghetto formation' follows a similar pattern to that experienced by American cities in the 1950s and 1960s, but it was on a much more modest scale and it did not become anywhere near as rigid and racially exclusive as in the US.

For example, Peach shows that at the beginning of the 1990s the 'ghettoization' that Ruth Glass feared following the American experience had not materialised, at least not to the same extent in Britain. Her findings showed that Caribbean average levels of segregation are about half the level for African Americans in the US. In fact, the story in the context of Black London is one of progressive de-segregation (in contrast to New York where segregation has increased since the 1960s). In 1961 ward level data for Caribbean-born London residents showed ID concentrations of 56, but by 1991 this figure had declined to 41. As we might expect at smaller enumeration (census tract) levels the ID figures were higher (65 in 1971 and 50 in 1991), while at borough level the figures were 38 and 34 respectively. Thus, although spatial scale produces considerable variation in degrees of segregation, the overall trend is clear, and it is towards greater assimilation of the Caribbean community among the general London population (Peach, 1996: 227).

The uniqueness of the American model is also clear from a contemporary analysis of the French *banlieue* (working class or low-income public housing projects that are mostly located on the outskirts of the major cities) where despite a significant ethnic minority presence (predominantly of North, Central and West African origin or descent) 'the mechanism of segregation and aggregation from which they result' are quite different from the black American ghetto (Wacquant, 1999a: 131). For Wacquant,

> the American ghetto gives us a realistic vision of the kind of social relationships liable to develop when the State jettisons its essential mission to sustain the organizational infrastructure indispensable to the functioning of any complex urban society . . .
>
> (ibid.: 132)

According to Wacquant, each of the factors that we have discussed above are relevant to the persistence of the American ghetto (racism, the culture of poverty, the 'pathogenic' behaviour of elements of the 'underclass', de-industrialisation and middle-class black flight), but the crucial determinant in his analysis is 'the policy of *deliberate urban abandonment* of these neighbourhoods pursued by the American government in the wake of the riots of the sixties' (ibid., original emphasis). In other words, the explanation for why a broadly welfare capitalist Europe has nothing like the scale of urban poverty, unemployment, ill health, drug use and violent crime as urban America is the result of the peculiar and unique character of a political system that responds only to the taxation and spending priorities of the largely white, mostly affluent and predominantly non-metropolitan majority.

QUESTIONS TO DISCUSS

1 What distinguishes the historic or traditional ghetto from the 'hyperghetto' of contemporary America?

2 Why, unlike previous generations of ethnic minorities, have black Americans found it harder to escape the poorest inner city neighbourhoods?

3 What in your view accounts for the fact that African Americans suffer disproportionately from violent crime?

4 Why is residential segregation by ethnic status so much more prevalent in American cities than in Europe?

CONCLUSION

> The race, economic class, and ethnicity of communities follow the highway arteries – with disparate groups clustering near off-ramps the way nineteenth-century communities clustered near ports and along rivers. The challenge of the twenty-first century is to harmonize the factions – creating communities with blended races, classes, ages, and economic groups.
>
> (Partners for Livable Communities, 2000: 41)

The first point to make about Anglo-American urban studies before and after the Second World War is that there were no sudden shifts in direction, nor any major theoretical innovations from the 1920s until the 1970s, not least because many social scientists believed that there was no longer any value in pursuing a specifically 'urban' sociology, since 'the urban' had lost its geographical specificity. The sociologist Ruth Glass argued that with such an extensively urbanised population as Britain's there was no need for a separate urban branch of sociology, especially since so much of it was preoccupied with 'the bric-à-brac of our own parlours' (Glass, 1955 in Pahl, 1970: 210). This view was also expounded in Peter Saunders' introduction to the second edition of his influential reassessment of urban sociology, *Social Theory and the Urban Question* where he observed that '[m]ost of what we find going on in cities can, in a society like Britain, be found going on outside them too, and this makes it virtually impossible to identify any specific aspect of social life which is distinctive to cities' (Saunders, 1995: 7).

Meanwhile, in the US, Riesman's call to abandon the over-simplified ecological models of the Chicago School encouraged urbanists to shift their attention to the history of urbanisation and to study the emerging cities of the developing world. As a consequence, the study of the city as a subject in its own right all but disappeared from the syllabus of most social science disciplines in the late 1960s (Pahl, 1970). Those researchers such as Pahl, Rex and Moore and Dennis in the United Kingdom and Webber, Greer and Whyte who persisted with the study of their own urban systems unquestionably made a lasting contribution to the study of urban society. But was this enough, as the young Manuel Castells asked, to prove the existence of an urban sociology – a sociology capable of thinking beyond its own locale to consider the structures and processes that often remain invisible to the casual observer (Castells, 1968)? This question was answered by a shift in urban research at the beginning of the 1970s to what we might call 'abstract theory' of a mostly Marxist vintage that focused on the 'political economy' of cities (see Chapters 6 and 7). At the same time in the US, there was a renewed focus on the life of communities that downplayed the urban/rural distinction in favour of other identifying categories such as the degree of 'social integration' or 'fragmentation' as a way of providing historical and geographical comparators for the measurement of social change.

Before the 1970s, ethnography, reportage and human ecology continued to feature strongly in studies of old and new urban communities, while the accessibility of this medium of investigation to an increasingly numerous and educated general public on both sides of the Atlantic made household names of Michael Young, Peter Willmott, Jane Jacobs and Holly Whyte. During this period 'the urban problem' became the province of the public intellectual to a far greater extent than in continental Europe, not least because as both the University of Chicago and the

LSE's history of urban scholarship confirms, the focus has always been on the policy implications of generally empirically oriented research. By contrast, as we shall see in the next chapter, continental European theorists tended both to study and see the city as 'the urban question'. Therefore, as a phenomenon that needed to be critically understood (rather than described or explained), one required a wider conceptual vocabulary than was commonly found in the editorial pages of an architectural magazine or between the covers of a work of popular sociology.

The second continuity is the persistent concern for 'community' – what it is, and especially what it ought to be – that stretches back to the earliest urban investigations that we encountered in Chapter 3, and which emerges in the proposals for ideal cities that we encountered in the previous chapter. From Booth's counting and classification of the urban populace, to the kinship networks of Bethnal Green, and Robert Putnam's 'bonding' and 'bridging' social capital – getting the measure of how people live and behave in cities, and increasingly in what Alvin Webber (1964) described as 'the non-place urban realm' or Garreau's (1991) edge cities, has been tied to questions of how we might live better. In particular, how the urban environment might be altered or refashioned to better stimulate the sort of *Gemeinschaft* relations that most urban observers in the Anglo-American world of urbanism believe is desirable.

In the following chapter we explore how the analysis of the capitalist city – a perfect example of the conjuncture of the universalism of the market with the particularity of place – has enabled empirical and 'grand theory' approaches to converge, and through the dialogue between urban theory and the urban experience, has produced some of the most significant and far-reaching studies of the past and present city.

FURTHER READING

Community studies in Britain

Asa Briggs (2001) has written an engaging biography of Michael Young, the late Lord Young of

Dartington, that deals extensively with his pioneering studies of inner-city and suburban communities in post-war London and also his founding of the Institute for Community Studies. Key texts in the British community studies school include Glass *et al.* (1964), Bell and Newby (1971), Willmott and Young (1960/1971) and Young and Willmott (1986/1992).

Jane Jacobs

Max Allen's edited collection of articles, correspondence, reviews and features by and about Jane Jacobs *Ideas That Matter. The Worlds of Jane Jacobs* is a wonderful evocation of her life and career (Allen, 1997). It contains a fascinating chapter on how *The Death and Life of Great American Cities* came to be written and the controversies that have been stirred since its publications. Other chapters also deal with Jacobs' life in New York, her views on European cities, and her later very active support of the 'city sovereignty' movement in her adoptive home of Toronto. Jane Jacobs' *Cities and the Wealth of Nations* (1986) makes an argument in favour of cities from the perspective of world trade and economic development. It offers a characteristically readable tour through the history of urban growth and decline without developing any firm theoretical conclusions. A review of Jane Jacobs', *The Death and Life of Great American Cities*, by Herbert Gans is reproduced in the Allen volume.

Social capital and civic empowerment

The main research hypotheses in Robert Putnam's epic *Bowling Alone* study can be found in two issues of *The American Prospect* (Putnam, 1993b, 1996). Other related articles on social capital and urban affairs include Briggs and de Souza (1997) and Berry *et al.* (1993). For a more critical view of social capital see Alejandro Portes and Patricia Landolt (1996). The Civic Renewal Movement (www.cpn.org) contains a wealth of material and readings on social capital and social renewal related themes.

Suburbs, new towns and edge cities

Kenneth T. Jackson's *Crabgrass Frontier* (1985) is a classic study of the suburbanisation of modern America, and can be read alongside F.M.L. Thompson (1982) on the historical development of the English suburbs. One of the few contemporary studies of the English new towns is to be found in Orlans (1971). An account of the post-Second World War development and of British new towns and suburbs, including a study of social life in these new communities is to be found in Mark Clapson (1998). For those looking for a quick introduction to Garreau's edge city investigations, an extract from the book can be found in Judd and Kantor's reader, *The Politics of Urban America* (2001). An academic analysis of edge cities can be found in Robert Beauregard (1995). A counter-blast to the ''burb bashers' is offered by Rob Kling and Mark Poster in *Beyond the Edge: The Dynamism of Posturban Regions* (1995) in which the authors argue that many of the negative criticisms of suburbia (or their preferred term 'postsuburbia') are misinformed or misplaced – pointing for example to the fact that Orange County in California has higher population densities, a more thriving economy and a more dynamic cultural and recreational life than many major urban centres. The volume also challenges many of Garreau's criteria for edge cities arguing for a much more fluid relationship between residential hinterlands and retail, commercial and transit centres. Critical views of extra-urban development from a town planning and architectural perspective can be found in the further reading for Chapter 4 of this volume.

Ghettos and gentrification

Contemporary accounts of the American ghetto are provided by Wacquant (1999a), Vergara (1995), Gregory (1999) and Cutler *et al.* (1997). Bonney (1996) explores the ways that the black ghetto is perceived by mainstream America. William Julius Wilson's *The Truly Disadvantaged* (1987) is excerpted in Le Gates and Stout (various editions). Anthony Downs of the Brookings Institute prospects a more

BETWEEN THE SUBURB AND THE GHETTO   99

positive vision for the American ghetto in 'Alternative Futures for the American Ghetto' (Downs, 1968), which can also be found in Le Gates and Stout [various editions]). Goldsmith and Blakely (1992), while not denying the extent of ethnic minority poverty in the US, are more optimistic about how public policy can help to increase opportunities for the urban poor. Rex (1988) offers an account of ethnic housing segregation in Britain although the title suggests closer parallels with the American experience than there are in reality.

Ruth Glass's original development of the term 'gentrification' is taken from 'Aspects of Change' in the volume, *London: Aspects of Change*, (Glass *et al.*, 1964). More recent elaborations of the concept include Beauregard (1985), Smith and Williams (eds) (1986), Bondi (1991), Warde (1991), Abu-Lughod (1994) and Smith (1996). Van Weesep and Musterd (1991) discuss gentrification in Europe, and Taylor (1992) the emerging phenomenon of Black gentrification in the US. The 'rent gap' theory is developed by Smith in Smith and Williams (1986) and Smith (1996) and is critiqued in Hamnett (1991, 1992) with a rejoinder in Smith (1992).

6

URBAN FORTUNES

Making sense of the capitalist city

I wander thro' each charter'd street,
Near where the charter'd Thames does flow.
And mark in every face I meet
Marks of weakness, marks of woe.

William Blake, *London*

INTRODUCTION

For nearly 200 years, capitalism has been a fact of life for western metropolitans, and with the end of the Cold War there is barely a city in the former communist or developing worlds that has not been exposed to the stiff trade winds of the global market. But how can we make sense of the co-existence of these two great universal phenomena? How and in what ways are they related to each other? What accounts for the different aspects of urbanism in certain capitalist societies and the diversity of capitalisms in some urban societies? In this chapter we take up several themes that emerged from the discussion of classical approaches to urban society in Chapter 2. In particular, we shall focus on the capitalist metropolis that established itself in several European countries and in North America by the middle of the nineteenth century. The key factor in the rapid urbanisation of the western city in this era was industry and the great advances in long-distance communication made possible by the steam locomotive and steamship. By the end of the nineteenth century, the invention of the telegraph, the cinema and the internal combustion engine ushered in a new period

of mass consumption made possible by the mechanisation of production on a previously unimaginable scale.

Average city populations doubled or trebled every fifty years during the industrial revolution, and this resulted in tremendous overcrowding and often appalling living conditions for the urban workforces who had been lured or forced away from the countryside in search of a better life (Stedman-Jones [1971] 1976; Briggs, 1982). Nowhere were the contradictions of industrial capitalism more evident than in England, which by 1850 had earned the epithet 'the workshop of the world'. At around this time Marx and his collaborator Engels began to study the process of wealth accumulation that they were to term 'capitalism' in order to better understand a system of exploitation that they believed should and would be overturned by 'the immense majority' – the new proletarian class on whose labour the entire system depended.

This chapter begins by considering Engels and Marx's early work on the industrial city, before examining subsequent Marxist analyses of the urban question in the work of David Harvey, Doreen Massey and Manuel Castells. The contemporary capitalist city is increasingly viewed through the lens of globalisation, and here we consider how previous and new insights into the nature of urban structures and processes are reflected in the burgeoning literature on global cities. In the concluding section some reflections on the capitalist city in theory and practice since the beginning of the industrial era are offered, and links are made between the urban analyses in earlier and subsequent chapters.

THE CAPITALIST CITY IN THE WORK OF ENGELS AND MARX

It is the elementary precondition of bourgeois society that labour should directly produce exchange value, i.e. money; and, similarly, that money should directly purchase labour, and therefore the labourer, but only in so far as he alienates [*veräussert*] his activity in the exchange. *Wage labour* on one side, *capital* on the other, are therefore only other forms of developed exchange value and of money (as the incarnation of exchange value). Money thereby directly and simultaneously becomes the *real community* [*Gemeinwesen*], since it is the general substance of survival for all, and at the same time the social product of all . . . in money the community [*Gemeinwesen*] is at the same time a mere abstraction, a mere external, accidental thing for the individual, and at the same time merely a means for his satisfaction as an isolated individual. The community of antiquity presupposes a quite different relation to, and on the part of, the individual. The development of money in its third role therefore smashes this community. All production is an objectification [*Vergegenständlichung*] of the individual. In money (exchange value), however, the individual is not objectified in his natural quality, but in a social quality (relation) which is, at the same time, external to him.

(Marx, 1973: 225–6, original emphasis)

Unlike his close collaborator, Friedrich Engels, Marx had very little to say about the environment in which capitalism actually operated. This was because Marx preferred to concentrate his intellectual resources on the analysis of the process or mechanism of the capitalist mode of production rather than survey the consequences of capitalist relations of production in everyday life. However, this is not to say that Marx was unconcerned with what today we

would call the sociological dimension of human experience. Marx had a highly developed notion of 'community' that in many ways rivals the social analyses of Max Weber and Emile Durkheim, but he refused to abstract (or reify) social life from the conditions of its production and reproduction.

Hence, in the extract quoted above, Marx is positing a typology of communities that are differentiated by their relationship to capitalism as a means of production. In bourgeois (i.e. capitalist) society it is the abstraction of labour through the exchange value of money that gives rise to the *Gemeinwesen* (the real community), by which term Marx does not mean 'true', but the governing or functional social system. Antique society, which we might equate with Tönnies' notion of *Gemeinschaft* (or traditional community), has only a limited money economy and relies more on direct exchange systems such as barter. Commodities exist but they cannot be universalised as currency. Money is thus the 'universal commodity' in Marx's formulation. It is this abstract commodity form which 'smashes' the 'natural' or pre-capitalist community by allowing labour value to be converted into exchange value and thus removable from the place and people of its manufacture. However, it is only under capitalism that the commodification of value reaches its absolute condition through the monopolisation of all means of production by the bourgeoisie and the systematic conversion of all previously existing labour forms to wage-labour.

'Community' is therefore transformed from a system of social interaction based on law and custom and obligation, to a world governed by the power of 'things'. Marx, anticipating his later work in *Capital* appears to be suggesting that in the capitalist money economy, men and women undergo a double alienation; first, as individuals through their enforced self-commodification as labour, and second in their social relations (the relations of production) through which money as the abstract commodity form regulates all human exchange.

To say that Karl Marx paid little direct attention to the city is not to deny the importance of urban society in his analyses of capitalism. When Marx refers to industrial capitalist society, it is predominantly the urban rather than the rural world he is describing. Feudalism was essentially an agrarian mode of production and, of necessity, it concentrated social power in the countryside, even if the surplus accumulated from feudal exploitation often found its way to the markets, exchanges and banking houses of the cities. Capitalism, on the other hand, could only flourish in circumstances where surplus value could be exchanged for a profit and this required the existence of wage-labour. As Marx writes:

> The foundation of every division of labour which has attained a certain degree of development, and has been brought about by the exchange of commodities, is the separation of town from country. One might well say that the whole economic history of society is summed up in the movement of this antithesis.
> (Marx, 1976: 462 in Saunders, 1995: 22)

The commodification of labour in the form of wages had a number of significant consequences for the development of towns and cities. It created a market for labour, and with the abolition of serfdom, peasants could move more freely between farms, or in some cases move from agricultural activity to domestic service or manufacturing in the expanding towns and cities of Europe of the late-eighteenth and nineteenth centuries. Thus, for Marx, 'modern history is the urbanization of the countryside, not, as among the ancients, the ruralization of the city' (Marx, 1964: 77–8 in Saunders, 1995: 23). The single most important explanation for the growth in urban populations in this period was the development of commerce and industry, and therefore, for Marx and Engels, the industrial city contained within it all the contradictions of a class society built on a minority ownership of production and its ruthless exploitation of the propertyless majority through wage slavery.

As a young radical, Engels was horrified by the conditions of the workers he saw in his adoptive Lancashire, and with the encouragement of Karl Marx whom Engels had met in Cologne in 1842, he began to research a study that he was to publish in German in 1845 under the title *The Condition of the Working Class in England in 1844*,[36] – a study that Marx certainly regarded as a major contribution to the

cause of socialism. Although the book owed much to contemporary (and rather nostalgic) accounts of Britain's industrial development that painted rather too rosy a picture of the condition of the poor before the industrial revolution, Engels' survey still retains much of its documentary authority.[37] David McLellan describes it as, 'a classic of early social geography and a pioneering study of the effects of early uncontrolled industrialization'.[38] Though flawed, the influence of Engels' account of social life in the crowded slums of early Victorian England has not been surpassed by later studies of a more systematic type such as Booth's classificatory survey of social class in late nineteenth-century London.

Engels is far from unique in being at the same time repelled and attracted by the sheer vastness and dense activity of a large port city such as London. Commenting on the sight that greets the traveller arriving into London from the Thames, Engels declares, 'all this is so vast, so impressive, that a man cannot collect himself, but is lost in the marvel of England's greatness before he sets foot upon English soil'. Yet, the new visitor is likely soon to become disillusioned '[a]fter roaming the streets for a day or two . . . after visiting the slums of the endless metropolis, one realizes for the first time that these Londoners have been forced to sacrifice the best qualities of their human nature, to bring to pass all the marvels of civilization which crowd their city' (Engels, [1845] 1993: 36).[39]

Engels' sociological concern with the dehumanising impact of the capitalist city on the individual labourers and their families who make up the majority of its population is interesting for its attention to what Marx would later call 'the social relations of production' rather than the mechanics of the capitalist mode of production as such. In this respect, Engels' judgement is an entirely negative one, and it would appear that the context of exploitation matters even more than the means of its contrivance.

> The brutal indifference, the unfeeling isolation of each in his private interest becomes the more repellent and offensive, the more these individuals are crowded together, within a limited space. And, however much one may be aware that this isolation of the individual,

this narrow self-seeking is the fundamental principle of our society everywhere, it is nowhere so shamelessly barefaced, so self-conscious as just here in the crowding of the great city. The dissolution of man into monads, of which each one has a separate principle and a separate purpose, the world of atoms, is here carried out to its utmost extreme.

(ibid.: 37)

Spatial differentiation in the capitalist city is also determined by class status, 'Every great city has one or more slums, where the working class is crowded together.' True, poverty often dwells in hidden alleys close to the palaces of the rich; but, in general, a separate territory has been assigned to it, where, removed from the sight of the happier classes, it may struggle along as it can' (ibid.: 39). As Katznelson (1992) writes:

> Engels . . . showed how Marxism might incorporate the city into its social theory in order to create an account of the emergence of different kinds of working-class sub-jectivity. In pursuit of this aim, he also introduced urban space, albeit in a very condensed way, into the core of Marx's macroscopic historical materialism and into Marx's account of the logic of capitalist accumulation'.

However, 'in the subsequent elaboration of these schemes Marxists, including Engels, dropped the urban emphasis and the spatial elements from their work. As a result, they needlessly denied themselves important resources capable of contributing to the development of social theory' (Katznelson, 1992: 153).

However, as Saunders argues, in the work of Marx and Engels it is 'not the city that is held responsible for the poverty and squalor of the urban proletariat, but the capitalist mode of production' (Saunders, 1995: 25). The city is merely the site on which these class contradictions are played out. A point elaborated by Lefebvre when he writes:

> For Marx himself, industrialization contained its finality and meaning, later giving rise to the disassociation of Marxist thought into economism and philosophism. Marx did not show (and in his time he could not) that urbanization and the urban contain the meaning of industrialization.
>
> (Lefebvre, 'Around the Critical Point' in Kofman and Lebas, 1996: 130)

For Marx and Engels then, it is not the city that makes capitalism but capitalism that gives rise to the modern industrial city, while the logic of capitalist development pulls into the metropolis the vast mass of wage labourers together with a 'reserve army' of the unemployed and underemployed on which the affluence of the bourgeoisie must depend. For writers like Lefebvre, the idea of the city as a mere palimps-est, or magic slate, on which the great drama of the class struggle is daily re-inscribed appeared a remark-ably one-dimensional and undialectical account be-cause it assumes that space is a mere 'backcloth' in the historic drama of the class war (Lefebvre, 1972: 106 in Saunders, 1995: 25). Time, on the other hand is given a very privileged status in Marx's under-standing of class formation since history holds the key to the riddle of previous class conflict and future time provides the solution to the contradictions of present-day capitalism through the inevitable emergence of communist society. In the following sections we explore how the notion of commodification and the organisation of urban society through time and in space has been explored in more contemporary critiques of the capitalist city.

PAVEMENTS OF GOLD: THE COMMODIFICATION OF URBAN SPACE

> for bourgeois society, the commodity-form of the product of labour, or the value-form of the commodity is the economic cell form. To the superficial observer, the analysis of these forms seems to turn upon minu-tiae. It does in fact deal with minutiae, but so similarly does microscopic anatomy.
>
> (Karl Marx, 1976, Vol. 1: 90)[40]

As we saw in the discussion of Benjamin's city writings in Chapter 2, the interest in *objets de com-merce*, however trivial or bizarre, offers an anatomy of capitalism that more abstract or general analyses cannot rival. As Marx writes, the commodity that at first sight appears to be 'an extremely obvious, trivial thing', on closer inspection shows itself to be 'a very strange thing, abounding in metaphysical subtleties and theological niceties' (Marx, 1976: 163 in Frisby, 1988). Commodities need not be tangible, portable goods that can be stacked in a warehouse, since a commodity in the Marxian sense means any 'thing' that has been transformed by the value of labour into an exchangeable good, service or property. Hence, 'land' is a shorthand term for 'commodified space' wherein the owner of a given terrain or territory is able to rent, sell or derive other benefits relating to his or her title of ownership. The city is not entirely the product of commodified space, public roads and squares and markets exist alongside grand palaces or modest townhouses, and yet even 'public space' can be bought, sold or rented and as such is valued as a state asset in national accounts.[41]

Therefore, the urban system is necessarily based on relations of production that bring into combination labour in the form of resident and commuting wage-workers and capital employed to produce manufactured goods and services for profitable exchange. Each element – labour, land or space, and fixed and liquid capital – exists in its own temporal domain. Thus the working-day of the office commuter into which has to be figured journey time and mealtimes is part of a different cycle to that of the building in which she works that may be designed for a twenty- or thirty-year period of occupancy before it is refurbished or demolished for perhaps a different purpose entirely. The money that the employee is transacting will be part of a global finance network in which trillions of dollars are daily moved from one account to another at a keystroke. Hence each of these circuits, though closely inter-related, exists in its own logical domain, the routines and rhythms of which are often in conflict with the long-term interests of the other elements.

In other words, the capitalist city involves the transformation of the use-value of space into the exchange-value of land while at the same time providing the means to produce and exchange other types of commodities. If we think back to the City of London when it was a grazing pasture for early Saxon farmers and contrast it with 'the Square Mile' of today where the daily turnover of its exchanges and investment houses is worth many times the gross annual product of the wealthiest nations, we can begin to understand the importance of capitalism in transforming the physical world in general, but cities in particular, into dynamic productive systems (see Exhibit 6.1).

Urbanisation and capital

For urbanists of a Marxist persuasion it is this crucial dependence between the dominion of capital and the configuration of space in the modern metropolis that holds the key to understanding the logic of the city. The central features of this relationship between the capitalist mode of production and urban development are: (i) an ever more specialised division of labour, (ii) the existence of 'a second circuit' of capital in property and buildings and in the financial institutions that control and distribute liquid capital (banks, investment houses, stock exchanges, etc.), (iii) the concentration of collective means of consumption (schools, hospitals, theatres, etc.) and (iv) the intervention of the central and local state in both controlling the growth of class resistance to the domination of capital, and in ensuring that crises of over-production and/or declining profitability are managed as far as possible through the control of labour markets and consumer behaviour.

As with all structural analyses of capitalist society the recurring question must be – to what extent are these features general aspects of capitalism rather than specifically urban questions? The response usually contends that the difference between urban or urbanised capital and 'non-urban' capitalist society is not qualitative but rather one of degree. Put simply, cities have a much higher level of specialisation in production and services than rural or peripheral locales, they have greater concentrations and higher yields of commercial and finance capital as well as collectively provided public goods. It also follows that 'class resistance' (strikes, civil disorder, crime) is likely to be far higher and far more vigorously policed. As we have seen, other social theorists, especially Max Weber and Georg Simmel have drawn attention to the distinct features of the city in contrast to the rural world, but while both believed that trade and commerce are the key to the existence and persistence of urban form they did not accord a special status to the capitalist mode of production as the organising principle of the modern city. Thus, the task for Marxist urbanism is to explain how and why, following Lefebvre, the capitalist production of space is such a determining feature of the contemporary urban environment.

The Marxist geographer, David Harvey, has devoted a long and distinguished career to providing just such a critique of the modern capitalist city. Harvey's original prospectus was sketched out in his book *Social Justice and the City* published in 1973 in which he describes the city as 'a pivot around which a given mode of production is organized' (Harvey, 1973: 202 in Halpern, 1997: 225). *In Consciousness and the Urban Experience* (Harvey, 1985) Harvey refines his focus to consider the relationship between money, time and space because, as he argues, we can thereby 'clear away some of the clutter of detail and lay bare the frames of reference within which urbanization proceeds'. Of course, as Simmel has shown, money is elusive, being both a store of value, but also a token 'devoid of content "save that of possession"' (Harvey, 1985: 2). Its use also predates the full arrival of the capitalist economy by thousands of years, so it is important to distinguish, as Harvey reminds us, between the money economy, and the capitalist economy.

Money economies existed before the 'great cities' came into being, but by the arrival of industrial capitalism in the 1830s, Europe and several North American and Latin American cities contained populations of over a million. The circulation of people as labour commodities, and the circulation of capital as the abstraction of that labour surplus, therefore, become ever more closely connected. This is why, according to Harvey, '[t]he style of urban life necessarily reflects such conditions', and also why there exists, 'a deep tension between the individualism and equality that the possession of money implies and the class relations experienced in the making of that money' (Harvey, 1985: 5).

The link between money and space is particularly intensified in capitalist society because

money creates an enormous capacity to concentrate social power in space, for unlike other use values it can be accumulated at a particular place without restraint. And these immense concentrations of social power can be put to work to realize massive but localized transformations of nature, the construction of built environments, and the like.

(Harvey, 1985: 12)

The commodification of space in time was made possible by a combination of accurate scale maps and the introduction of the cadastral survey that 'permitted the unambiguous definition of property rights in land' (ibid.). But it took 'the buying and selling of space as commodity' to 'consolidate space as universal, homogeneous, objective, and abstract in most social practices' (ibid.: 13). In other words, as property rights became increasingly diffused and privatised (i.e. no longer the sole monopoly of the monarch, lord or bishop), civil law in association with the new profession of surveying gave rise to accurate, divisible and tradable parcels of land. The growth of the city was enormously facilitated by the ability of investors in urban land to extract rent (or unearned profit) from the rising value of city space. Proximity to administrative centres, markets and communication networks attracted premium rents and, hence, then as now, the surest way to add value to even productive arable land was to build on it or convert it to a road, canal, railway or runway.

In the second part of *Consciousness and the Urban Experience*, Harvey takes the idea of the commodification of space, which as we noted previously is simply the conversion of land or territory (although other natural resources such as waterways, subsoil, oceans and airspace can be subjected to the same process) into an asset that can be bought or sold, and applies it to the great scheme for the demolition and rebuilding of the city of Paris undertaken by Baron von Haussmann under the Second Empire of Napoleon III. The creation of the *grands places* and spectacular elite theatres such as the Opéra from the ruins of the slums of Montparnasse and Les Halles served the dual purpose of removing 350,000 members of the 'dangerous classes' who had for a time controlled the city during the period of the 1848 revolution, while provisioning the city's growing bourgeois elite with new sites of cultural consumption and ease of access from the wealthy residential zones to the newly rebuilt downtown (Donald, 1992: 438).

This had the intentional effect of displacing large numbers of workers to new out of town settlements where it was cheaper and easier for large-scale manufacturing industries to establish themselves. In Second Empire Paris we therefore have a perfect illustration of the features of capitalist urbanisation outlined above: the state regulation of class conflict, the provision of new collective amenities and facilities (albeit not for the urban poor), the geographical distribution of labour according to the specialised needs of the new manufacturing industries, and the creation of a lucrative market in land and property speculation allied to the growth of an international banking and finance district. Drawing special attention to the role of capitalist cities in organising labour markets on behalf of capital, Harvey writes that '[t]he history of the urbanization of capital is at least in part a history of its evolving labour market geography' (Harvey, 1985: 19).

The spatial division of labour

This is a point on which the economic geographer Doreen Massey also agrees when she writes that '[t]he spatial distribution of employment . . . can be interpreted as the outcome of the way in which production is organised over space' (Massey, 1995: 65). Referring in particular to the problem of de-industrialisation in inner cities, Massey notes how locational theory and theories of regional organisation in geography began to pay more attention to production factors, while acknowledging that '[p]roduction really is a social process' and that it is therefore necessary to 'embed that problem within the broader context of what is going on in society in general' (ibid.: 15).

Massey's book, *Spatial Divisions of Labour*, did for economic geography what Castells and Harvey had achieved in the fields of critical urban sociology and urban geography, for she explains how Marxist analysis can be applied in specific economic contexts

to show how the spatial distribution of industry, employment and unemployment works in advanced industrial societies. Massey's insights can best be summarised as follows. There are essentially three forms of spatial structure that organise divisions of labour: (i) locationally-concentrated spatial structures, (ii) the cloning branch-plant spatial structure, and (iii) the part-process spatial structure. In the first example the entire establishment of the firm is concentrated within a single geographic space (headquarters/administration and production/service processes). In the second, HQ and administration form a central, separate and distinct nucleus that controls a number of identical branches with local level administration control and total production/service processes at different locations. In the third example, the production process is broken down into separate phases, aspects of which may be linked physically to the administrative control headquarters, while others will be combined with branch administration and control.

If we think about how these three spatial organisations of production might be organised at the level of the city, one could imagine how the original Detroit Ford Motor plant could exemplify Type I locationally concentrated spatial structures with its world corporate headquarters located within a short distance of its manufacturing and distribution centres and with shop-floor workers, supervisors and senior management forming a single, mostly local labour market. An example of Type II would be the operation of an oil giant such as Royal Dutch Shell that maintains its international headquarters in the Netherlands, but at its different production, refining and distribution centres around the world requires an on-site administration and control staff. In Type III one could point to a modern advanced technology industry such as aerospace and the city of Toulouse in France. Toulouse is both the corporate headquarters of EADS, the major shareholder of the Airbus consortium which manufacture the Airbus passenger jets, and also (along with Hamburg) the final assembly site. Across the Channel at Broughton and Filton near Bristol, the wings and the under-carriage for each aircraft are produced – with each assembly process requiring its own dedicated technical and administrative support. This pattern is repeated in the other *filières* of the Airbus consortium in Spain, France and Germany.

These types of extended production space (Type II and Type III) are characteristic of an increasingly integrated international division of labour (Froebel *et al.*, 1980; Cohen, 1981). However, as Smith and Feagin point out, this process is hardly new since Marx was identifying a similar trend as far back as the mid-nineteenth century (Smith and Feagin, 1987: 4). The implications for urban economies are that a given city-region will wish to exploit its location and labour market advantage by specialising in a particular service or product, distribution, type of marketing and so on. But when demand changes, or the cost base of a particular industry becomes unsustainable under increasing global competition, 'mono-product' towns are particularly vulnerable to economic meltdown. This was, and remains, true of the 'rustbelt cities' of the northeast and mid-west US from the 1970s onwards, along with many primary and manufacturing industry dependent towns in Europe. While capital moves on to find new investment opportunities, the workforce of the post-industrial city are often left counting the cost of a non-diverse economy.

States can be important actors in mediating the process of economic reorganisation and can play a generative or reactive role in this process of global restructuring. But as Graham and Marvin argue (2001), it is the larger corporations, and particularly multinational firms rather than governments that are creating the demand for increased labour mobility, for enhanced communication and transportation infrastructures, for the easing of taxation and credit export regimes and the removal of controls over zoning and land-use. Thus, the new spatial divisions of labour are impacting on households and communities in terms of their internal composition and spatial distribution, on the growth of the informal economy, and immigration, and on the formation of different types of political, commercial and cultural networks.

However, as Massey observes, firms do not have it all their own way because local labour markets and 'skills clusters' are just as important pull factors in the location of particular enterprises as geographically specific industries such as ports, mining, forestry, fishing, or tourism are in attracting non-local labour (Massey, 1995). The fact that powerful corporations such as Microsoft need to outsource their software development to southern India or Cambridge, England demonstrates this pull feature of discrete labour markets, and the increasing integration between these different scales of corporate activity and complex yet spatially bounded labour markets.

Thus the 'goodness of fit' of any given spatial distribution of labour will depend on how adapted its skills base is to a rapidly changing global–local (or 'glocal') economy. For this reason the 'Regulation School' or 'Paris school' of political economists have argued that the capitalist regulatory state is inevitably implicated in getting the labour supply fit right for a global capitalist market that is no longer reliant on First World skills and expertise.[42] What is true for the nation-state is even more so for 'the entrepreneurial city' (Jessop, 2000), whose managers and marketers actively promote their citizens' 'vital statistics' as an incentive to outside investors, while key infrastructure features and research support (especially higher education facilities) are vaunted as part of a highly sophisticated global sales pitch (Short, 1999b and see Chapter 7 of this volume). Having considered the ways in which the city can be analysed in terms of the spatial means of production, it is now time to explore the ways in which the city can be analysed as an articulation of the relations of capitalist production. In other words, how do capitalist social and economic relations structure and determine the contours of urban life?

Capitalism and 'the urban question'

As we saw in Chapter 3, both the American and British traditions of urban studies tended to take the nature of the mode of production within which cities operated very much for granted. To the extent that class figured in such studies, it was considered as a marker of social identity that gave meaning to an often difficult and unpredictable life-world, rather than as a property of a structural inequality that shapes and is, in turn, shaped by capitalist relations of production.

Thus, it was not until the 1960s when, under the stimulation of the workers' and students' movements, social sciences in continental Europe began to re-examine Marx's more sociological writings, particularly the *Grundrisse* as being relevant to the new configurations of power that had emerged under the 'regulated capitalism' of what became known as the Keynesian Welfare State. Having re-engaged with Marx's critical thought on the nature of productive relations, as opposed to the productive process, two theoretical tendencies emerged. One based on the later Frankfurt School and particularly the work of Jürgen Habermas, saw in advanced capitalist society a transformation in the mode of production resulting in the mediation of class conflict through the 'means of communication', a process which is co-extensive with rapid urbanisation.[43]

Many French Marxists, particularly those who subscribed to the structuralist Marxism of the philosopher Louis Althusser, tended to regard the 'ideological formation' of the city, which is necessarily the nerve centre of the bourgeois state, as a product of the capitalist mode of production.[44] Chicago School type empirical explanations of 'urban evolution' were refuted as part of the same ideological offensive that seeks to present 'difference' as change (Castells, 1977). Since Marx argued that real change could only come about after the fall of capitalism as a means of production, consumption and exchange, bourgeois ideology will continue to exist as Marx wrote, 'in the very surface appearance of things' (Marx, 1974).

In Europe and in Latin America the early work of Manuel Castells and Jean Lojkine placed particular emphasis on the commodified space of the capitalist city as the key determinant of urban practices (Castells, 1972; Lojkine, 1977). From empirical studies such as *Monopolville* (Castells, 1974) to the more theoretically speculative *City, Class and Power* (Castells, 1978), Castells applied the critical framework of structural Marxism to a predominantly European and still recognisably **Fordist** urban environment. Unlike Althusser, however, Castells did

not accept that capitalist ideological forms were so pervasive as to deny the worker subject any possibility of resistance. Indeed, Castells contribution at this stage – and it was one that he was to build on in later works – was to emphasise the importance of the urban realm as a site of struggle and contestation (see Chapter 7 of this volume).

'The urban question' and urban theory

A particular feature of the French urban studies of the late 1960s and early 1970s was a belief that the theoretical provided the only 'scientific' antidote to the ideological contamination of research concepts that had characterised the American empiricist studies of the post-war era. This required the construction of elaborate explanatory models in advance of the evaluation of the case study. However, as Castells was later to admit, this deductive approach based on a fairly crude Marxist model of urban power produced very unsatisfactory results. As he confessed, 'we produced what we consider the only major fiasco we have had in empirical research' (Castells, 1983: 298).

Essentially, *The Urban Question* argued that even in its social aspects the city was rather like a giant capitalist factory where the hierarchies inherent at the workplace were reproduced in the location of workers' housing in the most polluted and over-crowded parts of the city, in the colonisation of public space and social amenities by the middle classes, and in the double exploitation inflicted on the proletariat by requiring them to give back a part of their meagre wages in the form of rent, which was a form of alienation that not even slaves were subject to. As Castells later admitted, the deficiencies of *The Urban Question* were due to an insufficient grasp of the limits of classical Marxist theory when applied too inflexibly to changing social structures – slums were being replaced by light, modern apartments, the white-collar service class was becoming more numerous than 'men in blue overalls', women were entering the workforce in larger numbers and demanding equal pay and employment rights. However, even in this early work, territory and space were seen as integral to the process of capital accumulation and hence constituted key sites of class resistance.[45] In this sense Castells maintained a belief in the inevitability of conflict in the city that held out the promise of a genuine social transformation. Nevertheless, Castells acknowledged that there were still theoretical limitations to his approach:

> for Marxist theory, which is the major framework for this strand, has no tradition of the treatment of an urban problem. Because there was an immediate need for a theory, it was applied too mechanically, by adapting general Marxist concepts to the processes observed, without identifying those new aspects posed by the urban problems which necessitated new concepts and new interpretations according to the historical content.
> (Castells, 1983: 11)

This measured self-criticism contrasts with detractors such as Ruth Glass, doyen of the British Community Studies school, who fulminated against the way '[t]his awful verbal pollution spreads: it saps sense and energy, (in Castells' code it is called ideological purity). In the products of that language shoddiness is built in'.[46] Rather like Lefebvre's *The Production of Space*, *The Urban Question* was not so much an instruction manual on how the city actually works, so much as a philosophical judgement on whom the city works for and against. The theory may have been prone to the structural determinism of Louis Althusser at his worst, but it did at least give some theoretical direction to a world of urban studies that, with the significant exception of Lefebvre himself, had been rather overtaken by the technocratic planning literature or the liberal reformist micro-sociology of the community studies tradition.

By the 1980s, Castells' research had moved away from a strict focus on the city as a means of production and consumption and has concentrated on the urban environment as a site of contestation, not just on the level of class conflict, but including social, economic and cultural conflicts of a more general nature (Castells, 1983: 327). Having abandoned a static, structuralist-Marxist account of the urban process, in the early 1980s Castells and his collaborators began to embrace a type of 'action research' based in several different developed and developing metropolitan regions that has traced what his group sees as a burgeoning growth in non-institutionalised protest movements around the world that are challenging the very premises upon which urban society has hitherto been based. Since we deal with the subject of urban social movements more fully in the following chapter, we will limit ourselves here to a brief consideration of how Castells' social movement analysis connects up to his developing critique of capitalism and urban life.

In *The City and The Grass Roots*, (1983) Castells' 'post-urban question' approach is well exemplified through some excellent case studies of urban social movements in history, such as the Glasgow Rent Strike of 1915, together with major evaluations of action research such as the squatter communities in Santiago, Chile. The aim throughout is to give a theoretical expression to live political conflicts with the aim of making a positive intervention in the development of such struggles. Castells' more recent research in his monumental three-volume study, *The Information Age*, (Castells, 1996–2000a), has continued to engage with the impact of modern capitalism on human (and especially urban) society, but the project has become much more wide-ranging than his earlier work not just spatially (Castells' compass has now become global society) but also in terms of the research agenda (including analyses of sexuality and identity or the impact of new information technologies on the organisation of capital and labour).

At the same time Castells is much less optimistic about the potential of social movements (whether in the town or the countryside) to liberate the poor and the socially excluded from their unenviable fate. Whereas in *The City and the Grass Roots*, Castells talked of a burgeoning, particularly of urban social movements throughout the world, in *The Network Society*, social movements are no longer seen as the harbinger of a new radical politics but, 'fragmented, localistic, single-issue oriented, and ephemeral, either retrenched in their inner worlds, or flaring up for just an instant around a media symbol' (Castells, 2000a: 3).

The city, informationalism and the space of flows

Capitalism still looms large in Castells' more recent work, but he no longer ascribes a determining importance to capitalism as a mode of production, not least because there are so many varieties of capitalism that its territorial impacts are far from uniform. For all the transformations in the world economy Castells concludes that 'capitalism is alive and well in spite of its social contradictions'. But within the logic of capitalist organisation he detects a new mode of development that he describes as 'informationalism' (Castells, 2000a: 211). In homage to Max Weber, Castells describes the 'spirit of informationalism' as an amalgam of discrete but inter-linked processes. The first is found within business networks (from family based enterprises such as those found in northern Italy and in parts of Asia), entrepreneurial networks such as those concentrated in Silicon Valley, post-Fordist type 'unbundled' enterprises made up a series of subcontracted processes and functions and hierarchical communal networks such as those found in Japan. The second includes technological tools such as new telecommunications networks, personal computers and wireless communication devices, self-evolving software, and also a new caste of specialist managers and employees who can communicate effortlessly using the protocols and advances of the latest digital media. The third aspect is global competition that continues to drive innovation, connectivity and the development of intelligent systems capable of exactly matching consumer demand to products and services. The fourth factor is the state, which plays a different role depending on the nature of the economy and political culture. In less developed countries the state is likely to be interventionist and to control and manage key sectors of the economy, in Europe it focuses on regulation, in the US it tends to be much more of an advocate for economic liberalism within the context of global corporate supremacy. By linking these different elements together, a super evolved form of capitalism that Castells calls 'the network enterprise' is able to circumvent many of the problems that beset earlier manifestations of

capitalism from production bottlenecks to over-stocking, to 'labour rigidities' through 'just-in-time' ordering, the outsourcing of production and, in many countries, by weakened employment rights (ibid.: 213–14).

Having briefly sketched the component elements of the informationalist economy we now need to consider how Castells believes they link together in what he calls 'the space of flows' as opposed to the 'space of places' of the traditional economy. Castells points to a concentration of advanced services and high employment growth in the leading metropolitan centres of the world, with the 'upper tier of such activities' being concentrated 'in a few nodal centres of a few countries' (Castells, 2000a: 410). In particular, global business centres constitute 'networks of production and management, whose flexibility needs not to internalise workers and suppliers, but to be able to access them when it fits, and in the time and quantities that are required in each particular instance' (ibid.: 415).

As we noted previously, the international division of labour resulting from these technological advances is structuring business locations across the world in order to take maximum advantage of labour force characteristics, favourable trading opportunities and access to markets. Castells cites the example of the American micro-electronics industry where research and development, innovation and prototype development tend to be concentrated in the core areas of the company's operations. Skilled fabrication takes place in the branch plants of the company, typically in the medium-sized towns of the western United States, whereas large-scale, high-volume manufacture, because of the need to reduce labour costs in a highly competitive market, will usually be located in the newly industrialising economies of East Asia (Singapore, Malaysia, Taiwan, Indonesia). Sales centres, maintenance and aftercare are concentrated in the major regional centres of North America and Western Europe, along with the rapidly growing consumer market in Asia itself (ibid.: 418).

What defines these most successful new technology-based global enterprises is their situation within what Peter Hall and Manuel Castells have called 'milieux of innovation' (Castells and Hall, 1994). A milieu of innovation is characterised as 'a specific set of relationships of production and management, based on a social organization that by and large shares a work culture and instrumental goals aimed at generating new knowledge, new processes, and new products'. While such a milieu could exist as a distance-based network, Castells contends that 'spatial proximity is a necessary material condition for the existence of such milieux because of the nature of the interaction in the innovation process' (Castells, 2000a: 419). He then adds, '[w]hat defines the specificity of a milieu of innovation is its capacity to generate synergy' and, as we shall see in the next section, this notion of spatial synergy is similar to that of 'agglomeration economies' or 'untraded interdependencies' and 'neo-Marshallian nodes' put forward by Sassen, Storper and Amin and Thrift to describe the unique features of global cities.

QUESTIONS TO DISCUSS

1 How do writers such as Harvey, Castells and Massey believe that the city operates according to capitalist principles?

2 What evidence do Marxist theorists produce to support the claim that the city is divided by class conflict?

THE CAPITALIST CITY AND GLOBALISATION

Over the past decade and more there has been an 'exponential growth of publications dealing with globalization', and although the effects of globalisation on human settlements have received less attention than issues such as development and trade (Habitat, 2001: xxx–xxxi), within urban studies as a whole the study of globalisation continues to be a major growth area (Yeoh, 1999; Marcuse and van Kempen, 2000: 2). As UN Secretary-General, Kofi Annan, writes in his introduction to the Habitat volume, '[a]lthough globalization certainly affects rural areas, the impact of global economic change is largely centred on cities . . . At the same time, cities and their surrounding regions are themselves shaping and promoting globalization by providing the infrastructure and labour upon which globalization depends, as well as the ideas and innovation that have always emerged from the intensity of urban life' (Annan in Habitat, 2001: v).

While there is scant agreement on what the 'chaotic concept' of globalisation actually means (Jessop, 2000: 81), it is generally assumed that globalisation is a process whereby everything from capital, labour and goods to communications, culture and pathogens can be rapidly exchanged or transmitted from any one part of the world without the obstruction of national borders.[47] Keywords that are associated with globalisation thus include – connectivity, interdependency, 'time-space compression', integration, competition, fragmentation, complexity and chaos. Undoubtedly, technological advances have helped to drive the process of globalisation, but we should remember that technology can also be used to limit, block and censor communications, and therefore the integration of national economies within a global market and the removal or imposition of limits on investment, migration, trade, cultural expression and so on, are the work of human agents, albeit they are often powerful and privileged directors of economic, political and military forces.

Most writers on globalisation point to the dominance of large-scale multinational companies (MNCs) around the world as indicative of a new phase of global capitalism. But since the sixteenth century, the control of global markets, commodities and trade routes by government-sponsored enterprise has been associated with the expansion of mercantilism, and from the middle of the eighteenth century, with the growth of capitalism (Wolf, 1997: 232–66). Most economic historians also agree that, especially in the case of Latin America and Africa, western capitalism has produced profits for European and North American investors at the expense of the native populations who have not received a fair share of the value of their natural resources and labour. Thus, 'uneven development' in which a wealthy and increasingly avaricious core of advanced economies maintains an exploitative relationship with a territorially much larger and more populated periphery is a direct consequence of a geography originally imposed by imperialism and colonialism, but which is now maintained by 'the invisible hand' of the global market, and especially the most powerful world corporations.

Some indication of the scale of the disparity between large MNCs and the developing world can be deduced by the fact that in 2001 the world's largest company, Wal-Mart earned nearly $220 billion in revenue, which was $30.8 billion dollars greater than the entire income of the continent of Africa in 1997.[48] Cities, and especially the larger cities in the advanced capitalist economies have become nodal points for this expanding global network of corporate and financial enterprise such that in order to remain successful, these 'world cities' or 'global cities' must provide the facilities, infrastructure, labour supply, and tax regimes that are most favourable to the most powerful banks, investment houses and MNCs. At the same time, the effect of this concentration of globally oriented capitalist enterprises has been to hasten the process of de-industrialisation within established urban centres, and to de-territorialise a considerable amount of economic activity that used to take place within the confines of the traditionally defined city (Paquin, 2001: 337).

Meanwhile, in the developing world, and especially Sub-Saharan Africa, structural adjustment programmes that were supposed to make the developing world more competitive in world markets have failed to halt a reduction in agricultural employment and the migration of millions of poor people into the already overcrowded cities of the Third World, where jobs, basic amenities, and public services are scarce or non-existent (Stein and Nafzier, 1990; Castells, 1998: 116). Yet, despite the desperate plight of the urban poor in the developing world, the power of the 'cultural pull' of western-style consumption by MNCs creates a demand for products and services where previously self-sufficiency or local consumption

patterns predominated, thus intensifying the balance of payments crisis and adding to the 'high human costs and sacrifices' that 'are rending the fabric of African society' (Adepoju, 1993: 3–4 in ibid.: 115).

While there is no denying the scale of the problems afflicting developing cities, the United Nations own Human Development Report for 1999 provides a hint of why, despite all the risks inherent in migration, rural populations continue to abandon the land in search of a better future in the town. In Botswana, which is ranked 122nd out of 174 in the UN Human Development Index, its Human Poverty Index (HPI-1) dropped from 32.2 per cent to 22 per cent between 1991 and 1996. Yet this aggregate figure for the nation as a whole masks a stark urban–rural divide with an HPI-1 poverty rating of 11.7 per cent in urban areas compared to 27 per cent in the countryside. Although a small country, Botswana's urban population has skyrocketed from 12 per cent in 1975 to 66 per cent in 1997, and is expected to rise to 88.7 per cent by 2015 (United Nations, 1999: 131). For the desperately poor of Botswana and millions like them around the world, such a trend is to be welcomed, for even the most marginal stake in the city is a better option than the drought, disease and famine that afflicts so much of the global south's rural poor.

Urbanisation, as we have already noted, is associated with the rise of capitalism and with the spread of capitalist markets. With the demise of Soviet-style communism and the effective abandonment of 'import substituting' endogenous economic development in the non-western world, it would no longer be an exaggeration to say that development and capitalist development have become synonymous. Therefore we would expect the rate of urbanisation to have some direct, positive correlation with the index of human development (see Table 6.1).

Table 6.1 confirms this hypothesis, because although there are some variations between the top and bottom groups (the most 'urban' societies are not those with the highest HDI scores, and the most 'rural' are not the least developed), we can see that in 1997 none of the ten most developed nations had urbanisation rates of less than 70 per cent and none

Table 6.1 Urbanisation rates of the ten highest and ten lowest Human Development Index ranked countries, 1975–2015 (projected)

HDI rank	Country	Urban population as a percentage of the total national population		
		1975	1997	2015
1	Canada	75.6	76.8	79.8
2	Norway	68.2	73.6	78
3	USA	73.7	76.6	81
4	Japan	75.7	78.4	82
5	Belgium	94.9	97.1	98
6	Sweden	82.7	83.2	85.2
7	Australia	85.9	84.6	86
8	Netherlands	88.4	89.1	90.9
9	Iceland	86.6	91.9	93.8
10	United Kingdom	88.7	89.3	90.8
164	Rwanda	4	5.8	8.9
165	Central African Republic	33.7	39.9	49.7
166	Mali	16.2	28.1	40.1
167	Eritrea	12.2	17.7	26.2
168	Guinea-Bissau	16	22.5	31.7
169	Mozambique	8.6	36.5	51.5
170	Burundi	3.2	8.1	14.5
171	Burkina Faso	6.3	16.9	27.4
172	Ethiopia	9.5	16.3	25.8
173	Niger	10.6	19.1	29.1
174	Sierra Leone	21.4	34.6	46.7

Source: United Nations

of the ten least developed nations had urbanisation rates above 40 per cent.

The dual city

The international division of labour under globalisation is producing both regional polarisation on a world scale with certain cities in the global south (e.g. Metro Manila, Mumbai, Dakar) performing essentially a sweatshop function for global as well as local manufacturing industry, while in the developed world, a two-tier city, what Mollenkopf and Castells describe as 'the dual city' (Mollenkopf and Castells, 1991) of highly paid executives, professionals and technocrats work (but do not reside) alongside a low paid, low skilled, and generally ethnically diverse service class (Friedmann, 1986). Largely missing from this new urban 'space of flows' (Castells, 2000a) are the intermediate strata in mostly traditional occupations, and living in nuclear family-based households that constitute the predominantly white suburbs and small towns of the developed world. As we saw in the previous chapter, globalisation also has profound social consequences in terms of stimulating the demand, both positively and negatively, for living space in the metropolitan city resulting in some districts in gentrification (the acquisition of residential and commercial space by wealthy incomers), and in others in ghettoisation (where those on low incomes constitute a significant proportion of the resident population).

World cities and global cities

'World cities' as Peter Hall (1966) reminds us have always existed in the sense that some of the earliest cities were ports or settlements that had grown up along vast trans-continental trade routes. However, Saskia Sassen wants to identify the process of globalisation with a different type of metropolis that she calls 'the global city' (Sassen, 1991, 2000 and 2001a). Sassen defines global cities as 'strategic sites for the management of the global economy and the production of the most advanced services and financial operations. They are key sites for the advanced services and telecommunications facilities necessary for the implementation and management of global economic operations', and, '[t]hey also tend to concentrate the headquarters of firms, especially firms that operate globally' (Sassen, 2000: 21). The notion that global cities are defined by their 'command function' as sites of global capitalist enterprise and international management is strongly rooted in the urban globalisation literature (Hymer, 1972; Cohen, 1981; Friedmann and Wolff, 1982; Friedmann, 1986, 1995; Feagin and Smith, 1987; Brotchie *et al.*, 1995; Knox and Taylor, 1995). But another approach, while not denying the importance of the command and control aspects of global cities, stresses international connectivity as another important test of 'globalness' (Smith and Timberlake, 1995). World city marketers also identify a wider range of indicators in addition to those mentioned – including concentration of government, higher education, culture and tourism (Mayor of London, 2002: 300).

In the revised version of *The Global City*, Sassen (1991, 2001a) develops a more extensive checklist consisting of seven essential features. The first is the geographical dispersal of a firm's activities across different countries. As these activities become more diverse and non-dependent, coordination and management of the different operations increases in scale and complexity. As a consequence, many firms decide to outsource their administrative and financial operations to specialist service providers (information technology, legal, finance and management consultancy firms). This is also increasingly happening with core services such as payroll and telecommunications that would previously have been provided in-house. Also following from this trend is the global firm's increasing reliance on 'agglomeration economies' where the particular mix of specialist services and skilled personnel creates an 'information centre' which is specific to a certain type of urban environment that allows greater speed and flexibility in response to market demands than would be possible

---

EXHIBIT 6.1   The City of London

The City of London is one of the world's leading financial centres and each day it turns over more than $500 billion of foreign exchange.

An average of $543 billion a day was traded on the London International Financial Futures Exchange (LIFFE) in 2001 (more than any other futures exchange in the world).

London also accounted for 56 per cent of the global equity market in 2001 (worth $2,651 billion), and 70 per cent of the Eurobond market, with City firms managing $2,850 billion worth of institutional and private assets (in 2000).

The world city status of London is further confirmed by the fact that 480 foreign banks have offices in London, and 448 foreign companies are listed on the London Stock Exchange, while 75 per cent of the Fortune 500 companies have offices in the UK capital.

(International Financial Services, London; Corporation of London)

in-house. Sassen's fourth point is that global firms are freer to move their headquarters because the more advanced companies have outsourced their specialist functions to the agglomeration economy and hence no longer need to be near their service providers and suppliers (ibid.: xix–xxi).

Building on her companion volume, *Cities in a World Economy* (Sassen, 2000), Sassen also asserts that the growth of specialist business-to-business services is leading to the creation of transnational urban systems where 'the economic fortunes of these cities become increasingly disconnected from their broader hinterlands or even their national economies'. Also

> the existence of major growth sectors, notably the producer services, generates low-wage jobs directly, through the structure of the work process, and indirectly, through the structure of the high-income lifestyles of those therein employed and through the consumption needs of the low-wage workforce.
>
> (Sassen, 2001a: 286)

It is the latter two aspects of Sassen's approach that make it more than an economic model of globalisation per se because, as she remarks elsewhere, the narrow focus on the upper circuits of capital that typifies the economic globalisation literature often 'excludes a whole array of activities and types of workers from the story of globalisation that are in their own way as vital to it as international finance and global telecommunications are' (Sassen, 2000: 7). Because large transnational corporations need cleaners and postal clerks as much as they require top executives to function effectively, office location managers have to bear in mind the availability of cheap and plentiful unskilled labour. This is why metropolitan cities continue to attract international companies because global cities are more likely to provide adequate supplies of cheap (often immigrant labour) that can be utilised with few of the costs or regulations that apply to the native (and more likely unionised) workforce (Friedmann, 1986: 77). Sassen is therefore right to emphasise the point that the increasing cosmopolitanism of the city is largely a function of structural income disparities that are trenchantly maintained by stockholders and managers of the larger international companies.

What this debate begins to hint at is that globalisation is not simply an economic process, it is, above all, a political process whereby market access, building permissions, tax breaks, labour laws, transportation infrastructure and so forth are all entirely dependent on the decisions of local, national and occasionally supranational government. However, in their framing and negotiation such opportunity structures for globalisation are overwhelmingly local (Clarke and Gaile, 1998: 2). As Amin and Thrift argue (2002), this emphasis on the importance of local markets and infrastructure for the viability of cities in an increasingly competitive global market can also be found in the work of economists such as Paul Krugman (Krugman, 1991; Fujita *et al.*, 1999) who, along with Sassen, highlights how local agglomerations of economic actors provide economies of scale and international trade advantage. Both Allen Scott (1981) and Sabel and Piore (1984) and Sabel (1994) draw attention to the ways in which **post-Fordism**, by moving from vertical to horizontal integration, places an even greater premium on the urban and regional economy as a source of suppliers and clients. Thus, paradoxically, these indirect or untraded interdependencies (Storper, 1997: 222), 'neo-Marshallian nodes' (Amin and Thrift, 1992) or 'service clusters' (Taylor *et al.*, 2003) have reinforced the need for spatial proximity at a time when the integration of the global financial and trading systems continues at an accelerating pace. Thus, the emergence of urban-regional economic networks should be seen as a complementary process of local economic agglomeration nested within a wider regional and national system of global integration.

Storper, in contrast to Sassen, sees the development of specialist agglomerations of high-value, high-specialisation services as a far more diffuse trend than the conventional triumvirate of global cities (London, Tokyo and New York). If one defines 'globalness' in terms of dependence on international transactions into and out of the city's economic base, it can also be shown that globalness is not specific to the Big Three (Llewelyn-Davies Planning, 1996). In fact, globalness is a truly worldwide phenomenon (Storper, 1997: 224–5; Taylor and Walker, 2001).

At the same time, it is also undoubtedly true that the major league world cities control a disproportionate share of telecommunications traffic (compared to their resident populations). Graham and Marvin cite a study which reveals that no less than 55 per cent of all international private telecommunication circuits into the United Kingdom terminate in London, while over three-quarters of all advanced data traffic generated in France originates within the Paris region (Graham and Marvin, 2001: 316).

Just as the growth of towns led to a speculative market in urban land, so the new globalised metropolis is absorbing the vast majority of new infrastructural investment in the form of teleports and fibre optic cable networks, along with more traditional forms of communications such as airports, road networks, subways, surface rail and other high-speed inter-hub links. In the aftermath of the Second World War, national governments had seen the development of transport and communication networks as part of a national economic development strategy in which the priority was seen to be reducing the wealth gap between core and periphery (Brenner, 1998b: 445 in Graham and Marvin, 2001: 309). But in the late twentieth and early twenty-first centuries the trend has been towards an 'unbundling' of national infrastructures in order to concentrate resources on the most successful urban regions because these are held to be the key to national prosperity (Graham and Marvin, ibid.).

At the same time, the traditional hostility of American capital towards big government (except as a provider of lucrative contracts) swept through the policy-making process of conservative governments in other parts of the English-speaking world – especially the United Kingdom and Australasia – to such an extent that the mantra 'public bad, private good' became an unquestioned orthodoxy in US type urban economies (see the following chapter). Although a pivotal role for the state was preserved in more *dirigiste* regional development strategies such as in France (under the aegis of the state controlled territorial planning authority, DATAR) where state investment continued to provide most of the capital expenditure

for new infrastructural projects in urban agglomerations (Savitch, 1988).

In the Far East, and especially Japan it is also important to note that, traditionally, private capital had less influence on urban development plans than firms enjoyed in the US, although even in these more state-directed economies, private capital has become pivotal in, for example, creating the world's richest market in property speculation in the Tokyo Metropolitan district with major consequences not just for the regional and national economy but for the investment environment of the entire world given the pre-eminence of Japanese banks as the major source of investment capital in the international economy (Waley, 2000: 136–8).

While we can, therefore, point to distinct processes and trends that are associated with an increasingly open world economy, the idea of globalisation as a successful conspiracy of leading multinationals bent on world domination and the further emiseration of the developing world has come under criticism from a number of quarters. As the chief implementer of structural adjustment programmes, it should come as no surprise that the World Bank sees global integration through the capitalist market as 'a good thing', and an antidote to, rather than a cause of, world poverty. In a 2000 report, the World Bank claimed that '[t]here is compelling evidence that globalisation has played an important catalytic role in accelerating growth and reducing poverty in developing countries' (World Bank, 2000: 1).

Radical urbanists such as Harvey, Castells and Sassen share the view that globalisation is creating a great deal of suffering and exploitation, increasingly concentrated in the marginal spaces of cities around the world. But if we take long-term trends into account, poverty measured by the official poverty level fell in India from 57 per cent in 1973 to 35 per cent in 1998, infant mortality rates have been cut in half around the world between 1970 and 1997, and average life expectancy has increased from 55 years to 67 years (World Bank, 2000: 3). Yet, as the World Bank admits, 'despite impressive growth performance in many large developing countries, absolute poverty worldwide is still increasing' (ibid.). Where countries are urbanising fastest their economies are recording higher rates of economic growth (see Table 6.1 above); in some cases this is leading to increased inequalities between rich and poor, but in others it is not. Nevertheless, and this is significant in the context of an increasingly urbanised poor population, the integration of financial markets and the increased volatility and scale of private capital flows means that governments and institutions are either unable or unwilling to protect the poor from the effects of economic crises. Additionally, workers in particular industries (and there are few now who enjoy robust employment protection) face increased insecurity including the need to change jobs and even city of residence at regular intervals (ibid.: 7; Sennett, 1999).

Such judgements cannot be resolved at the level of the empirical evidence, since for many liberal advocates the triumph of global capitalism is about creating opportunities for people to better themselves. According to this perspective there are no 'losers', just those who haven't won yet. The more pessimistic view asserts that globalisation makes every city, not just those in the developing world, a haven for sweatshop manufacturers and low-wage service industries because suddenly 'the reserve army of labour' can be measured in millions rather than thousands. Such firms tend to cluster in the more rundown parts of the city for the same reason that high-tech firms concentrate in Santa Clara County, California – because that is where the right labour force is to be found – unprotected low-wage migrants in the case of textiles and manufacturing, highly qualified local and foreign young college graduates in the case of Silicon Valley. Rather than proving or disproving the myths of globalisation, a considered analysis reveals that the effects of globalisation on the city are not sudden and clear cut, but halting and blurred with, as Goldsmith and Blakely argue, an often neglected role for critical urban spaces themselves in shaping the extent and nature of the new global economy (Goldsmith and Blakely, 1992).

QUESTIONS TO DISCUSS

1 **What in your view makes a world city?**

2 **Why are cities particularly important for the global economy?**

3 **What are the main costs and benefits of urbanisation?**

CONCLUSION

Although urban society has changed considerably since the mid-nineteenth century, the themes we have explored in this chapter all connect back in some way to the interpretation of the capitalist city provided by Engels and Marx. The common themes we can identify are: (i) the importance of commodities and commodification for urban development and urban capital markets, (ii) the intensification of the division of labour under capitalism, (iii) the need to manage and control class conflict as a result of the concentration of wage-labour in the city, and (iv) the increasing centrality of the city as a command and control centre for the globalised capital economy.

The geographer David Harvey's emphasis on the city as a key site for the reproduction of capital can be seen as part of the same Marxian analysis that led Manuel Castells to conclude that cities were also designed to reproduce and concentrate labour power. Later, however, Castells broadened his focus to consider collective actors in the city and to show how their political behaviour was strongly related to their dependence on public goods and services (Castells, [1972] 1977, 1983). This 'consumption location' approach (see Chapter 7) has strong Weberian overtones since it stressed market position and consumption status more than the relationship of urban actors to the means of production. Harvey, on the other hand even in his later work such as *The Condition of Postmodernity* (1989) has stuck to a fairly orthodox Marxist position in which contestatory politics and cultures are understood as the articulation of the relations of production (or class forces) that exist in a given urban locale at a determinate moment in time.

This has led critics of Harvey's urban theory, such as Michael Peter Smith to accuse the former of being a functionalist and of ignoring the ways in which urban society is being shaped and developed by counter-hegemonic groups and actors around the world (Smith, 2001: 24). Meanwhile, the global–local dichotomy that is consistently articulated by Castells, Harvey, Sassen, Zukin and other leading urban theorists, Smith argues, serves to simplify the complexities of 'transnational urbanism' and misses the real story of a highly functional globalism that (paradoxically) is at its highest stage of development within the 'marginal spaces' of the developing world and in the poor, migrant communities of the western metropolis (ibid.: 3–5). But without modern capitalist technologies, communication networks, and the push factors of global south destitution and the pull of western prosperity, the marginal spaces Smith describes would remain unconnected.[49]

Although Jessop also accepts that globalisation is not 'a coherent causal mechanism', this does not mean that it is insignificant or that the claims for it are necessarily overblown. The emergence of the 'entrepreneurial city', not only in its natural home of the US, but in many countries around the world points to the renewed importance of the networked metropolis as the key site of the informational economy and the most adapted spatial scale for a transnational capitalism that is less and less constrained by national borders (Jessop, 2000: 97). This trend is part of a process that has been given significant impetus by the collapse of communism in the Soviet Union and Eastern Europe, leading to what Graham and Marvin describe as the privatisation of splintered infrastructures and urban spaces by transnational corporate capital (Graham and Marvin, 2001). Global cities are the key locations for this process of fracturing and reconfiguration of economies, polities and cultures, leading some writers to conclude that many of the features we associated with postmodernism are to be found in the contemporary globalised metropolis (Magnusson, 1996: 281–2) (for a fuller discussion see Chapter 8).

As we saw in Chapter 5, this process is but one element in a worldwide trend towards a privatised urban realm that has seen the conversion of public streets into private malls (Gottdiener et al., 1999), the emergence of resident-only gated communities (Blakely and Snyder, 1999), and the increasing use of private companies to provide security and surveillance in commercial and residential areas (Davis, 1992).

However, history teaches us that conflicts between wealthier city dwellers and the urban poor have always been a feature of 'improvement' strategies by civic administrations. One need only think of Haussmann's 'class cleansing' of central Paris or the charming Victorian gardens of London's Belgravia and Chelsea where access is only granted to the square's wealthy residents. All this points to the continuing relevance of Lefebvre's seminal essay on 'the right to the city' since, as we shall see in the following chapter, the critique of the city as a site of social and economic exclusion is a salutary reminder that the city walls, curfews and ghettos of the middle ages are far from being historic relics.

FURTHER READING

Marxism/Marxist theory

Ira Katznelson's *Marxism and the City* (1992) is one of the best treatments I've found of Marx and Engels' contribution to urban theory that is at the same time accessible to a non-expert audience. On the other hand, David Harvey's 1982 classic, *The Limits to Capital*, remains unparalleled in its rigorous Marxian analysis of the capitalist city, but it repays a prior knowledge of Marxian economics. Other relevant collections in this tradition include Feagin and Smith (1987) and Tabb and Sawers (1984).

Ida Susser's *The Castells Reader* (2002) includes essays and extracts from the earlier period of Castells' career that are of a more explicitly Marxist mien (along with his more wide-ranging recent work). The two volumes Castells wrote in the 1970s and early 1980s, *The Urban Question* (1977) and *The City and the Grass Roots* (1983), are still required reading for anyone who wishes to understand the re-invention of urban theory after decades of epistemological stagnation.

The informational city

Castells has also made a major contribution to the literature on the information economy and the city. Particularly important are *The Informational City* (1989) and (co-authored with Peter Hall) *Technopoles of the World* (1994), and the *The Rise of the Network Society* (2000a). Castells' essay on the 'space of flows' is also to be found in Le Gates and Stout, *The City Reader* (various editions). Graham and Marvin's *Telecommunications and the City* (1996) also deals with many themes related to globalisation, urban restructuring and the new communication technologies – especially in regard to their social impact.

Globalisation and the capitalist city

There are far too many relevant volumes on the political economy of globalisation to list here, but from an urban theory perspective a list of key references ought to include Friedmann and Wolff (1982), Friedmann (1986), Sassen (1994), Knox (1995), Knox and Taylor (1995), Smith and Timberlake (1995) and Short (1999a). An abridged version of Sassen's *The Global City* appears in Fainstein and Campbell (1996) along with an excellent chapter by Fainstein herself on the changing world economy and urban restructuring.

Spatial divisions of labour/post-Fordism

In addition to Massey's signature text (1995), Ash Amin's edited collection *Post-Fordism: A Reader* (1990) while not exclusively dedicated to spatial considerations of post-Fordism, provides an excellent selection of essays that examine the causes and consequences of new production processes and their implications for work and settlement patterns. The chapters contained in Part I of John Rennie Short's *The Urban Order* (1996) are a clearly written and authoritative introduction to many of the themes and theorists discussed in this chapter.

7

THE CONTESTED CITY

Politics, people and power

All politics is local.

> Thomas P. 'Tip' O'Neill

The homily that 'all politics is local' is fatuous and self-defeating in the face of global neo-liberalism.

> Neil Smith

INTRODUCTION

All discussion of urban formations, the organisation of urban space and the behaviour and aspirations of city dwellers can be said to depend on relations of power to some extent. In previous chapters I have tried to show how writers and researchers have interpreted such phenomena sometimes using deductive (i.e. preconceived) and at other times inductive (or 'open ended') methods of investigation. In this chapter, I concentrate on power in its political form, especially in terms of its relationship to the management and organisation of urban space, the populations contained within it; the social, economic, cultural and political activities that urban government is charged with regulating or sanctioning; and the ways in which social actors have sought to contest and resist forms of authority or the measures that authorities have tried to implement in different urban contexts.

Of course the first point to acknowledge is that no two cities have an identical politico-administrative profile, even within the same national territory, and this makes the task of comparative generalisation particularly difficult. A no less cumbersome problem is the fact that the autonomy and powers of city governments vary widely from place to place, so that some urban governments could be said to be mere adjuncts of the national administration whereas others enjoy considerable fiscal, legislative and juridical independence vis-à-vis the national state. Furthermore, the relationship between levels of government extends upwards from the national state to encompass supranational authorities such as the European Union and, often, down to the district or neighbourhood level.

To this model of hierarchical and vertical *government* must be grafted the increasingly complex and extensive world of *governance*, which describes an ensemble of what have been termed 'spheres of authority', 'jurisdictions' or simply 'networks' that are all involved in decision-making or decision-taking. If government is monocentric, hierarchical and prescriptive, governance is usually contrasted as polycentric, non-hierarchical and non-directive. Government is characterised by structures of command, whereas governance is associated with partnership and consensus building (see Table 7.1, p. 129).

Finally, as we saw in the previous chapter, we cannot imagine the metropolitan world outside of the conditions necessary for its material production and reproduction. Urban change has always been driven by the concentration and movement of people and wealth, but today it is fashionable to condense these observations under the umbrella heading of 'globalisation'. Cities, especially the larger cities, are at the forefront of processes of globalisation that are altering the urban political order in profound and novel ways. We therefore try to identify and make sense of these processes in the contemporary metropolis and attempt to relate such developments to the insights we have encountered in other accounts of modernisation and modernity.

The chapter begins with a discussion of elite accounts of urban power before turning to look at the classic pluralist approach of political scientists such as R.A. Dahl, after which we consider the rational choice perspective on urban political behaviour, followed by an examination of the urban political economy approach, and concluding with an examination of more recent theories of urban politics, including urban regime theory, the growth machine model and institutionalism.

In the second section, following on from our discussion of how globalisation impacts on urban economies in the previous chapter, we look at this process in terms of its impact on the form and scale of urban and regional governance. Having learned that globalisation is not a process that begins and ends with the operation of the global market and that it is important to understand in which precise ways the new configurations of the local and regional state are developing in the context of demands for a better politico-spatial 'fit' for these emergent urban formations, we move beyond the arena of urban governance to consider broader societal dimensions of urban power. In the third section we focus on 'movements from below' and, in particular, on how poor and marginal groups are able to assert their rights to the city. We consider how differential outcomes in terms of the life chances and expectations of urban populations and their advocates have been articulated via movements for change and demands for a greater

voice in the management and allocation of urban resources. Finally, in the conclusion we attempt a confrontation between urban theory and theories of urban politics in order to highlight the issues and preoccupations that have traditionally distinguished the two schools of urban analysis. We then consider whether the gap between the two positions is closing and assess the contribution of urban political science to a resurgent concern with processes of power within urban theory.

APPROACHES TO THE STUDY OF URBAN POLITICS AND URBAN GOVERNANCE

Historically it has been – and in some parts of the world it is still – possible to talk of an urban community as a clearly bounded habitat, densely populated, politically and economically integrated, and distinct from its rural or maritime surroundings. In modern western societies such a net division between the urban and rural is far more difficult to detect. Urbanisation has resulted in the settlement of more than nine tenths of the population of the United Kingdom in towns and cities with somewhat lower but still high levels of urbanisation for Northern America, Japan and Western Europe (Fainstein and Fainstein, 1982: 145).

Thus, to talk of 'urban politics' as distinct from politics in general may seem spurious to western political analysts. But, in reality, if the territorial distinctions in the rural–urban continuum have become blurred, political scientists with an interest in urban enquiry have tried to avoid the problem by referring to 'the local' or 'sub-national' as a plane of analysis that is distinct and separate from the central and national. But here we encounter an immediate difficulty because the concept of the urban does not bear with it the limitations of the 'centre–local' dichotomy. This is because the centre–local or centre–periphery distinction is a political rather than a territorial demarcation and, as such, is bound up with articulations of the modern state. The urban, on the other hand, is a spatial and cultural category that encompasses and compresses every manifestation of state power.

Susan Fainstein has argued that in the UK urban politics emerged more slowly as a field of study than it did in the US. The much weaker position of British local government (as compared to the US), as well as the lesser prominence of urban social movements there, had much to do with this relative obscurity (Fainstein, 1993: 257). It is also the case that research on urban politics in Britain was rooted in the study of public administration which, historically, has lacked a strong theoretical tradition so that 'local government studies' have often been characterised by a strict empiricism with occasional forays into comparative analysis. Cities were, therefore, regarded as the places where local administration happens and the idea that urban politics involved more than the machinations of Town Hall committees took some time to penetrate the academic literature. In the US, however, the constitutional autonomy enjoyed by civic authorities stimulated interest in these miniature republics as polities in their own right. From the writings of Tocqueville in the early nineteenth century, city government in America has been seen as central, not only to the development of urban society but to the character of the nation as a whole. In the following sections we explore how different political perspectives on urban governance have sought to explain the organisation and outcomes of socio-institutional relations in the city in the past, and how contemporary approaches to urban politics are redefining the study of the urban polity in new and significant ways.

Elite theories of urban politics

Since the time of Aristotle the government of the city has been at the centre of political discourse. The ancient Athenians did not believe that the city should be governed with the consent of all its inhabitants since the *demos* represented only the non-slave male minority, while women, slaves and foreigners (i.e. non-Athenians) were excluded from decision-making. Aristotle and his followers believed that certain qualities were required of rulers that were not held

in common, and only those who had achieved a certain level of education and who were governed by reason rather than passion or the desire for popularity were fit to govern.

This classical view was further elaborated by the work of political theorists such as Niccolò Machiavelli in the sixteenth century, who gave his name to the art of political cunning and intrigue, and in the nineteenth and early twentieth centuries by Vilfredo Pareto and Gaetano Mosca. These later writers believed that all representations of decision-making as democratic were essentially false, and they built on the notion of 'the iron law of oligarchy' formulated by Robert Michels which contended that all organisations and bureaucracies (especially political parties) concentrate authority at the top of the decision-making tree, while the voters and the rank and file members of the organisation play an essentially supporting role.

How far were these claims for the universality of the elite model of government applicable to the management of cities? In particular, could the most self-avowedly democratic system in the world – that of the US – also fall prey to these oligarchic tendencies in its decision-making processes? In the 1920s Robert and Helen Lynd set out to find the answer in their famous study of 'Middletown' – their fictional name for the archetypal Midwest community of Muncie, Indiana (Lynd and Lynd, 1929). The Lynds concluded that a social and economic elite did, in fact, run the affairs of Middletown, and they confirmed their findings in a follow-up study published eight years later (Lynd and Lynd, 1937). Floyd Hunter's studies of Atlanta in the 1940s and 1950s saw elite rule as being structured into the very fabric of the city's social and economic life where those with economic wealth and political authority get to decide what policies and decisions should be taken on behalf of the largely voiceless and powerless majority (Hunter, 1953, 1959). Domhoff (1970, 1978) argued that elites shared common identities and a collective purpose through their membership of voluntary associations and policy groups (Logan and Molotch in Fainstein and Campbell, 1996: 291), a view that contradicted the dominant model of civic governance associated with the pluralist model developed in the 1960s by Robert Dahl.

Classic pluralism and the riddle of *Who Governs?*

As the author of over 40 books and 200 research papers, the American political scientist Robert Dahl is, without doubt, one of the most influential writers on government to have emerged since the Second World War. But despite his voluminous output on theories of democracy, it is his 1961 study of New Haven, Connecticut that is still cited as the definitive text for what has become known as the 'classic pluralist' account of urban government (Dahl, 1961).

Dahl's intuition was that the conventional view that elite groups within city hall and the business community held sway over the rest of the population, with only inter-factional struggles for power disturbing the smooth fabric of urban government, was wide of the mark. He set out to test the limitations of the elitist view through looking at the historical evolution of political leadership in New Haven, by examining official and unofficial sources of office-holding and key leadership roles in the community, and by the use of detailed questionnaires and in-depth interviews.

Dahl did, in fact, discover an oligarchic and elitist polity in the New Haven of the eighteenth and nineteenth centuries, but into the twentieth century he argued that the leadership of the city had become increasingly pluralistic and diverse. The answer to the question *Who Governs?* can be answered quite succinctly as 'no one in particular' or better 'no one group in particular'. It was Dahl's claim that while inequality at the level of the individual certainly existed and carried over to the political level, he also believed that group mobilisation and the competition and coalition-forming behaviour this engendered ensured no one group could monopolise power for long. Dahl's book emerged around the same time as Edward Banfield's *Political Influence* (Banfield, 1961), which painted a similar and surprising picture of Daly-era Chicago and Nelson Polsby's *Community Power and Polilitical Theory* (1963), which also became

an instant classic, and helped to establish behavioural political science, and particularly the study of group political behaviour in the front row of the discipline.

However, critics of Dahlian pluralism have highlighted the way in which (in US cities) it is not so much what is discussed but what is kept off the decision-making agenda that is key to understanding the structure of power in any given community. Bachrach and Baratz in their highly influential study 'Two Faces of Power' (Bachrach and Baratz, 1962) revealed how powerful local interests were able to dominate the policy agenda to such an extent that non-elite interests were never represented. Similarly, Paul Peterson has argued that redistributive measures are kept off the policy agenda because urban managers are compelled to attract and retain businesses that tend to avoid high tax environments. This argument put forward in his book *City Limits* (Peterson, 1981), nevertheless has its detractors who contest the view that cities have little control over the economic forces that bear down on them, even in the US (Swanstrom, 1988: 123; Clavel and Kleniewski in Logan and Swanstrom, 1990).

Public choice and urban politics

Public choice (sometimes referred to as 'rational choice' (Almond, 1990) or 'social choice' (Arrow, 1951; Buchanan, 1957)) is a theory of political behaviour inspired by neo-classical economics and, in particular, by the formal models that have been derived from utility theory and game theory (Von Neumann and Morgenstern, 1944; Riker, 1992). Its principal assumptions are that human agents are utility maximisers who will seek to make choices that are optimal in terms of the net payback to the individual or the group where the group is acting in concert. Also, following Schumpeter (1987) and Downs (1957), this assumption holds that voters' political preferences are set in relation to the political offers (or policies) put forward by political parties in a pluralist democracy in much the same way consumers decide over the relative merits of 'brand y' or 'brand x' in a supermarket. Just as with a competitive market in goods, politicians' policy offers are considered to be highly responsive to public opinion (consumer demand) and will be targeted as far as possible to match the preferences of the median voter (Downs, 1957).

One of the earliest models of public choice theory, as applied to urban decision-making, was provided by Tiebout in the 1950s (Tiebout, 1956). In Tiebout's analysis, residents of the city can be parcelled into groups of consumers with different sets of tax (price) and public goods (product) preferences. In large metropolitan authorities, those who hold minority preferences will be saddled with the local tax rates and level and type of services that suit the majority. For this reason, Tiebout argued that market responsiveness worked best in societies where municipal administrations are smaller, because where there is a plurality of local suppliers of public services, it is easier to match tax and service levels with the demands of discrete populations (Gottdiener, 1987: 67–8).

According to Dowding,

> [t]his market-analogue is thought to encourage produc-tive efficiency as governments are forced to produce goods at competitive prices; encourage allocative effi-ciency as local governments will be forced to provide the goods that (potential) residents desire; and reduce tax burdens as governments attempt to encourage rich residents to locate in the area in order to receive the higher marginal tax payments.
>
> (Dowding, 1996: 53)

However, Rusk believes the evidence points in the other direction, and that 'elastic cities' which expand to encompass their suburbs and the wider growth region perform better across a range of social and economic indicators than those whose municipal boundaries have remained unchanged since incorpor-ation. As a consequence, single authority government brings advantages to all inhabitants of the metropol-itan area, not just the less affluent central neigh-bourhoods (Rusk, 1993).

Whatever the merits and demerits of public choice outcomes may be, does this theory explain real behav-iour in residential location decisions and in voter choice in the metropolis? This is hard to answer for the simple reason that very little research has been done on the attractiveness of larger scale authorities as compared to smaller municipalities for businesses and new residents – no doubt because the 'white noise' of so many other context-dependent variables would drown out any meaningful statistical associa-tions. At the same time, evidence of the growth in secessionist movements wishing to acquire self-governing status for their neighbourhoods and to 'divorce the city' would support the impression that wherever possible, affluent voters will seek to avoid paying the 'external' costs of those less fortunate than themselves. That said, opposition to annexation proposals by big city authorities in the 1970s and 1980s also came from minority groups who saw it as a ploy to undermine their political influence at city hall (Rusk, 1993).

Thus, although public choice may have something to say about strategic political considerations, it fails to explain why, in the US, the national government is such a weak regulator of urban administrations – other than via fiscal carrot and stick measures – but such a strong enforcer of fair competition rules in the free market. In contrast to most other OECD coun-tries, in the US the taxation collected by the federal government is mostly channelled to costly national policy programmes such as defence and agriculture rather than to support public housing or education. In Europe, central governments will typically provide three-quarters or more of local government revenue, and will, consequently, require common eligibility rules and non-discriminatory allocations policy (although with the increasing success of centre-right parties this is beginning to change). This arrange-ment reduces the propensity for authorities to bid for the location of important companies and affluent residents by providing low tax regimes. It also means that 'shopping around' by prospective residents of the city is only likely to produce marginal bene-fits in terms of lower local tax rates, and these are likely to be offset by the reduced provision of public goods that middle-class urbanites disproportionately consume (art galleries, opera houses, theatres, concert halls, etc.).

For these and other reasons, Michael Keating believes that public choice theory (along with urban regime theory, which we discuss below) tends 'to generalize excessively from the American experience, rooted in a specific set of assumptions and historical traditions' (Keating in Judge *et al.*, 1995: 127). In other words, it tells us something about how the urban political market place works, but not how and why the market came to exist, and why it seems to vary in its behaviour from city to city. For answers to these questions we have to venture further into the dense undergrowth of urban political economy.

Urban political economy

John Walton describes 'urban political economy' as an interdisciplinary and intercontinental 'joint undertaking of neo-Marxists and left liberals'. The neo-Marxists are predominantly European and include figures such as Manuel Castells and David Harvey, while the 'left liberals' are predominantly North American urbanists who were influenced 'less by Marx than by C. Wright Mills and Floyd Hunter' (Walton, 1993: 302). However, several urban analysts claim to work within a political economy approach without necessarily subscribing to the ideological precepts mentioned by Walton. As we shall see in the following sections, urban regime theorists such as Clarence Stone describe their work as 'a political economy approach' (Stone, 1993) as do public choice theorists such as Bish and Ostrom (1973). Stephens and Wikstrom (2000: 107) cite as political economists, Robert Dahl, Charles Lindblom, Anthony Downs, James M. Buchanan, Gordon Tullock, William H. Riker, R.L. Curry and L.L. Wade, few if any of whom, I suspect, would identify themselves as left liberal or neo-Marxist.

A variation of public choice theory, local public economy, sees public entrepreneurs and citizens acting in order to obtain the best possible services as they would in a private market (Parks and Oakerson, 1989; Keating in Judge *et al.*, 1995: 126–7). Thus, it would appear that urban political economy as an area of urban research includes radical, liberal and conservative analysts along with structuralist, elitist and pluralist perspectives on urban power processes. What all (urban) political economy scholars agree

on is the importance of studying human behaviour (be it individual or collective) in order to better understand the workings of governments and states. Following from this, a political economy perspective would tend to see government, or 'the state', as an articulation of wider social and economic processes and forces rather than as an actor in its own right.

In parallel with developments in Marxist urban theory in continental Europe, in the US an interest in urban political economy 'emerged in response to a crisis in the reigning mode of urban inquiry' as urbanisation brought to bear considerable pressure on the already creaking infrastructure of many cities around the world. This process led, in turn, to intensifying social conflicts over economic resources and access to basic services such as health and education (Hill, 1984: 123–4). Urban sociologists at that time were mostly interested in social integrationist analyses that emphasised organisational approaches to the study of urban phenomena. Urban economists remained for the most part attached to neo-classical theories of economic behaviour, while most urban political scientists (at least in the US), as we have seen, were staunchly committed to pluralist accounts of urban government that stressed the competitive and autonomous nature of local political processes (ibid.: 124).

As we saw in the previous chapter, this was to change in the 1970s and 1980s with the rising influence of Marxism on the social sciences in Europe and in the developing world where, in urban studies, interest was being directed towards the work of Friedrich Engels on 'the great towns' and the writings of Karl Marx that dealt with relations of production and the commodity form. Rather than seeing cities as precursors of advanced industrial society as Weber implies, along with Manuel Castells and David Harvey, urban political economy writers such as David Gordon, Michael Storper and Richard Walker began to argue that the configuration of the urban region results from 'the capitalist imperative' (Storper and Walker, 1989, and see Chapter 6). In particular, they argued that the labour market, and the locational and infrastructural needs of industrial production rather than market allocations give rise to the changing dynamics of regional growth. Storper and Walker's 'labor theory of location' claims to account for the shift of business away from the relatively unionised north and east of the US to the less unionised Sunbelt cities of the south and west (ibid., see also Gordon 1977 and 1984). Although, as we saw in the previous chapter, some capital is more sticky than others, and there are limits to the freedom of industrial relocation even in the US.

Sandercock and Berry utilise a political economy approach that focuses on the Australian real estate market in an effort to explain 'who's getting what out of the urban economy and the urban planning system, and why?' (Sandercock and Berry, 1983). Much like Castells in *The Urban Question* and Lefebvre in *The Production of Space*, Sandercock and Berry argue that the market in urban real estate and land speculation has led to marked uneven development in Australia's cities, and an increased economic burden for the poor and working class who have struggled to keep a foothold in the city as a result of the capitalist induced housing scarcity (Sandercock and Berry, 1983). These two versions of Marxist political economy may differ in their emphasis of the determining factors in the configuration of urban space, but they are unanimous in their conclusion that it is the forces of capital (or business) that determine the urban landscape rather than political or social factors. In the following sections we explore the ways in which the political economy approach has seen 'a return to politics' through the study of urban growth machines and urban regimes in the US and beyond.

The urban growth machine

The idea of the city as 'a growth machine' has antecedents not only in the field of political economy but also in elite theory because it emphasises the importance of having access to major financial and political resources in order to influence urban policy-making. Methodologically the growth machine model is also associated with what has been termed 'the new urban politics' in its emphasis on case-by-case analysis, a strong focus on local context, and 'bottom-

EXHIBIT 7.1 City slogans	
Boston	Progress Through Partnerships; America's Working City
Dallas	The City of Choice for Business
Milwaukee	The City that Works for Your Business
San Jose	Capital of Silicon Valley
Norfolk	Where Business is a Pleasure
Phoenix	Moving Business in the Right Direction
New York	The Business City that Never Sleeps
	(Adapted from Short, 1999b: 49)

up' approaches to the study of policy formation (Jessop *et al.*, 1999: 144).

The signature texts of the urban growth machine literature are the essay by Harvey Molotch, 'The City as a Growth Machine' (Molotch, 1976) and the *Urban Fortunes* volume that he and John Logan produced a decade later, Logan and Molotch (1987). The aim of this new school of urban policy research was to redirect the pluralist urban politics agenda from 'Who Governs?' to 'For What?' (Logan and Molotch in Fainstein and Campbell, 1996: 291). Growth coalition analysts have more of a foot in the political economy camp than urban regime theory (discussed below) because they argue that the main force behind pro-growth initiatives is the business community – especially the rentier class who have an interest in boosting the value of property and land holdings.

Since the nineteenth century, boosterism has been an essential feature of city growth and development in the US, and the successful evolution of cities such as Chicago, Atlanta and Las Vegas would not have been possible without the active cooperation of ambitious mayors, property owners, railway directors and entrepreneurs. However, what Molotch demonstrated was that even established cities such as New York need to compete for real estate investment, prestige sports venues, universities, and qualified labour in order to sustain a vibrant capitalist market economy. Thus, city managers have to 'imagineer' (Rutheiser, 1996) their cities by selling their virtues and attractions to potential businesses and residents while at the same time keeping their exisiting political and business constituencies on board (see Exhibit 7.1).

However, Logan and Molotch emphasise that these growth coalitions take time and energy to construct, they require cooperation between different sectors and levels of capital and, above all, without a receptive political leadership such coalitions are unlikely to succeed. In this context, it has been argued that the role of 'the political' has been under-theorised for we do not have a satisfactory explanation as to why some city leaders are pro-growth, while others are neutral or even anti-growth (Harding, 1999: 679). It has also been argued that the seeming independence of political leaders in US urban politics, which is a unique feature of an unusually decentred polity, masks the important role that the state performs at every level in the management of the urban system.

Urban regime theory

Urban regime theory represents an attempt at responding to some of these criticisms of the growth machine model by focusing more on the nature of informal coalition behaviour (regimes) and acknowledging the lead role played by city administrators in establishing such urban partnerships. Urban regime theory is most closely associated with the work of the American political scientists, Stephen Elkin and Clarence Stone (Elkin, 1987; Stone and Sanders, 1987; Stone 1989).[50] Stone offers a succinct definition of an urban regime as 'an informal yet relatively stable group with access to institutional resources that enable it to have a sustained role in making governing decisions' (Stone in Judd and Kantor, 2001: 26). In the long tradition of American urban political science, both writers have developed their ideas through the study of a particular city government. In Stone's case, Atlanta provides the testing ground for a model that has been applied with some

utility to other American cities, though with rather less success in the British and European context (Harding, 1997).

Urban regime theory takes from pluralism the idea that no one group monopolises power at local level, but it also backs elitist accounts of urban power by pointing to the existence of tight-knit coalitions of urban 'movers and shakers'. Stone's 'pro-growth regime' also resembles Logan and Molotch's 'growth coalition' because the chief objective of such an expansion-oriented regime is to build its economy by attracting business and real estate development, and hence the local tax base. Where urban regime theory differs from the growth coalition approach is in the identification of non-business directed regimes such as progressive regimes (or community-based political coalitions), and caretaker (or maintenance) regimes that essentially try to maintain the status quo (Sites in Judd and Kantor, 2001: 215).

One of the leading proponents of urban regime theory in British political science has argued that its chief advantage lies in its offering 'a new perspective on the issue of power' because '[i]t directs attention away from a narrow focus on power as an issue of social control towards an understanding of power expressed through social production' (Stoker in Judge *et al.*, 1995: 54–5). The approach has also been welcomed for its recognition of the division of labour between the (local) state and business or capital, where both have different remote interests while being dependent on one other for their long-term survival (Kantor and Savitch, 2002: 277–8).

However, critics have questioned whether it would not be better to describe the regime approach as a model rather than a theory in its own right because it lacks a universal application. As Stoker has observed, 'regime theory needs to escape the "localist" trap and place its analysis in the context of the broader political environment' (Stoker in Judge *et al.*, 1995: 55). There is some encouraging research in this direction on comparative regime theory by Alan Harding and others (Harding, 1994; Stoker and Mossberger, 1994) that suggests a wider application may be possible beyond the metropolitan level, but it is doubtful whether the early enthusiasm for urban

regime theory will be sustained in the face of traditional and emerging urban political analyses. In the
next section we consider an approach that has seen a
revival in fortune in recent years as political scientists
have begun to re-examine the institutional nexus of
urban government in order to help explain contemporary political developments in cities.

The institutional perspective

In both the growth coalition and urban regime
versions of urban political economy there is an
assumption that the state, and especially the local
state, is a relatively weak actor when it comes to the
organisation of spatial hierarchies in the city. But
while conservative political economists see the market
as enabling individual and business choice and, therefore, as an instrument of political freedom, radical
political economists see the state in general as prey to
class struggles and a necessary and essentially passive
mechanism for reproducing the productive forces of
capitalism (Skocpol, 1985: 5).

Much of the critique of societal and economy-
centred models derives from an institutional or statist
perspective that seeks to restore formal political
institutions and their 'logics of operation' to the
centre stage of (urban) political theory. Institutional
theory, in Jon Pierre's words, 'highlights the overarching values that give meaning and understanding
to political processes' while 'institution' refers to the
'overarching system of values, traditions, norms and
practices that shape or constrain political behaviour'
(Pierre, 1999: 373). Institutions differ from organisations in the sense that although institutions do
possess organisational logics, the ultimate source of
legitimacy for institutions is the sovereignty of the
nation-state. Institutions therefore imply authority,
accountability and accessibility, while organisations
need possess none of these features.

The institutional perspective has strong antecedents in the sociological theory of Max Weber,
which is why in the British literature the emphasis on state centred accounts of urban politics
are referred to as neo-Weberianism while, in the US,
this approach is often called 'state managerialism'

(Gottdiener, 1987: 66). In essence, institutionalist
analyses support the idea of 'the relative autonomy of
the state', believing (contra Marxist and pluralist
interpretations) that political institutions do not
merely react to social and economic processes in wider
society, they often direct and give rise to them
(North, 1990: 112). Thus, Gurr and King write,

> the purposes pursued by the modern state with respect
> to cities are not primarily those of private capital or
> social movements or other interests inside the state, nor
> of the 'general welfare'. Rather they are the interests of
> the state, which is to say the primary interests of the
> state in maintaining public order and authority in urban
> populations, in securing public revenues, and the inter-
> ests of officials in the pursuit of *their* programmatic goals
> with respect to urban welfare.
>
> (Gurr and King, 1987: 9,
> original emphasis)

Matthew Crenson, in his work on air pollution,
develops an institutionalist perspective that shows
how governments and parties (particularly 'machine
type' municipal parties) work hand in glove to set
and maintain the political agenda without recourse
to wider publics or interests (Crenson, 1971). City
administrations are major stakeholders in the local
economy, often the largest employer in the city, and
almost invariably the biggest consumer of services
and goods. Added to these economic capacities are a
raft of legal and political privileges that private
capital certainly does not enjoy – such as powers of
eminent domain or compulsory purchase, the right to
impose land-use restrictions according to particular
zones, the right to impose taxes, the right to control
or ban certain types of traffic, and so forth. The 'relative autonomy' of the state relates in large measure
to how far these powers can be exercised according to
the priorities of the city administration rather than
those of voters and organised interests. Where the
voice of public opinion is weak or disunited, urban
governments find it much easier to pursue their
own policy agenda and to effect 'institutional capture'
of local party representatives (Cockburn, 1977).

But however central the state may be to what
happens within cities in every sphere of human
activity, 'the state' is not a static actor that behaves

in a consistent way in every urban context. Globalisation as we noted in the previous chapter is putting not only the urban economy under enormous pressures, it is also changing the internal organisation of government and the nature of the relationship between different levels of government. In the following section we explore how the 're-scaling' of government is changing the nature of politics and government at the urban level and prompting a re-evaluation of the spatial contexts in which political decisions affecting the lives of cities are taken.

QUESTION TO DISCUSS

1 Which of the approaches to the study of urban politics discussed above do you find the most convincing and why?

LOCAL AND GLOBAL: THE RE-SCALING OF URBAN GOVERNMENT

The rise of 'glocal' governance

The UN Development Report in 1999 emphasised that human settlements and, in particular, cities, are important nodes in the new forms of governance that are currently emerging around the world. However, as globalisation requires cities to act increasingly 'as territorial units in competitive processes', cities are becoming at the same time, 'more and more fragmented: socially, economically, physically and politically' (Sassen, 2001b: 57). This disparity between the city's economic and political mission and the administrative-institutional form it assumes in many locales requires what has been called 'a repertoire of governance strategies' which need to involve responses to, or the development of, 'markets, hierarchies, and networks' (ibid.). In order to formulate its response to the myriad world of public, semi-public, private and non-governmental actors that have a stake in the urban complex, governments need to address three tasks – coordinating, steering and integrating (see Table 7.1).

Table 7.1 Key features of governance

Coordination	Not just vertical, but also lateral and inter-agency management of services, facilities, infrastructure and the urban environment
Steering	A 'command and control' function that reserves to the political authority only the key strategic decisions, leaving to outside agencies (often private or non-profit) the task of delivering services (or 'rowing')
Integrating	Also referred to as 'joined-up government' – making different internal and external policy operations and networks compatible and complementary

Governments need to be involved in markets because, as we saw in the previous chapter, the regulation of capitalist relations of production is essential to the stability and reproduction of capital itself. As the state becomes increasingly 'hollow' (i.e. no longer a direct provider of goods and services), the routines and assumptions of competitive market economics have become increasingly evident within the institutions of the state itself. Thus, urban governments refer to themselves, local businesses and the local community as 'stakeholders', while urban managers engage in 'place marketing' to attract new investment and visitors to their cities. Markets matter at the local level in deciding whether the council's own workforce or an outside contractor will collect and empty the city's refuse. Simultaneously, at the global level, cities can be involved in bidding to stage an Earth Summit or in persuading a multinational oil company to build a new refinery on its doorstep.

Combined with these market-centred pressures on urban government to become more open and 'entrepreneurial' there are evolutions in the form of the state itself aimed at improving 'the spatial fix' (Harvey, 1989) of government for a variety of reasons. The first and most compelling reason for the 're-scaling' of urban government is in order to manage the changing environment of the local, regional, national, national-regional and international economy better. However, improving global competitiveness is not the only reason for re-scaling and re-organising urban-regional government. In an increasing number

of countries, demand for political and cultural autonomy from an over-powerful and directing central
state explains the push for devolution and decentralisation better than the frustrations at the loss of
economic growth opportunities. By the same token,
urban administrations that could improve the overall
economic performance of their region by merging
with their metropolitan neighbours have chosen not
do so because voter antipathy towards 'metropolitan-
isation' is a greater disincentive to action than
improved growth potential.

As we noted earlier, this is a particular feature of
municipal government in the US and helps to explain
why the contested and heterogeneous phenomenon
of the 'new regionalism' has been limited to a handful
of well-known urban agglomerations such as Portland, Oregon; Minneapolis-St Paul; Chicago; Philadelphia; Pittsburgh; Boston and Miami/Dade County
(Carr and Feiock, 1999: 477; Dreier *et al.*, 2001;
Brenner, 2002). This is surprising if the evidence
suggests that 'elastic cities that can expand their
boundaries are in better fiscal shape, have fewer social
problems and have more prosperous economies'
(Rusk, 1993 in Carr and Feiock, 1999: 476). However, Swanstrom points out that the economic
growth arguments for new regionalism are far from
compelling (Swanstrom, 2001), while the political
and institutional obstacles for Mumford-type city
regions are no less daunting than they were during
the 1930s. A third, and more theoretically contested
factor has to do with the changing nature of the
nation-state itself. Many writers sustain that the
nation-state in its old guise as the chief helmsman
of the Fordist economy is no more (see especially
Storper 1997; Brenner, 1998b, 2002; Jessop, 2000).
In its place we have an increasingly fragmented
polity where, under neo-liberalist pressures, power is
being re-distributed vertically from international
treaty organisations (NATO, the European Union,
the WTO, etc.) through national governments,
and regional governments all the way down to
district or neighbourhood councils, and horizontally
between levels of governance as interspatial competition for capital and resources intensifies (Brenner,
2002).

In many policy areas it is now possible to identify
interactions between different levels of government
that involve 'scale jumping'. For example, the Rio
Earth Summit in 1992 introduced a local environmental action plan known as Local Agenda 21, and
many municipal authorities around the world have
adopted these targets for environmental sustainability
without recourse to national governments (Castells
and Borja, 1997). In the European Union, the Committee of the Regions develops recommendations
for EU regional policy, in many (though not all)
cases with representatives exclusively drawn from
subnational government, while in other policy arenas
there is evidence that regions and cities are emerging
as influential political actors at a supranational level
(Hooghe and Keating, 1994; Hooghe, 1995).

In some countries, pressure for the re-scaling of
urban-regional government emanates from political
actors, business leaders and representatives of labour
at the city or local level, in others the re-scaling
process may be more 'top-down' with national or
(in the case of the European Union) even supranational executives seeking to impose new administrative boundaries or to add or subtract functions
from different layers of subnational government. In
other countries we see a mixture of both 'bottom-up'
and 'top-down' initiatives aimed at reconfiguring
existing territorial dimensions and functions, but
not necessarily in the same direction (Keil, 1998b).
Negotiations between the different scales of policy
communities can produce either a 'one size fits all'
variation on the existing subnational government
architecture (such as with the French regional reforms
of the 1980s) or a 'variable geometry' approach where
individual regions can bargain for enhanced autonomy from the centre (such as Catalonia, Galicia,
Andalucia and the Basque Country in Spain)
(Guerrero, 1997; Martínez-Herrera, 2002).

Although one might expect globalisation to promote a reconfiguration of the urban polity according
to the spatial fix best suited to global city marketers
and their powerful and demanding corporate
clients, the reality is rather different. Governments
that have championed neo-liberal market-oriented,
pro-capitalist policies, such as the Thatcher govern-

ments in Britain in the 1980s used their powers to eliminate metropolitan government. But there is little evidence to suggest that this move was intended to 'circumvent the territorialized opposition of both industrial capital and manufacturing workers' (Duncan and Goodwin, 1988 in Brenner, 1998b: 24) so much as to frustrate the political ambitions of the government's local political opponents (i.e. the 'new urban left' politicians who were in control of several of Britain's largest cities) (Livingstone, 1987).

Ironically, the former 'new left' leader of the Greater London Council, Ken Livingstone, who is now the elected Mayor of a pared down version of his former authority, has become a strong advocate of London's global city role, and is known to favour high rise office developments so that London will not lose out to its global competitors in the international service economy.[51] Almost any other example that one could choose would involve similar tales of political intrigue, bizarre alliances and conspiracies involving politicians, bureaucrats, business interests, and lobbyists from every intersection of the 'glocal' realm all vying for their version of the institutional-political fix that is best likely to promote or defend their position (Swyngedouw 1997; Brenner, 1998a; Swyngedouw and Baeten, 2001). Although much of the political rhetoric of urban boosters will be about global competitiveness, as a casual glance at the Metro pages of the New York Times or Chicago Herald Tribune reveals, city politics remains resolutely local – especially where local elites are susceptible to the caprices of the local electorate.

Contrary to Brenner's claim that the national state provided a static container of 'timeless space' prior to the 1960s (Brenner, 1998b: 28), 'glocal' spaces have a long history of subsisting within the framework of the national state. The City of London from its first beginnings has attracted international capital and capitalists precisely because it was able to function as an 'off-shore island' within the national capital of Britain. Other privileged (and ancient) globalised command and control centres also fail to conform to the 'pre-globalisation' hierarchy of scales myth – most notably Vatican City – which outdoes its secular rival, Rome on almost every count of 'global reach'.

At the end of a long list of exhortations for cities to be more competitive, to become more efficient in managing resources, and better able to bring private, public and citizen interests together, the United Nations Habitat Centre's report ends on an almost Lefebvrian note by asserting its 'support for citizenship in the sense of rights to the city' (Habitat, 2001: 55). But in keeping with the increasingly neo-liberal cast of international policy-making, the UN report argues (contra Lefebvre) that

> [t]he withdrawal of the state and limitations on institutional demand making have combined to create new spaces for political contestation. This development signals emerging opportunities for civil society to engage government and the private sector in new forms of cooperation that enable the low-income communities to participate as empowered partners. More broadly, this development is about authentic citizenship, meaning the rights and responsibilities of the urban citizenry.
>
> (ibid.: 56)

How we identify an 'authentic citizen' is not made clear, but what is manifest is the increasingly communitarian tone of international agencies and governments in stressing 'responsibilities' alongside rights, even where as its own report shows, the urban poor are experiencing a reduction in entitlements and therefore reduced opportunities to be 'authentic citizens'. In the following section we look at how and why these 'new spaces for political contestation' are emerging in cities around the world, and what these urban social movements have to teach us about citizenship and the nature of urban power relations.

POWER FROM BELOW? THE CHANGING FACE OF URBAN SOCIAL MOVEMENTS

As the authors of a recent study of urban social movements note, '[t]he study of urban social movements and the specificity of their collective actions have traditionally occupied a limited and marginal status in social movement theories and deliberations' (Hamel et al., 2000: 1). The reasons for this, the authors believe, are to do with the fact that such movements are seen as operating on the micro as opposed to the macro scale, and are therefore best studied as components of the local political process and local political economy. As previously noted, if we consider urban movements to be organised interests of one kind or another, political scientists have often sought to incorporate such collective actors within a Schumpeterian framework where mobilisation and organisation strategies are seen to be organised around a competition for influence and resources at the micro and occasionally meso level, but rarely if ever on a national or international scale. A consequence of this research agenda has been the neglect of urban movements as 'representative of broader values, ideals or emancipatory possibilities' (ibid.). However, such a micro focus may itself have been associated with the early phase of social movement activity that subsequently loses its localist bias as the struggles and conflicts (such as anti-capitalism) are articulated in increasingly wider imaginative universes. In other words, as urban activists begin to act locally they also start to think globally, and these global reflections frequently yield the insight that local activism is not enough (Keil and Ronneberger in Marcuse and van Kempen, 2000).

The social movements of the late 1960s had privileged the street and public space as a terrain of conflict that had enabled workers, students and other city dwellers to share the experience of mass protest. Of course 1968 was not the first occasion that such social forces had been brought together in order to resist and challenge the dominant political order. What was interesting and important from the perspective of a critical sociology, however, was the 'insertion' of these new social actors within a rapidly emerging metropolitan culture. In the past, such bohemian avant-gardes had been confined to the artist quarters of Paris and some of the larger European metropolises, but after 1968 the avant-garde took on a 'mass' aspect, spawning an alternative culture that was rapidly exploited by commercial interests, but which nevertheless retained a strong kernel of anti-materialism. Squatting took on ever more organised forms in many European cities (see Exhibit 7.2), rent

strikes and tenants' unions became a commonplace, campaigns for free public transport, and feminist marches to 'reclaim the streets' all pointed to the birth of an alternative civil society in the larger urban cores (Castells, 1983). The increasing social diversity of the city accompanied by the economic inequality which many new migrants experienced, also promoted the emergence of racial politics which found its apogee in the US in the 1960s and early 1970s (see Chapter 5).

These events were to have a significant impact on social movement theory because several of the most influential social theorists including Alain Touraine and Manuel Castells were actively involved in the student and workers' protests themselves, and were therefore able to gain a direct insight into the new contestatory politics. It is not that this experience marked an immediate departure for Castells at least from classical Marxist analyses, but it certainly persuaded him that an adequate theory of urban social movements needed to take into account other subordinate subjects who, in Lefebvre's terms, were denied their rights to the city, but were finding new and inventive ways to make their demands heard.

In *The City and the Grass Roots* (1983), Castells builds on his direct experience of squatter groups and citizens' organisations in France, Spain, Chile and also California where he moved to a professorship in urban sociology in 1979. The case studies of urban social movements from around the world, and in different historical time periods, emphasise the non-class exclusivity of progressive movements and the importance of social movements as collective voices for 'new' social identities, or the claims of existing communities that have become marginalised through the actions of governments and planners. What unites disparate groups and actors in the city is their collective consumption of resources, be it water, housing, transport, or in the case of gay minorities, the right to equal treatment by the law and legal authorities. A point that Castells illustrates in *City, Class and Power* (Castells, 1978) by referring to general 'urban problems' such as traffic congestion, pollution and housing shortages as having a generic impact on the

EXHIBIT 7.2   'Squatting in Amsterdam', from Ed Soja, 'The Contested Streetscape in Amsterdam' in Borden *et al.* (eds) *The Unknown City*, 2000

The contemporary residential rejuvenation of Amsterdam's Centrum, more effectively than any other place I know, illustrates the power of popular control over the social production of urban space in general and, in particular, over the ongoing process of urban restructuring. It has been perhaps the most successful enactment of the anarcho-socialist-environmentalist intentions that inspired the urban social movements of the 1960s to recover their 'right to the city', *le droit à la ville*, as it was termed by Henri Lefebvre, who visited Amsterdam many times and whose earlier work on everyday life inspired the Amsterdam movements [. . .] In 1965, while Watts was burning in Los Angeles, a small group of Amsterdammers called the Provo (after their published and pamphleted 'provocations') sparked an urban uprising of radical expectations and demands that continues to be played out on Spuistraat and elsewhere in Amsterdam's 'magical centre' of the world. [. . .]

The Provos concentrated their eventful happenings in both Dam and Spui squares and managed to win a seat on the city council, indicative of their success in arousing wider public sympathies. Their artful challenges to hierarchy and authority lasted for only a few years, but they set in motion a generational revolution of the 'twentysomethings' . . . that would dominate the renewal of the Centrum over the next two and half decades. In no other major world city today are young householders, whether students or young professionals, in such command of the city centre.

[. . .]

After 1980, the movement did not decline so much as become a more generalized radical pressure group protesting against all forms of oppression contained within what might be called the specific geography of capitalism from the local to the global scale. Squatters, for example, merged into the women's movement, the antinuclear and peace movements, and the protests against apartheid (a particularly sensitive issue for the Dutch) and environmental degradation (keeping Amsterdam one of the world's major centers for radical Green politics), as well as against urban speculation, gentrification, factory closures, tourism, and the siting of the Olympic games in Amsterdam.

urban population even though the effects of each problem will be felt differently according to one's class and status position (ibid.).

Conventional Marxist accounts failed to acknowledge the radical potential for social change of actors who are not directly involved in the process of production, whereas Castells saw these marginalised groups as the advance guard of a new identity politics that would fundamentally change our ideas about urban power and its uneven distribution in the modern city (Susser, 2002: 7). But can we draw distinctions between different types of urban movement and what criteria should we adopt?

One of the leading proponents of pluralist urban political theory, Christopher Pickvance, has developed a 'linked sub-models' approach to urban (social) movements based on four main categories of contestation – movements for the provision of housing and urban services, movements over access to housing and urban services, movements for the control and management of the urban environment, and defensive movements around environmental or social threats (Pickvance, 1985). There will inevitably be a degree of overlap between these various categories, as social actors will have more than one field of interest. But unlike Castells, Pickvance is scrupulous in avoiding any progressive claim for the existence of such movements and details the long history of urban campaign groups dedicated to reducing local government spending, while also cautioning against an elision between 'grass-roots' movements and campaigns for social justice (ibid.).

In Pickvance's view, the presence of political parties is but one among many possible conditions for urban political activity. However, unlike (the earlier) Castells he does argue that the existence of forms of collective consumption is not sufficient in itself to give rise to social movements. In addition to collective consumption, two other criteria must exist, there must be local-level political processes, and there must be spatial proximity. Without these factors present it is almost impossible to distinguish between a geographically non-specific social movement and an urban social movement (ibid.).

Spatial proximity is seen as a key determinant of urban political culture both in terms of relations to the means of collective consumption (transport, social services, housing, health, etc.) and in terms of the functioning of social networks. But to simply state that close proximity to social provision produces a different form of consumption to that of more diffuse communities is an act of description not explanation, and does little to expand our knowledge of the urban political process. Instead, we need to explain why certain constellations of social and economic forces develop around particular sites of conflict. For example, the competition for space between residential and commercial uses, the requirements of finance capital as against that of manufacturing capital, or the exigencies of comprehensive urban planning versus the preservation of community identities.

Community resource mobilisations in defence of economic advantage or 'quality of life' among affluent urban stakeholders contrast in significant ways to the repertoires of protest and resistance associated with the urban poor, and especially the black and Latino communities of cities such as New York because the welfare and coercive arms of the state, and the vagaries of the global economy constrain the lives and activities of the urban poor to a far greater extent than they do the middle classes. At the same time, these 'communities of resistance' (Castells, 1997) as Robin D.G. Kelley has shown in relation to black grass roots movements in America's cities, are capable of challenging orthodox views of how social protest

is meant to be conducted among the so-called 'under class' (Kelley, 1998).

The increasing interest in the dynamics of coalition building by policy-makers, interest groups and local communities in the urban context has given rise to some interesting theoretical frameworks that have attempted to go beyond the limited perspectives of structuralist, pluralist and functionalist approaches. For example, more recent work looks at how once traditional urban social movements have become community resource organisations that are directly dependent on the state for some or all of their funding (Shragge and Church, 1998). But does this mean that urban movements have been 'transformed' (to use Gramsci's phrase) into organs of the capitalist welfare state, or does this trend betoken, instead, an invasion of the state's welfare agenda by activists and campaigners (Della Porta and Andretta, 2002)?

The Italian urban movement known as Le Vicine di Case (loosely translated as 'Women in the Neighbourhood') revealed another 'autonomous' face of urban social movements as 'self help groups' that were at the same time directed towards the political transformation of their environment.

> '[E]vents such as street parties were organised, and regular political discussions were held. An informal governance of space slowly took shape. In a significant testimony: Neighbours began to learn from those who cared about our area and our city. This process took place outside the institutional political arena and without the forms of traditional representation, which in fact may hamper true political activity. Large organisations or official committees chaired by specialists were not necessary for our purpose. What was necessary was an exchange of opinions regarding what would improve the quality of our lives, and an awareness of our freedom and possibilities. The task was to upset the balance of power in the given urban order. Women can challenge the 'rationality' of power by introducing in the official political equilibrium a degree of imperfection, humanity, unpredictability. This capacity gives women the 'keys to the city', a city which therefore can become more comfortable, modifiable and safer.
>
> (Interview, Ruggiero, 2001: 157–8)

This and other examples of urban social movements that Manuel Castells documents in Glasgow, Madrid, Santiago and San Francisco reveal that in making their claims to a more equal urban future, such grass roots activists are not representative of the 'normal politics' of Dahl-style political pluralism or the collective consumption model favoured by writers such as Pickvance. Neither is the 'logic of collective action' (Olson, 1965) an adequate description of the huge variety of citizen campaigns ranging from 'no shopping days' to 'reclaim the streets' demonstrations to marches in defence of the 'sans papiers' (unofficial migrants) in cities around the world. Rather, such urban movements are distinguished from the typical resource mobilisation actions of instrumental protests by their refusal to bargain with a 'steady state' social system. This is why structural-functional social science in the US and beyond has found it so hard to accommodate anti-systemic movements within its one-dimensional view of social power – and as a consequence it has lacked the theoretical resources with which to explain events such as the Los Angeles Watts riots in the 1960s and the 'burn baby, burn' protests that followed in the wake of the assassination of Martin Luther King in cities across the US.

If, on the other hand, as Elizabeth Lebas rightly argues, we take Lefebvre's notion of the 'right to the city' as a revolutionary demand – a demand that inspired the slogan of the May 1968 movement in Paris to 'be realistic, demand the impossible' – the performative function of riots, civil disturbances and street protests becomes clearer in the opportunity such mobilisations provide to give voice to the voiceless and power to the powerless.

CONCLUSION: THEORIES OF URBAN POLITICS AND URBAN THEORY

More than twenty years ago the British political scientist Patrick Dunleavy argued that,

> [o]ver the last thirty years there has been remarkably little discussion of the theoretical basis for studying urban politics. While the debate over empirical theory has grown more vigorous in many other fields of political science and sociology, urban political theory has been only rarely discussed in a systematic way.
> (Dunleavy, 1980: 22)

In the intervening period these charges still remain unanswered despite a proliferation in the number of empirical studies on local and urban politics throughout the world and a growing recognition within the profession, research funding bodies and government of the importance of urban economies, and urban politics and policy-making (Thomas et al., 1984; Hausner et al., 1987; Young and Mills, 1993; King and Stoker, 1996). Other authors take a less pessimistic view and argue that 'the sheer eclecticism of theories and the variety of approaches to theoretical issues . . . makes the study of urban politics so vibrant in the 1990s' (Judge et al., 1995: 1). While writers such as Dunleavy lament the lack of systematic theory, Judge, Stoker and Wolman celebrate theoretical pluralism and eclecticism. It is fair to say that the latter group regard the interplay of urban political theories (a term which is used in a loose sense to describe any form of urban political analysis) as a positive and enlightening one:

> In the process of questioning and criticism, inconsistencies or omissions have been discovered, commonalities identified, and refinements and modifications generated within and between the diverse theories . . .
> (Judge et al., 1995: 4)

These writers are not arguing that urban politics has seen a theoretical convergence, although this claim has been made by some for urban regime theory, but that criticism from rival theories has refined the premises on which the key schools of urban political analysis have been based – in particular pluralism, elite theory, and (neo) Marxism. However, each of these theories exists as part of the broader repertoire of political science, therefore one could argue that developments within urban political studies have remained firmly within the confines of the 'master discipline'.

Nevertheless, within the study of urban politics and government there are signs of theoretical innovation outside the traditional disciplinary boundaries of political science. For example, Neil Brenner's examination of globalisation and the re-scaling of government is partly inspired by the spatial theories of Henri Lefebvre (Molotch, 1993; Brenner, 1997), while Bob Jessop has sought to develop and extend the ideas of contemporary German social theory and structuralist Marxism to the study of local, regional, national and supranational state formations (Jessop, 2000; Jessop and Sum, 2000). An appreciation of the cultural dimension of urban political economy and the social construction of institutional priorities is evident in the work of Muller and Surel (1998) among others (Le Galès, 1999: 300).

The status of 'the city' has also been raised by central government's use of the metropolis as a site for policy experimentation and innovation over the past decade and more. Many of the most interesting and contested reforms of the post-war British state under the Thatcher and Major administrations were directed at urban government and the network of services and resources that the local Keynesian Welfare State had universally provided since the Second World War. This has been echoed in the US by the Clinton administration's support for 'new public management' style approaches to 'empowering local communities' based on the principles of 'steering' rather than 'rowing' put forward by Osborne and Gaebler (1992).

The axis around which the two worlds of organised/institutional politics and what we might broadly term 'civil society' centres is 'who gets, what, when and how?'. The difficulty, as we have seen, is that the urban polity in many cases is not the monopoly provider of goods and services, but is integrated vertically and horizontally into decision-making chains of which 'the city' is only a part. The study of local political processes and movements is important in understanding how power in the city is organised and resisted but, as we saw in the previous chapter, endogenous political and economic factors need to be situated within a broad spectrum of interlocking structures and processes.

Such a critique also requires us to 'think outside the box' of conventional accounts of power and, in particular, to move beyond the functional paradigms that continue to bedevil political science research, especially in the English-speaking world. Over the past three decades there has been a growing interest within the humanities and social sciences in what might be best described as the 'ontology of power' –

how power relates to us as human subjects in the totality of our experiences, and not just as voters, or demonstrators. Urban studies have been particularly receptive to ideas of power that relate to culture, representation and difference because it is in the city that these expressions of the contradictions and potentials of the modern and postmodern experience are most in evidence, as we shall see in the following chapter.

QUESTIONS TO DISCUSS

1 **Why are issues of scale relevant to the discussion of urban politics in a globalising world?**

2 **How would you define an urban social movement and what would distinguish such a movement from generic social movements?**

FURTHER READING

Approaches to urban politics

A good collection of material relating to the variety of approaches to urban politics we have encountered in this chapter is to be found in Judge *et al. Theories of Urban Politics* (1995). For those who would prefer to sample Clarence Stone's, *Regime Politics: Governing Atlanta, 1964–1988* (1989), an excerpt from this volume is included in Judd and Kantor (eds) *The Politics of Urban America* (2001), which also features a more critical article on the limits of urban regime theory by William Sites. Fainstein and Campbell's volume (1996) contains an extract from Logan and Molotch's 'Urban Fortunes', but those who wish to begin at the beginning should consult Harvey Molotch's (1976) seminal essay in the *American Journal of Sociology*, which is essential reading for those who wish to understand the growth machine hypothesis as a new political economy of the American city.

The politics of scale

On 'new regionalism' Swanstrom (2001) provides an excellent synthesis of the debate on the pros and cons of Mumford's long-prospected 'city region' concept in the US, while Brenner (2002) insists that the re-structuring of the state is not so much a response to local political exigencies as an effect of the re-scaling of global capital. Cox (1997) is a good reference for material on global/local restructuring and issues of territory and scale.

Urban movements

In addition to classic studies of urban social move-ments such as Fainstein and Fainstein (1974), Mingione (1981) and Castells (1983), more recent studies such as Ruggiero (2001) and Hamel *et al.* (2000), have extended the theoretical range and research orientation of urban social movement studies. Studies of specific urban protest organisations such as those around HIV and AIDS and the anti-globalisation/anti-capitalism movement can be found in Shepard and Hayduk (eds) (2002), while Jeff Ferrell's *Tearing Down the Streets* (2003) looks at a whole range of counter-cultural movements from skateboarders to graffiti artists, to pirate radio broad-casters and 'reclaim the streets' activists.

8

FROM PILLAR TO POST

Culture, representation and difference in the urban world

Every culture has its characteristic drama. It chooses from the sum total of human possibilities certain acts and interests, certain processes and values, and endows them with special significance . . . The stage on which this drama is enacted, with the most skilled actors and a full supporting company and specially designed scenery, is the city: it is here that it reaches its highest pitch of intensity.

Lewis Mumford, *The Culture of Cities*

INTRODUCTION

Urban writers have long used the metaphor of the mirror to emphasise the city's unique capacity to project back at us the image we have of human subjectivity and human society. In nature we may be far from the madding crowd, but our 'species being' to quote from Marx is only recognisable insofar as we interact with other human subjects who are also capable of distinguishing man and woman from nature. Nowhere is this self-awareness more acute and extensive than in the modern city, but at the same time as Simmel remarks, this detachment from the natural world can lead to a certain indifference to those that share the urban environment (the blasé attitude) and make us take refuge in traditional forms of belonging such as religion or ethnicity. At the same time, new forms of identity develop as subjectivity becomes freed from the constraints of surveillance and conformity associated with many traditional societies.

The emergence of the private realm and the enormous increase in geographical mobility in the nineteenth and twentieth centuries meant that city dwellers had the possibility of constructing or recon-

structing their identities in new and sometimes radical ways. With few exceptions, all the great avant-garde movements in art, literature and music were concentrated in the larger cities. The same was true for architecture, design and fashion as it was for that quintessentially urban art form, the cinema.

This chapter takes these themes of the interaction between private and public space, the manner in which the city constitutes a language with its own distinct cadences, grammars and vocabularies, the city as 'dominant' and 'other' space where identities are manifested and contested in the very form and use of the urban landscape, and finally we consider how the pattern of urban culture established by the process of modernity is being rewritten in and through the hybridising forces of the new information technologies, globalisation and cultural difference.

URBAN CULTURES

In the introduction to his unique account of urban civilisation, *The Culture of Cities*, Lewis Mumford writes,

[t]he city as, as one finds it in history, is the point of maximum concentration for the power and culture of a community. . . . Here in the city the goods of civilization are multiplied and manifolded; here is where human experience is transformed into viable signs, symbols, patterns of conduct, systems of order. Here is where the issues of civilization are focussed: here, too, ritual passes on occasion into the active drama of a fully differentiated and self-conscious society.

(Mumford, 1938: 3)

In other words, cities are both the locus and the focus of civilisation – they allow society to reach its greatest potential and concentrate its greatest contradictions. At the same time, cities are 'mirrors of modernity', they allow us to engage in self-reflection both as individual subjects and as members of discrete groups and tribes. The way that we act in urban society as Simmel, Benjamin and Lefebvre would argue is performative, and following Rousseau it is easy to see why the city was regarded by post-Enlightenment philosophers as a stage or a theatre on which the human drama was played out. If art is the imitation of nature, then that most unnatural of physical constructions – the city, in its signs, symbols and rituals is also a perpetual cultural work in progress.

In defining culture as 'a system for producing symbols' (Zukin, 1995: 12) Sharon Zukin highlights the way in which 'culture' as an explanatory term has been put to routine and generic use by the social sciences. Culture no longer necessarily refers to art or the aesthetic but to 'systems of representation' or even 'ways of being'. Thus, we now speak of 'the advertising culture' or the 'drugs culture'. Politicians talk of 'dependency cultures' or 'welfare cultures', business leaders refer to 'entrepreneurial culture', and even museum directors will describe their activities as being part of 'the heritage culture'. Culture has thus become the equivalent of what Lefebvre calls a 'milieu' where social actors communicate through shared knowledge systems that may or may not be exclusive to its habitués.

For Sharon Zukin, the imprinting of the urban landscape by capitalism means that there is no respite, no place to hide from the commodified space of the market. She writes:

Today, urban places respond to market pressures, with public dreams defined by private development projects and public pleasures restricted to private entry. Liminality in the landscape thus resembles the creative destruction that Schumpeter described, reflecting an institutionalised reorientation of cultural patrons, producers, and consumers.

(Zukin, 1991: 41)

Following on from the work of David Harvey (see Chapter 6), Zukin is interested in the commodification of places as sites of cultural consumption and in how the city, especially such an icon of modernist excess as New York City has begun to consume itself. This process is at the heart of her study of 'Loft Living' (Zukin, 1988) which showed how an ostensibly bohemian artists' quarter that had grown up among the abandoned garment warehouses of Manhattan could be destroyed by property redevelopers who have made millions by expropriating a cultural 'scene' which is then used to price-out the very artists who have created it.

In her more recent work, *The Cultures of Cities*, (Zukin, 1995) she explicitly builds on Mumford's interest in urban, design, democracy, and the market economy by exploring what she calls 'the symbolic economy' of Manhattan. She moves beyond Harvey's original formulation of seeing the urban landscape as being shaped by circuits of capital by arguing that such circuits also interact with circuits of culture. Zukin identifies certain key locations of the symbolic economy as exemplars of the inter-marriage between culture, public space and commerce. So for example, she identifies Bryant Park in Manhattan as an example of how local businesses had collaborated in turning what was a haunt of 'vagrants and drug dealers' into an arena for cultural events, fashion shows and outdoor restaurants.

The local office workers began to return to Bryant Park because their security was assured by the presence of private security guards paid for by the Bryant Park Restoration Corporation. The principles adopted by the restoration corporation were straight out of the Holly Whyte textbook on how to restore cities for 'normal users' (Whyte, 1980) and the 'defensible space' prognostications of Newman (1972). Remove congregation sites for 'undesirables', fence off children's playgrounds, provide utility spaces that will attract middle-class users (i.e. jogging tracks and tennis courts rather than basket ball hoops), and maintain a visible security presence. This 'pacification by cappucino' (Zukin, 1995: 28) model certainly worked for the returning office workers, but it meant that the socially despised former users were simply displaced into other less policed, less surveilled 'public spaces'.

This example of the cultural re-appropriation of public space by capital also plays out at the more macro level in terms of city growth coalitions' strategies of place marketing (see the previous chapter). Conventional artistic institutions such as museums, galleries, opera houses, theatres, cinemas and concert halls also contribute to this 'symbolic economy' by attracting what Pierre Bourdieu calls the consumers of cultural capital (see below), who mostly coincide with the affluent economic and political elites that city governments believe they need to woo in order to secure their economic future. No less important as consumers of 'public culture' are the millions of tourists and visitors who pass through the world's major cities each year who are sold 'a coherent visual representation' of the city as symbolic commodity (Zukin, 1995: 271). This coherent vision does not go uncontested however and, as we saw in Chapter 5, attempts at gentrifying public space, such as the battle over Tomkins Square Park have met with resistance (Smith, 1996 and see Chapter 5 this volume), while alternative, subaltern visions of the metropolis in their various forms combine to under-cut and disrupt such dominant cultural strategies – from street art (graffiti), to rap, to skateboarding – even though each 'counter-culture' is constantly prey to cooptation and commodification.

To understand the contested ground of urban cultures better it is important to see how the notion of private and public space developed in European and later North American urban society from the later middle ages to the present day using as our guides three writers whose ideas have inspired a good deal of subsequent reflection on the cultural production of modernity.

QUESTION TO DISCUSS

1 What does Sharon Zukin mean by the commodification of cultural space and how does it relate to the commodification of space we discussed in Chapter 2 and Chapter 6?

Public space and the private realm: Habermas, Sennett and Bourdieu

This section deals with an important dimension of the urban experience, namely how the city came to be both a site of public culture and association at the same time as preserving and even intensifying the 'closed world' of the family and the private household. The American sociologist Richard Sennett and the German philosopher and social theorist Jürgen Habermas both see this as a historic process coupled to profound changes in the structural organisation of the society and economy of European states. The late Pierre Bourdieu, whom we encountered previously in relation to the development of the concept of 'social capital', adds to this picture by exploring how group affinities are structured culturally through mechanisms of belonging.

Jürgen Habermas' treatise on national development and urbanisation in Europe and America, *The Structural Transformation of the Public Sphere* (Habermas, 1989), although not an enquiry into the urban condition per se does, I think, offer some particularly helpful insights into the culture of cities. Although Habermas' main task is to show how concepts such as 'public opinion' develop with the growth of bourgeois society from the eighteenth century onwards, he is also (like Sennett) interested in exploring the interface between the private and public sphere. Habermas believes that as the institution of the family loses monopoly control over its privatised domain – state schooling, taxation, building and sanitary inspections all begin to encroach on its territory from the nineteenth century onwards – the 'surreptitious hollowing out of the family's intimate sphere', by the twentieth century, 'received its architectural expression in the layout of homes and cities'. The result of this process meant, '[t]he loss of the private sphere and of ensured access to the public sphere' and 'is characteristic of today's urban mode of dwelling and living' (ibid.: 157).

Taking his cue from William H. Whyte's landmark study of American mass society, *The Organization Man* (1956), Habermas sees the layout of the suburban landscape as a physical expression of the pressure to conform, a kind of civic version of 'army post life' as Whyte puts it. In suburbia, the intimate sphere begins to dissolve as the neighbourhood group takes on the role of surveillance and solidarity previously associated with elder generations of the extended family as described by Whyte in his later study of the 'Kaffee Klatschers' of Park Forest, Illinois (see Chapter 5). 'Discussion as a form of sociability gave way to the fetishism of community involvement as such', notes Habermas before quoting Whyte again – 'Not in solitary and selfish contemplation . . . does one fulfil oneself, but in doing things with other people . . . even watching television together makes one more of a real person' (Whyte, 1956: 353 in Habermas, 1989: 158). Following H.P. Bahrdt's observation that urban society requires a careful balance between the public and private spheres so that the one does not overwhelm the other, resulting in either excessive fragmented individualism or the totalitarianism of the 'mass society', Habermas concludes that the self-educated urban 'public' of the eighteenth century that formed the basis of our modern understanding of 'civil society' has all but disappeared. Two centuries ago, 'the people were brought up to the level of culture, culture was not lowered to the level of the masses' (ibid.: 166). Thus, it would seem for Habermas that the blame implicitly lies in the disturbance of that perfection of urban form, the late medieval European city, which allowed the two spheres of public and private to co-exist harmoniously in a world where the possession of 'learning' and 'culture' was *de rigueur*. Like Robert Putnam he seems to be arguing that suburbanisation, TV culture and the automobile are threatening the reproduction of the very social capital on which civilisation has depended for hundreds of years.

Richard Sennett would certainly concur with this critique of suburbia and, like Habermas, he has consistently argued in favour of 'dense and dirty cities' as a rich source of creativity, learning and human values. Sennett has called himself 'an urban anarchist' and still admits to being a socialist scholar in an age of 'post-socialism'. Like Lewis Mumford, whom he much admires, in the *Fall of Public Man*, Sennett sets out to provide a social history of the city, focusing particularly on the nineteenth and early twentieth centuries, but he also tries to show how 'public man' – and the gender specificity is unavoidable given the historical context – becomes increasingly privatised and passive in a mass urbanised society where 'feeling' is relegated to the intimate realm of the nuclear family and where political discourse is polluted by homespun psychologisms (Cox in Sennett, 1993: xvii–xix).

Sennett makes no apologies for declaring his antipathy towards this process, just as in his earlier work he condemned urban planning in America for taking the conflict out of cities by flattening them into disaggregated, low density communities where the retreat behind the white picket fence denotes only an arrested psychological development at the adolescent stage. Suburban America is symptomatic, therefore, of a society that in its anxiety to control or repress perceived threats, has locked itself into an anti-metropolitan identity of an *ersatz* and consequently dysfunctional *Gemeinschaft* (Smith, 1980: 153–61).

Sennett believes cities should be cosmopolitan, dense, and even dangerous places where anything can happen, as in the Dionysiac City where 'dislocation, deconstruction [and] disorientation' prevail (Sennett, 1990: 238). He contrasts this with the ordered and regulated world of Apollo, which finds its apotheosis in the planned rationalism of the *ville radieuse* and the manicured lawns of suburbia. The Apollonian ideal is what drives architects and planners to achieve a perfect, universal human form. But Sennett reminds us that the way of Apollo is also an acceptance of the need for balance and a respect for the limits of human ingenuity and accomplishment. Dionysius teaches us not to accept the banal and the routine as our guiding principle, Apollo calls for humbleness in pursuit of the ideal. Sennett's re-articulation of these ancient truths is at the same time a hymn to the anarchy of the modern metropolis, and also a recognition that difference and otherness, while they have their aesthetic consolations, also breed disengagement and withdrawal (ibid.: 129). Like Simmel, Sennett is suspicious of *Gemeinschaft* tyrannies, but he worries whether the metropolis will win the battle in favour of an 'outer directed' subjectivity against the anti-city values of the suburbs (Riesman, 1950).

The urban landscape forms what Robert Park called a 'moral order' or what Sharon Zukin terms 'a landscape of power' (Zukin, 1991). But Sennett urges us not to see urban culture as determined by money and power alone, since there must always be a cultural dimension to culture itself. In other words, it is not that urban culture is free from the determinations of power and money, but that the complexity and contradictions of urban civilisation are its wellspring. Thus, an artist such as Fernand Léger is both an interpreter and a critic of the urban complex – at the same time as being a part of what we have come to call a producer in the 'cultural industry'. As Walter Benjamin showed, the integrity of the artistic gesture is not destroyed by the means of its consumption. But in order to understand its wider significance we have to be alive to its inner contradictions.

This concern with culture as both an **ontological** and strategic aspect of the human condition also runs through the work of Pierre Bourdieu. Considered by many to be the most important sociologist to emerge from France since Emile Durkheim, Bourdieu has contributed a significant body of work to many aspects of sociology and anthropology. The key text as far as Bourdieu's potential importance for urban research is concerned is his book *Distinction* (1986) in which he develops a number of concepts such as 'social space', 'economic capital', 'cultural capital' and 'social capital' that have interesting applications to the study of urban communities. Bourdieu also extends ideas that have been introduced in earlier studies concerning the notion of 'habitus' and 'field'.

Bourdieu is much more in the tradition of sociologists such as Simmel, Veblen and Tönnies than the American structural-functionalists in his emphasis on the cultural dimension of modern society (Rosenlund, 2000: 222). He is also an empiricist's theoretician, for although Bourdieu's is a theoretically rich sociology, it is based as he says, on a

belief that the deepest logic of the social world can be grasped only if one plunges into the particularity of an empirical reality, historically located and dated, but with the objective of constructing it as a 'special case of what is possible' . . . as an exemplary case in a finite world of possible configurations.

(Bourdieu, 2000b: 4)

This principle is applied to Bourdieu's concept of 'habitus', which corresponds somewhat to Habermas' notion of *Lebenswelt* (or Lifeworld) (Bourdieu, 2000a). The habitus is constructed from a series of dispositions or a repertoire of (mostly unconscious) thoughts, feelings, tastes and preferences that shape our social practice (Jenkins, 2002: 77–84). Our dispositions also serve as social and cultural markers that position us within what Bourdieu calls 'social space'. Social space, in Bourdieu's schema is based on two principles of differentiation – economic capital and cultural capital. For our purposes we need not go too deeply into how these types of capital are relationally distributed, suffice to say that what Bourdieu is attempting to construct is an alternative social stratification that stresses *relational* factors more than one's direct connection to the means of production (Marx) or one's consumption location or status group (Weber). Thus, one can be poor in economic capital on the salary of a state schoolteacher, but rich in cultural capital by being proficient in the playing of a musical instrument or in one's knowledge of art history. By contrast, an oil millionaire will be rich in economic capital but may not possess a single book. Bourdieu's observation in *Distinction* is that people with roughly similar stocks of economic and cultural capital tend to have the same dispositions – i.e. they tend to like and consume the same things. This social space also maps onto 'actually inhabited space' following my observation in a previous work that where circumstances and resources allow 'location follows vocation' (Parker, 2001a).

How does this social space that Bourdieu refers to conform to individual and group behaviour in the physical space of cities? Social actors in the city (or anywhere else for that matter) use what Bourdieu called their practical sense or practical knowledge to navigate around their social world. Most people, most of the time, do not think of themselves as reflexively responding to complex signals and cues (as I mentioned in the introduction to this volume), they mostly operate on 'autopilot'. This socially skilled but unreflective reproduction of practices, norms and values Bourdieu refers to as the Doxa or 'doxic experience'.

The 'doxic city' I would argue is very similar to the helter-skelter, clock driven metropolis that Simmel describes where urbanites respond automaton-like to warnings at underground stations to 'mind the gap' and 'move along right down the carriage please!'. The habitus that Bourdieu describes is, in its physical spatial aspect, a domain of interaction with significant locations, symbolic landmarks and special time-space markers (meeting points at particular times of the day) that cognitive urban researchers such as Kevin Lynch have shown to be fundamental to our 'image of the city' (Lynch, 1960). At the bodily level, Bourdieu's notion of 'hexis' is 'a mediating link between individuals' subjective worlds and the cultural world into which they are born and which they share with others' (Jenkins, ibid.). This then is our own customised way of being in society, how we stand, walk, talk, dress, flirt, argue and so on.

Urban subcultures are as profoundly shaped by these hexic routines as were the Kabylia tribe from whom Bourdieu developed the concept. One only has to look at the importance of a fashionable brand of trainer or sneaker, the latest street slang, or the greeting rituals of teenage urbanites to appreciate how language and the use of the body helps us to locate ourselves within a given urban milieu. But surely the notion that this culture is 'self generated' is rather naïve because, as Zukin argues, there is precious little culture that has not been or cannot be commodified. Valuable though Bourdieu's insights are as a way into understanding urban cultures, the

relationship between what he calls 'the field of power' (or politics) and other 'social fields'[52] is not made explicit enough in terms of how power determines social space – and by extension cultural practice.

In the end, we appear to have come full circle to re-phrase Marx's dictum in concluding that 'men make (urban) culture, but not in the conditions of their own choosing'. In the following sections we take these themes further by exploring the manifestations of gender and sexuality in the urban experience, before looking in more detail at ethnic and cultural identity and difference both as an incubator of cultural creativity and as a catalyst of conflict in the contemporary metropolis.

QUESTIONS TO DISCUSS

1 **Do you agree or disagree with the negative judgements that Habermas and Sennett pass on the culture of 'suburbia'?**

2 **Why do cities seem to be such rich sites for the accumulation and display of what Bourdieu refers to as 'cultural capital'.**

SEX IN THE CITY: GENDER AND SEXUALITY IN THE URBAN EXPERIENCE

The gendered city

In her book, *The Sphinx in the City*, Elizabeth Wilson writes that: 'Woman is present in cities as temptress, as whore, as fallen woman, as lesbian, but also as virtuous womanhood in danger, as heroic womanhood who triumphs over temptation and tribulation' (Wilson, 1992: 6). In stressing the sensual pleasures that the city offers those with the leisure and opportunity to indulge their desires, Wilson makes use of the Benjaminian figure of the *flâneur* to suggest that 'perhaps . . . the prostitute could be said to be the female *flâneur*'. But as she admits:

In recent years feminists have argued that there could never be a female *flâneur*. They have gone further, to suggest that the urban scene was at all times represented from the point of view of the male gaze: in paintings and photographs men voyeuristically stare, women are passively subjected to the gaze.

(ibid.: 55–6)

Wilson's heroines, women that felt free to roam and experience the pleasures of the city unchaperoned were, as she herself admits, exceptional and few in number. Working women did not have the leisure or the resources for such urban encounters and were largely confined to their immediate neighbourhood. Middle-class women until late into the twentieth century rarely appeared in public unaccompanied by a maid, a husband or a female friend or relative. Thus, according to Wolff, '[t]he experience of anonymity in the city, the fleeting impersonal contacts described by social commentators like Georg Simmel, the possibility of unmolested strolling and observation first seen by Baudelaire, and then analysed by Walter Benjamin, were entirely the experiences of men' (Wolff, 1990: 58 in Wilson, 1995: 66).

This view is supported by Ferguson's analysis of *flânerie* in nineteenth-century Paris because in the eyes of the *flâneur*'s depictors:

No woman is able to attain the aesthetic distance so crucial to the *flâneur*'s superiority. She is unfit for *flânerie* because she desires the objects spread before her and acts upon that desire. The *flâneur*, on the other hand, desires the city as a whole, not a particular part of it.

(Ferguson, 1994: 27)

Women in nineteenth-century Paris were expected to have a more immediate and practical relationship to the city, which limited their role to that of shopping. Women from higher social classes could escape this social determinism by forming *salons*, but these were not public forums and the idea of a respectable woman sitting alone at a pavement café would have been unthinkable until well into the twentieth century. Even in the period immediately following the Second World War, pioneering feminists such as Simone de Beauvoir were all too conscious of the deviant act that an unaccompanied woman committed by writing or reading in a public place. For working-class women, or women from the less affluent social classes, the street was much more of a communal space, or even a feminised space, to the extent that adult males were either at work or enjoying the closed world of the public house or bar. This left public amenities and markets as well as the street itself as a mostly female domain during the working week. Therefore, it is important when discussing the 'gendering of the urban landscape' not to assume that this means 'the masculinisation' of city spaces, and to remember that the urban experience varies quite markedly depending on women's social and ethnic/cultural identity as much as it does for men.

A number of feminist urbanists argue that women's presence in space is highly constrained by gender roles and that urban policy is oriented to the needs and routines of male city users in every sphere (Watson and Gibson, 1995; McDowell, 1999). Often public transport is geared to the needs of the male commuter and urban transport policies in many cities privilege the private motor vehicle, making streets unsafe for children and degrading the environment that women, as frequent users of streets and local amenities, rely on more than men (Brownill, 2000: 116). The lack of adequate street lighting, badly designed subways, unstaffed public transport facilities, and the diminishing visibility of police foot patrols, limit all but the criminal's use of the city. The right of women to 'reclaim the streets' – as feminist campaigners demanded in night-time marches around the world in the 1970s – was a response to the fact that many women regarded many parts of the city as being off limits after dark because they feared for their personal safety (McDowell, 1999: 150). But while there is a considerable literature on the oppressive character of the urban environment, as McDowell points out, 'the public spaces of the city have [also] been significant locations in women's escape from male dominance and from the bourgeois norms of modern society' (McDowell, 1999: 149) – an observation that one could equally apply to gay and lesbian city users and residents.

Queer spaces

If one thinks of the novels of Radclyffe Hall, Jean Genet, Christopher Isherwood and James Baldwin, the city features centrally in the exploration, assertion and repression of homosexual life and love. But only in recent years has the study of urbanity from the perspective of the gay and lesbian experience become a subject within its own right in the social sciences and humanities. For writers such as Gill Valentine, although gender studies provides a useful starting point for understanding the spatialisation of social power, 'the use of space by other groups is not only the product of gender; heterosexuality is also powerfully expressed in space' (Valentine, 1997: 284). Nancy Duncan makes a similar point when she argues that 'the private/public distinction is gendered' and that '[t]he public/private dichotomy . . . is frequently employed to construct, control, discipline, confine, exclude and suppress gender and sexual difference preserving traditional patriarchal and heterosexist power structures' (Duncan, 1996: 128).

Yet, as we have noted in the context of women's experience of urban life, the city can both oppress and liberate, and this dichotomy applies even more so to gays and lesbians where the fear of physical violence has to be balanced against the advantages of privacy and tolerance that the anonymity of the city offers and the opportunities for networking and socialising with one's peers (Valentine, 1997: 287). The concentration of 'gay communities' or 'gay colonies', even if such terms are notoriously imprecise, in districts such as New York's Greenwich Village occurred because, as the city's artistic or 'bohemian quarter', it had a reputation for tolerance, a proliferation of cheap rooming houses for single persons, the possibility of creative (although not well paid) employment and a lively social scene (Chauncey, 1994: 229). On a reduced scale Lillian Faderman (1991) identified bars and 'tea-shops' in San Francisco and Chicago as places where lesbian women could associate but, as Bailey points out, they were certainly not safe spaces (Bailey, 1999). It is hard to demonstrate that this formula for the establishment of an urban gay presence applies equally well to other cities in North America and

Europe, but the statistical evidence certainly confirms the positive correlation between an expressed gay identity and urbanisation in the US (ibid.: 54).

The visibility of urban gay and lesbian culture has undoubtedly increased since the 1960s and is now marked by annual events such as Gay Pride marches and Mardi Gras festivals from London to Sydney and Berlin to San Francisco. San Francisco has one of the most visible gay minorities in the world, but as Castells showed in his study of the city's gay neighbourhoods, the evolution of an identifiably 'gay space' was a stepwise development that was closely tied to broader social movements and political watersheds – the anti-war and civil rights movements in the 1960s, and especially the Stonewall riots at the end of that decade (Castells, 1983, 1997).[53] News of San Francisco's tolerant lifestyle soon spread, and the city became a popular destination for many gay men from all over America, and with the advent of cheap air travel, from around the world (Castells in Susser, 2002: 202).

The diffusion of 'queer spaces' from the 1970s onwards in all the world's major cities was partly the result of a more confident and affluent gay and lesbian community, although as Green (1997: 51) points out in the context of the London lesbian and gay scene in the 1980s, the lower earning power of lesbian women meant that the opportunities for socialising were more restricted than for gay men, and lesbian businesses found it harder to survive (ibid.: 58). Therefore, it is important to remember the gender imbalances in the experience of gay men and lesbian women's access to the city, and also to realise that just as in the straight community, status and class act as entrance points and barriers to the urban experience.

Nevertheless, among gay and lesbian city users, there does appear to be a greater appreciation of the symbolic significance of place, and a sensitivity to the changing rhythms of times and places when discrete milieux go from being 'straight' to 'mixed' to 'gay' spaces. This movement may, therefore, be less a sign of instability and more an indication of the sophistication of the urban repertoire of the gay world. Having the cultural knowledge about where and when to see and be seen is also important for many gay and lesbian city dwellers, and this is why queer theorists have been so interested in exploring the psychology of the *flâneur* or *flâneuse* as a performative feature of modern gay identity (Wilson, 1992, 1995; Butler, 1999; Munt 2001).

QUESTION TO DISCUSS

1 How does one's gender, cultural or sexual identity shape the way we encounter the city and the way the city relates to us?

HETEROTOPIA OR BUBBLING CAULDRON? CULTURAL AND ETHNIC IDENTITIES IN THE MODERN METROPOLIS

The elaboration of heterotopia as a concept in social theory is often ascribed to two interventions by Michel Foucault (Genocchio, 1995; Hook and Vrdoljak, 2002). In the introduction to *The Order of Things* (1997a) Foucault sees language and text as potential spaces of heterotopia (see the following section), while in a 1967 lecture Foucault talks of utopias as 'sites with no real place' whereas heterotopias are real places – 'something like counter-sites, a kind of effectively enacted utopia in which the real sites, all the other real sites that can be found within the culture, are simultaneously represented, contested, and inverted'. Foucault goes on to state that such places 'are outside of all places, even though it may be possible to indicate their location in reality' (Foucault, 1997b).

In his 1967 lecture, Foucault refers to two types of heterotopia – 'heterotopias of crisis' where for example certain rites of passage take place, or where sick or 'unclean' members of society are banished, and 'heterotopias of deviation' where we house the insane, or criminals, or the dependent elderly; or the dead. Heterotopias also combine sites 'that are in themselves incompatible' such as the contradictory site of the urban garden, which in ancient Persia symbolised the four corners of the world in a microcosm of 'denatured' nature. Finally, heterotopias are 'linked to slices in time' – they open onto 'heterochronies' – where we can find assemblages of time such as museums and cemeteries (the storage facilities of past lives). But there are also fleeting, transient, time defying heterotopias such as carnivals, festivals and fairs that disrupt and upset traditional temporal routines and uses of space (ibid.).

Implicit in Foucault's heterotopology is the backdrop of the city, and the existence of heterotopias in mythical and cultural history is exemplified by zones of ethnic and cultural confluence such as the ancient city of Babel, or the imagined city of Xanadu celebrated in Samuel Taylor Coleridge's eponymous poem, where

> twice five miles of fertile ground
> with walls and towers were girdled round:
> And there were gardens bright with sinuous
>   rills
> where blossomed many an incense-bearing
>   tree

which carries strong resonances of the planned heterotopia of the Garden City (see Chapter 4). Other more recent examples of heterotopia would include spaces of colonial-indigenous interface such as the Indian railway colonies of the British Raj (Gbah Bear, 1994) and all other 'sites of crossing between peoples, cultures and knowledge systems' (Tavakoli-Targhi, 2001).

In the contemporary city we do not have to look far to find such heterotopic spaces – the shop doorway, pedestrian subway or public park that doubles as a dormitory for the homeless, the curtilage of a concert hall that does service as a skate park, or the disused warehouse that has become an artists' commune. Thus, heterotopias do not have to be isolated places that are somehow 'foreign' to everyday experience; they also exist as 'locations of culture' within cities (Bhabha, 1994). In what the performance artist Guillermo Gomez-Peña terms 'the new world (b)order' exemplified in the transnational urbanism of New York City we find a 'new society

... characterized by mass migrations and bizarre interracial relations'.

> As a result new hybrid and transitional identities are emerging. . . . Such is the case of the crazy *Chica-riricuas*, who are the products of the Puertorican-mullato and Chicano-mestizo parents . . . When a *Chica-riricua* marries a Hassidic Jew their child is called *Hassidic vato loco* . . .
>
> The bankrupt notion of the melting pot has been replaced by a model that is more germane to the times, that of the *menudo chowder*. According to this model, most of the ingredients do melt, but some stubborn chunks are condemned merely to float. Vergigratia!
>
> (Gomez-Peña, 1992–3 in Bhabha, 1994: 218–19)

This cultural hybridity is not without its contradictions however, since cultures of difference can more often be a bubbling cauldron than the appealing *menudo chowder* Gomez-Peña describes – as exemplified in the mordant comedy of Spike Lee's *Do the Right Thing* (1989) or the disturbing portrait of alienation in extra-urban France in Matthieu Kassovitz's movie *La Haine* (Hate, 1995). Heterotopias can also be heterodox dystopias – the badlands of modernity as Hetherington (1997) describes them. Rodney King's plaintive appeal for us all 'to get along' highlights the fact that too often cultural difference is a source of division and enmity rather than an opportunity for engagement and discovery of other ways of being and seeing.[54]

Yet, in this same city of Hispanic barrios, black ghettos and white-only gated communities there are, as the journalist Richard Rodriguez notes, 'children who are Jewish Filipinos with Iranian cousins who are married to Guatemalans'. 'No wonder', he adds that, 'LA has become the true capital of America'. But ironically it was only through the terrible events of late April and early May 1992 that 'the city entire – was born during those dark nights, while the sirens wailed and old women realized that they shared the same city as teenagers in Compton'.[55] At the same time, the knowledge of connectivity makes for an apprehensive co-existence and a greater determination among some white residents to put more distance between them and what Roger Keil has called the 'Bantustan' of South Central Los Angeles

– a white-managed heterotopic space into which the city vents its excess highways, refuse, campus extensions, offices and prisons (Keil, 1998a: 218).

Globalisation compresses and compacts these diverse communities within the interstices of the commodified urban realm – the ghetto and the barrio – from which new cultural identities emerge, such as the Oakland daughter of a Mexican mother and black father who calls herself 'blaxican'.[56] Yet, at the same time there is a 'culture of anxiety' around the diffusion of diversity that continues to preserve the characteristic economic inequalities and cultural segregation of the dual city. In this sense, culture retains its function as a marker of social and economic differences. In order to understand why certain cultural experiences and claims are privileged over others we need to look more closely at how systems of meaning are created, maintained and displaced in the urban context.

QUESTION TO DISCUSS

1 **Can you think of any examples of 'heterotopic space' in cities you have observed? What gives them this quality?**

THE CITY AS TEXT: READING THE URBAN CONDITION

We share knowledge of our social worlds using verbal and non-verbal communication. But rather than perceiving the means of communication as merely a neutral conductive device, such as a telephone line, socio-linguists from de Saussure onwards have generally agreed that language is not merely a meaningful array of shapes and sounds, but that the structure as much as the content of communication matters to what is said and how it is said (de Saussure, 1974). Before the Second World War this theoretical innovation did not, however, extend far beyond certain specialist fields of linguistics and linguistic philosophy, as well as social anthropology where

Claude Lévi-Strauss pioneered the use of structuralist analysis in ethnography (Giddens, 1979). However, it was only when the ideas implicit in structural linguistics were combined with the radical philosophy of Nietzsche and Heidegger and the psychoanalytic theory of Jacques Lacan that, stiffened by a generous dose of critical Marxism, the heady cocktail of 'post-structuralism' could begin to intoxicate first the world of literary studies, and then in swift succession the remaining humanities and social science disciplines (with the possible exceptions of the traditionally temperate disciplines of economics and experimental psychology).

Under the influence of critical philosophers such as Jacques Derrida and the less easily categorised writings of Michel Foucault, the post-structuralist or 'cultural turn' (Jameson, 1998) has been one of the most controversial and far-reaching developments in social theory since the Second World War.[57] Urban theory has been particularly receptive to such developments at its intersections with human geography, sociology, social psychology, cultural studies, planning theory, architectural theory, film theory, feminism, black studies, and gay and lesbian studies. But where once urban theorists sought to provide universal theories of the urban condition or even more convincing theories for urban development and variation, post-structuralist and post-modernist analyses of the contemporary city are dismissive of totalising epistemology and argue instead for 'open' readings of urban phenomena that contest and debate positions rather than assert the primacy of certain deductive models.[58]

For the sake of clarity, and in order not to conflate some quite distinct conceptualisations of urban analysis from the penumbra of postmodernisms, I will use the term 'discourse theory' to refer to those writers and theorists who are interested in narrativity and representation (Wetherell *et al.*, 2001). A narrative is a story, but a story pre-supposes an author or authors, and a desire to communicate or represent some event or happening. Usually, but not always the events described will feature characters, and these characters will generally be recognisable as identity bearing subjects that share at least some minimal character-istics with the reader. Narrative strategies are also attempts to portray the world in terms that will reinforce or enhance the power and position of the narrator.

Some examples might include the discourse of patriotism, a narrative that serves the interest of the nation-state; or the doctrine of papal infallibility that aimed at recovering the loss of temporal power by the papacy through the assertion of theological absolutism, or the 'code of silence' or omertà in the world of the Sicilian mafia that serves the interests of illicit power. In each example it can be seen that the narratives serve the interests of the power holders – in other words they are 'dominant discourses' that allow for no negotiation, debate or discussion. This is what discourse theorists intend by 'narrative closure' – it is the means by which authority erases or seals off the production of its own value-system by presenting it as 'God given' or 'the way we do things here', or 'the work of market forces', thus making the particular interests of (usually) a powerful minority appear to be the shared values of the multitude.

As we will recall from our previous examination of Marxist social theory, this understanding of the construction of dominant discourses, which can be understood as a 'meta-narrative' or general worldview, derives from Marx's earlier writings on ideology (Marx, 1974 [1846]) and especially the development of the concept of hegemony by the Italian Marxist, Antonio Gramsci (1971). It was Gramsci who, through the use of historical and contemporary examples, showed how dominant ideologies had to be created, maintained and adapted in order to achieve consensus over the general population. Force alone could not sustain a dominant class in power for long, which was why the collaboration of what Gramsci called the subaltern class (i.e. those who were forced to sell their labour in order to live) was needed in order to make the values of the bourgeoisie those of society as a whole (Marx, 1974 [1846]). This task was achieved by the deployment of intellectuals (newspaper reporters, school teachers, priests, artists, actors, musicians, poets, etc.) in the service of the dominant power holders in order to forge a particular moral value system, (e.g. love of nation, respect for the rule of law), or to endorse the organisation of society according to a particular economic and political system (such as 'the invisible hand' of the free market working unconsciously in the interests of the world as a whole).

In order to combat the ideological offensive of the ruling class, Gramsci believed that a 'counter-hegemony' based on the ideas of Marxism and the radical collectivism of peasants and workers could challenge the 'false consciousness' of capitalist society with the 'true' values of a genuinely socialist alternative. However, as Foucault shows, every discourse (including Marxism itself) will exhaust itself as the context of its production shifts and changes; producing new, competing discourses and a fresh set of myths on which to build another narrative. The alleged failure of Marxism as a meta-narrative, or an account of human society that pretends to the status of universal truth, is at the heart of many of the disputes between what we might call orthodox or traditional radical social theory and the variety of 'anti-essentialist' or 'anti-foundationalist' critiques of contemporary society associated with postmodernism (for a summary see Dear (2000) chapter 15). But although the 'deconstruction' of dominant discourses may not result in the utopian finale claimed by classical Marxism, many cultural theorists from Benjamin to Barthes and Baudrillard have appreciated the value of the critique as a tool for explaining the modern, and indeed the postmodern condition.

What, then, does the study of cities have to do with the world of narratives and representations? First, as I argued in the introductory chapter, all cities are products of the human imagination, and our imagination works best when it is able to link discrete elements of information to form a unified pattern. Indeed, neuropsychology appears to support the idea that good cognitive function is closely related to our ability to piece together tiny fragments of data to form a coherent picture. By reflexively monitoring the image we have of our environment, our social relations and habitat we can gain (or lose) what Giddens calls 'ontological security', or more prosaically our sense of well-being in the world (Giddens, 1994).

Second, 'stories' or narratives do not have to be written down and contained within the pages of a book. Some narratives take this form, but narrative discourses can assume a variety of forms – they can exist in the architecture of buildings, in the contours of a racing car, in the tune of a folk ballad, or even in the tilt of a hat or the lacing of a pair of sneakers. Third, the combination of these different narrative events gives rise to particular cultures that are distinguished through their difference from one other.

Social identity is a product of these cultures of difference but, as such, it is always being defined and redefined against 'the other'. Cultures of difference are to be found in even the remotest village of the Siberian tundra or in the depths of the Amazonian rainforest, but only in cities, as Simmel argued, does the process of **sociation** allow the formation of new identities, new narratives and new cultures. Urbanity is the laboratory for the configurations and significations of modernity, and the study of the urban condition therefore affords myriad opportunities for exploring the ways in which the city operates as a site of representation, contestation and identification.

Deconstructing the city

In a series of fascinating studies on psychiatry and the clinic, prisons and sexuality Michel Foucault developed a 'genealogy of history' in which changing modes of discourse (or 'value systems' or 'ideologies') are explored through their relation to changing configurations of social power at different times and in different places (Rabinow, 1984). With the exception of 'heterotopic space' that we discussed earlier in this chapter, Foucault's ideas have been relatively under-explored in urban theory. This is a pity, because Foucault's deconstruction of traditional readings of time and space offers the prospect of a non-evolutionary view of urbanity in which the city can be understood as the site of multiple urban forms, narratives and identities. It is in this last arena of representation that some of the most interesting encounters between the rapidly changing urban cosmos and the theoretical developments associated with discourse theory are most in evidence.

Here Foucault's distinction between the four key **tropes** of representation – metaphor, metonymy, synecdoche and irony is important because it offers the prospect for a non-historicist reading of urban form that can accommodate narratives of social change such as 'globalisation', 'postmodernism', 'splintered urbanism', 'neo-traditionalism', and so forth without placing them in an evolutionary sequence. Consistent with this rejection of the linear view of history is Michel Foucault's notion of genealogy (Parker, 2000b). In Foucault's *History of Madness and Civilisation* (Foucault, [1960] 2001), following the scheme proposed by Hayden White it is possible to translate each of the four narrative tropes in the following ways (White, 1987: 119):

metaphor (resemblance)
metonymy (adjacency)
synecdoche (essentiality)
irony (doubling)

Because Foucault's 'methodology' resists any form of executive summary, it is best to illustrate the concept of 'tropology' in Foucault's own account of the history of madness and civilisation. In Foucault's genealogy of madness, for example, each trope corresponds to a distinct representation of 'madness' associated with a definite period in time. This genealogy occurs as each discursive formation undergoes a finite number of shifts before reaching the limits of the episteme (a notion similar to that of ideology or *forma mentis*) that sanctions its operations.

So for example, while the hold of the Catholic Church remained supreme up until the sixteenth century in Europe, madness is seen almost as a quality of the divine or the demonic (as it still is in many tribal cultures and in the more evangelical Christian churches). But as literacy and education spread beyond the priesthood, a more sceptical, rational discourse emerged which was to develop into the 'meta-narratives' of philosophy (metaphysics) and the natural sciences. From this point on, the idea of the insane being touched by God or by the devil could

no longer resist the weight of the discourse of scientific rationalism, not least because the latter was becoming the credo of an economically powerful urban bourgeoisie. Consequently, we see a shift in attitudes towards the insane where previously their behaviour was seen as beyond human understanding, and therefore something that one must tolerate, to the view that the insane were 'ill' and that they should be confined in order to receive 'treatment' and to protect the sane community from their unreason.

Using these insights from Foucault, it is also possible to conceive cities as 'epistemic formations' or containers of particular discursive strategies that sanction or prohibit means of individual and collective expression. What makes a city 'Baroque' or 'Gothic' therefore depends not only on a certain architectural style but a complex grammar of aesthetic and spatial inclusion and exclusion that structures the nature of being in the city and the ontological features of the city itself. Like languages, cities can therefore be read as power maps that are over-written with the deposits of once dominant but now defeated, yet faintly resonating epistemic systems (an approach adopted in Benjamin's readings of the poetry of Baudelaire, see Chapter 2). However, like language itself, urban forms are not the products of a unique time and place, and there is no historical inevitability about the transition from one urban form to another.

Let us first take the example of the classic or antique city, which I want to argue is most closely associated with the trope of metaphor, or resemblance. The ancient Athenian or Attican city is reproduced according to the same aesthetic principles and with the same functional elements in each new settlement with each city having an agora, a gymnasium, and a temple, all in broadly similar relation to one another. Similarly, in Roman city planning, the perpendicular street plan with a forum centring on the *'decumanus maximus/cardus maximus'* intersection can be found throughout the Roman Empire. This suggests the idea of an archetypal Athenian or Roman city that is deliberately and carefully reproduced at the site of each new outpost of empire – just as the Greek and Roman languages colonise and invade the vernacular, so Greco-Roman built forms displace those urban features that were peculiar to the native landscape. Here, it is important to emphasise Derrida's point that language, or rather discursive strategies are, above all, strategies of power (Derrida, 1972). The Doric column in the Syrian Desert is not just a roof support; it is a symbol that says 'we are the masters now'.

In medieval Europe, by contrast, the urban trope that characterises the city had changed to that of metonymy (or adjacency). The town loses its democratic geometry and becomes hierarchical and relational. The main thoroughfares follow the traffic of commerce and pilgrimage, but converge on a centre where the key institutions and markets are all tightly clustered. The divisions made by the major streets, and by natural divisions such as rivers, divide the city into 'quarters' (the old Roman term survives but no longer describes the topography) where place and trade or even ethnos are synonymous. Think for example of the City of London's Cheapside, Poultry

and Jewry and how such vocational locations are reproduced in much the same way in other European market towns. This is the beginning of the spatial division of labour, but it has a residential component too — one thinks particularly of the development of the merchant town house with the master and his family occupying the central floors, the servants the basement quarters, and artisans or journeymen as paying tenants in the attic rooms. Here we have a perfect micro-model of how space and living room and light are apportioned according to social position in the medieval city.

The synecdochal city refers to 'the part standing in for the whole' and here we are in the realm of 'the total city' of the industrial metropolis where all the factors of production are brought together within the confines of the city. Utopian city designs are also synecdochal in their ambition to reconcile humanity and nature and all the diversity of the human community within a 'model' settlement. Finally, the notion of irony as 'doubling' needs elaborating,

Figure 8.1 The Venetian Hotel–Casino complex, Las Vegas.
Copyright Gardner/Halls, courtesy of the Architectural Association Photo Library

because in Foucault's interpretation he is referring to discourse turning in on itself, or better, turning itself inside out. This is what fascinated Benjamin about Parisian high society in the nineteenth century where rose gardens would be recreated in ball rooms, while garden paths would be carpeted and trees dressed in muslin. So we see a reflection of language back on itself in order to disrupt or to 'play' with the established meanings embodied in the discourse. Although there are many examples of the ironic trope in urban culture, I would argue that it is really only after the deadly serious interval of the Modern Movement in architecture that the ironic city is given free reign in the unencumbered expanse of Southern California or the paved desert of Las Vegas where every architectural style has been begged, borrowed or stolen (see Figure 8.1).

The ironic city is the result of a discursive strategy that has become 'ungrounded', it has broken free from the moorings of production that kept it in a particular time and a particular place because the capital that sustains it exists outside time and outside space. The ironic discourse of postmodernity is therefore connected to the de-centring of strong urban narratives in favour of a weaker, polymorphous and therefore more 'flexible' urban form that matches perfectly the paroxysms of what Susan Strange has called 'casino capitalism' (Strange, 1997).

QUESTION TO DISCUSS

1 **How might we use concepts developed from the study of language to better understand the form and function of cities?**

HYBRIDITY, VIRTUALITY AND THE POSTMETROPOLIS

All around, the tinted glass facades of the buildings are like faces: frosted surfaces. It is as though there were no one inside the buildings, as if there were no one behind the faces. And there really is no one. This is what the ideal city is like.

(Jean Baudrillard, 1997: 221)

Postmodern urbanisms

So far we have restricted our discussion of the cultures of urbanity to the sphere of modernism and modernity – a world that in Marshall Berman's much quoted phrase is constantly striving to impose order on chaos (Berman, 1982). The point of postmodernity (if it can be described as a project rather than a critique) is, in the words of Peter Marcuse 'to cover with a cloak of visible (and visual) anarchy an increasingly pervasive and obtrusive order' (Marcuse, 1995: 243). But what is the nature of this pervasive and obtrusive order and how has it managed to achieve such a global reach? In this section we try to probe a little deeper into what has been called 'postmodern urbanism' (Ellin, 1996; Dear, 2000; Soja, 2000b). As we have come to expect with any -ism prefixed by the word 'postmodern' there are a number of competing definitions for what Michael Dear associates with a radical break in urban epistemologies and the onrush of 'multiple urban (ir)rationalities' (Dear, 2000). Let us start with the most straightforward aspect, that of the postmodern aesthetic. Arguably, postmodernism existed first as an architectural movement before it took on wider philosophical and literary/cultural pretensions (for commentaries see Anderson, 1998 and Jameson, 1998).

In particular, Denise Scott Brown, Robert Venturi and Steven Izenour are cited as the original proponents of postmodern architecture and urban design, although this is an accolade that Scott Brown and Venturi have attempted publicly to decline (Ellin, 1996: 57). In Venturi's signature volume, *Complexity and Contradiction in Architecture* ([1966] 1977) the case for a break with the dull unity of modernism in design was accompanied by an appeal for 'messy vitality over obvious unity. I include the non sequitur and proclaim the duality' (Venturi, 1977: 16). In the 1960s, at the height of the 'international style' of high modernism (associated with the architecture of Le Corbusier, Gropius, Mies van der Rohe and their followers – see Chapter 4), Venturi's intervention struck a chord with a public that no longer thrilled at the sight of summitless skyscrapers, but who instead saw endless concrete and glass monoliths crowding out the skyline and puncturing the urban landscape with a blatant disregard for the integrity of existing population densities and building elevations. There was, though, nothing particularly new or radical in Venturi's criticism, since modernism had never had a particular easy time with social scientists or popular journalism. Crude economics and the well-insulated world of city planning departments rather than a new aesthetic sensibility provide the key to explaining why high modernism took such a physical hold on the world's cities in the period after the Second World War.

However, it was Venturi and his partners' recommended antidote to modernist uniformity that aroused such controversy, especially with the publication in 1972 of *Learning from Las Vegas*, (Venturi *et al.*, [1972] 1993) a book that has earned the status of a manual for postmodern designers, architects and planners. Venturi and his co-authors took the example of Las Vegas as a demonstration of how 'decorated sheds' could be used to create contrasting exteriors and interiors, and where the use of space is contingent on the people that pass through it rather than ordered through some grand heroic design. This was best exemplified in the use of neon and high profile signage to promote the attractions of the hotels, bars and casinos that crammed 'the Strip' district of Las Vegas (Gottdiener *et al.*, 1999). Peter Hall described the publication of *Learning from Las Vegas* as 'cataclysmic', and marking 'the end of the modern architectural movement and its displacement by post-modernism' (Hall, 1988: 300). Charles Jencks has also seized on Venturi's intuition about the celebration of a populist aesthetic as the key to explaining the shift from modernist to postmodernist form and layout in the built environment (Jencks, 1980 in Jameson, 1998: 30).

If Le Corbusier and CIAM are associated with the 'master planned' communities of the *'ville radieuse'* or the totalitarian architecture known as neo-brutalism, Venturi and his contemporary architects Charles Moore, Michael Graves and Frank Gehry represent what one might call a 'post-rationalist' break with the international style (Jameson, 1998: 10). In post-rationalist or postmodern architecture, buildings are no longer intended as 'machines for living in' but as spaces in which encounters and events can take place in a non-programmed way. The architect Bernard Tschumi even describes this type of architecture as disprogramming or crossprogramming, which he defines as

> using a given spatial configuration for a programme not intended for it, that is, using a church building for bowling. Similar to topological displacement: a town hall inside the spatial configuration of a prison or a museum inside a car park structure . . .
> (Tschumi, 1994: 155)

With its explicit reference to cross-dressing and transvestism this architectural theory is heavily inflected with the vocabulary of postmodernism. An international arts centre in France is described as an 'in-between' space, 'made of accidents – the place of unexpected events' (Tschumi, 2000: 409) and this sums up neatly the idea of the postmodern urban environment as a space where the contingent, the accidental, the haphazard and the spontaneous are promoted as alternative design motifs to the 'form follows function' dogma of high modernism. In postmodern architecture, as Jameson also points out in the context of the Bonaventure Hotel in Los Angeles, there is a deliberate blurring of the entrance and exit points that divide the outside world from the commercial space of the building itself (Jameson, 1998: 11) (see Figure 8.2) in much the same way as the 'proto-modern' arcades of late nineteenth-century Europe lured the passer-by into a 'public' gallery of commodified space. Interestingly, according to Jameson the attempt to 'arcadise' the postmodern space of the Bonaventure by renting some of the floors to expensive boutiques has not met with commercial success, suggesting that accidental shopping is the province of the time-rich tourist or the Victorian *flâneur* rather than 'the grab it and go' contemporary hotel patron. Now the Lobby Court Coffee Bars 'proudly serve Starbucks Coffee', proving that even in a postmodern space, the brand is everything.

Now the Bonaventure is aimed very much at the leisure and business market, where the marketing blurb mentions the fact the building is 'one of the ten most photographed buildings of the world', and

less controversially that it is located at the inter-section of six converging freeways and 400 miles of subway and light track. The form may change but function continues to matter a great deal.

Thus it would seem, as we noted in Chapter 2, that Benjamin's cities of 'exteriorised interiors' is not a trope that we associate with the postmodern condition so much as the urban condition itself. What postmodern architecture is trying to achieve is a heightening of our sense of the ambiguities of the city by rendering visible the unstable identities of urban

form in much the same way as Benjamin and de Certeau celebrated in the fragmentary imaginary of the *flâneur* (de Certeau, 1984). At the same time, post-modern architecture is having fun at the expense of a modernist tradition that celebrates functionality and spatial segregation by literally 'dressing up' prisons to look like hotels and vice versa (Davis, 1992). However much this ironic narrative may serve reactionary political ends by imposing 'defensible space' (Newman, 1972) on the poor and the power-less, these disruptions of established notions of architectural practice remind us that the built environment, far from being a backdrop to the daily routines of the metropolis, actually forms our understanding of what cities are all about. This artificially constructed three-dimensional world is the product of a myriad of overlapping and conflicting narratives through which the individual city-user has to pick her way, sometimes naively, sometimes knowingly, but never unwittingly.

The city as spectacle

Most people will recognise the Bonaventure Hotel as a futuristic backdrop for the postmodern cult movie *Blade Runner* or in the Arnold Schwarzenneger vehicle *True Lies* in which Arnie gets to take a ride with a horse in one of the glass elevators. The notion of the 'spectacular city' or the city as a screen or theatre onto which, and upon which, images and representations are presented for the diversion and amusement of the public is as old as the city itself. Aeschylus and Sophocles wrote plays that parodied the urban order of their time and invited the audience to experience wars, riots, and sexual and political intrigues through the characters of their dramas. The dramatic repertoire of ancient Athens thus depended on a highly sophisticated and self-reflexive audience that was keenly aware of the political uses to which satire, comedy and irony could be put.

In more than 2,000 years the city has continued to inspire new dramatic forms such as opera and music theatre, but it is with the invention of cinema at the turn of the twentieth century that the city's potential as a drama in its own right was finally unlocked. The early films of the Lumière Brothers featured railway engines in order to demonstrate the versatility of ciné film in capturing true locomotion. The industrial city, so brilliantly evoked in Fritz Lang's 1926 movie, *Metropolis*, was all about movement, crowds, noise, smoke and a constant atmosphere of menace and danger. Perhaps Le Corbusier was inspired by Lang's bi-plane flying between the towers of a futuristic New York when he proposed citing an aerodrome for aerotaxis on the roof of the city's central station (Le Corbusier, 1929 in Le Gates and Stout,

1996: 372). It is hard to imagine the impact that Charlie Chaplin's film roles would have had without the ever-present backdrop of the modern metropolis. Implicit in *City Lights* (1931) and *Modern Times* (1936) is the metaphor of the city as an unforgiving, harsh and inhuman place where destitution and hopelessness are surrounded by the most magnificent architecture and opulence – Wall Street and the Bowery cheek by jowl as the twin sites of the American Dream torn apart by the ravages of the Great Depression.

Fritz Lang's dystopian fantasy reaches its modernist apotheosis in Stanley Kubrick's adaptation of the cult Anthony Burgess novel *A Clockwork Orange*, (1971) in which Le Corbusier's *unités d'habitation* have become a shorthand for an anonymous, inhuman, concrete jungle where the same social pathologies that were associated with the Victorian slums find their most terrible expression. The *banlieue* of Paris and South Central Los Angeles look very much the same to contemporary cinematographers as Salford appeared to Engels back in the 1840s. Movie films such as John Singleton's *Boyz n the Hood* (1991) and the uncompromising lyrics of rap artists such as the late Tupac Shakur furnish us with contemporary examples of the city's perpetual agonistic drama. In this sense urban culture is not so much the mirror of modernity as the crack in the glass of a modernity no longer able to bear the weight of its own contradictions.

CONCLUSION

In essaying the territory of urban culture I have tried to grapple with two aspects of the question. One we might broadly call 'the cultures of urbanity' as exemplified in the critical sociology of Sennett, Habermas and Bourdieu. In other words, what does it mean to be a social actor in the city? How do we make our social worlds, and recognise and interact with other members of our habitat? The second theme is connected with the idea of 'culture as representation'. How do we represent this habitat at the symbolic and aesthetic level? How do we inscribe it with values and meaning? These ontological and representational dimensions of culture clearly interact insofar as the built environment is the work of humankind and therefore a physical expression of imagination and creativity. At the same time, our imagination and identity are constructed in large part from the raw material of the cityscape, and the diversity of populations that make it resonate.

In analysing different cultural readings of the urban experience I have tried to emphasise the value of theory in trying to make sense of cultural variety in the city and in helping us to distinguish between the urban and the non-urban. Along with Sennett I think it is necessary to insist on cosmopolitanism as a key test for the urban condition. Without religious, ethnic and cultural difference the city lacks the ecological diversity to recombine in new and surprising ways. The cultural industries are keenly aware of this, which explains why we find a super concentration of artistic and creative talent in the world's most dynamic cities (London, New York, Paris, Chicago, Los Angeles, Berlin, Tokyo). As cultural production becomes increasingly globalised, cities stand to gain as repositories of cultural diversity and excellence that only such global central places can sustain. This is not only true for culture as performance or artefact, it is also true for culture as understood by Bourdieu – as the capital or resource of human dispositions. In no other place are there so many possible fields of interaction as in the urban habitus and this means that any specialist or different vocation, disposition or lifestyle will find its articulation somewhere in the city.

These themes link with the theoretical preoccupations of the foundational urban theorists we encountered in Chapter 2, all of whom sought in different ways to survey the impact of modernity through its reformulation of consumption habits and cultural praxes. Culture, consumption and the commodity form, as Sharon Zukin demonstrates, are drawn ever more tightly together as the public sphere becomes an increasingly discriminating and discriminatory place with the advent of the state-sponsored 'meta gentrification' that we encountered in Chapter 5, and the outright privatisation of thoroughfares and parks

that aim at removing the plebeian right to the city that even the emperors of Ancient Rome were forced to recognise.

As we saw in Chapter 4, modernism has an ambiguous legacy in terms of its democratic credentials, and not for nothing has the epic, monumental, steel and concrete architecture of Mussolini's Italy and Stalin's Russia earned the epithet 'neo-brutalist'. The search for aesthetic unity often betrayed a fundamentalism in the achievement of that objective that has diminished the cultural significance of the modern movement, especially in regard to its architectural influence. The emergence of postmodernism in urban design practice in the 1970s should thus be seen as a political response to the dictatorial prescriptions of CIAM and its acolytes in the planning and architecture professions and a reassertion of the subject, or the 'urban persona' as the authentic articulation of the urban experience over and against the anonymous tyranny of Brasilia-type 'villes radieuses'.

Although many have questioned whether postmodern architecture and design is anything more than the wealthy fantasy of a philistine global capitalism, insofar as the 'postmodern condition' described by Jean Francois Lyotard (1979, 1984) is about the search for self-realisation, the emergence of previously marginal subjects such as women, youth, gays and lesbians and migrants highlights the trialectical relationship between identity, representation and space that we associate with the late modern metropolis (Soja, 2000b). As Paul Virilio writes:

> The city is but a stopover path of a trajectory, the ancient military glacis, ridge road, frontier or riverbank, where the spectator's glance and the vehicle's speed of displacement were instrumentally linked.
> (Virilio, 1986: 5–6)

Virilio's highlighting of the sovereignty of the mobile gaze finds echoes in the cinematographic city – where the metropolis reappropriates its Renaissance role as stage and backdrop to the tragic drama of late modernity (*Blade Runner*) while its antinomy is celebrated in the atavistic postmodern nostalgia of Seaside, Florida (*The Truman Show*). Technology has democratised the right to the city in the many virtual worlds that exist suspended in cyberspace (see Chapter 9). William Gibson's vision of an unmistakably urban hyper-reality (Chiba City) in which cyborgs roam through vast data landscapes resonates with a new generation of academic scholarship that sees identity, representation and communication as part of the same mutating cybernetic thread (Haraway, 1997; Plant, 1998). It is in these new heterotopic spaces that the city is sowing the dragon's teeth of its future imaginings.

QUESTIONS TO DISCUSS

1 How does postmodern architecture differ from other traditions in architecture such as modernism?

2 In what ways can the urban experience take on the quality of cinema?

FURTHER READING

Public space and the private realm

The ideas of Sennett, Habermas, Bourdieu and Foucault are taken up by a variety of authors including (on the body and space) Seamon (1979), Ackerman (1990), Rodaway (1994), Sennett (1994) and Grosz (1995). Rabinow (1984) remains an excellent resource for those wishing to acquaint themselves with Foucault's ideas in one volume. Sorkin (ed.) (1992) and Zukin (1991) are both excellent on the relentless privatisation of the public realm.

Sex and the city: gender and sexuality in the urban experience

Sophie Watson contributes a considered feminist critique of the city – 'City A/genders' to the reader she has co-edited with Gary Marks (Bridge and Watson, 2002), which also contains Elizabeth Grosz's essay 'Bodies-Cities'. Daphne Spain (2002) argues that issues of gender and space have been inadequately dealt with by urban theorists from the Chicago School to the LA postmodernists. Joan Rothschild's *Design and Feminism: Revisioning Spaces, Places, and Everyday Things* (1999) explores the work of Dolores Hayden among other feminist architects. Hayden's influential 'What Would a Non-Sexist City be Like' may be found in Le Gates and Stout (eds) (various editions). Frank Mort's 'The Sexual Geography of the City' in Bridge and Watson (2000) provides a concise introduction to a range of writings concerned with sexual identity and urban space.

Heterotopia and cultural identities

Foucault's lecture on heterotopia can be found in various translations, for example, Leach, N. (ed.) *Rethinking Architecture* (Foucault, 1997b) and *Diacritics* 16, 1 (1986), 22–7. Another discussion of the theme is to be found in the introduction to *The Archaeology of Knowledge* (2002). Steve Pile and Nigel Thrift's *A City A–Z* is full of intriguing essays and analyses of urban cultures and the urban experience including poetry, photo-essays and cartoons. The editors provide introductions to each section that relate the excerpts to broader themes in urban theory.

A special issue of the journal *Ethnicities* (September 2002) edited by Amin and Thrift offers some thought provoking insights into themes such as multi-ethnicity and notions of Englishness in the British city, fashion and cultural commodification, whiteness and the construction of 'the dark city', and ethnic presence and discrimination in contemporary French cities. Helen Liggett brings her vision as an artist and photographer to bear on the city as site of critical analysis and discovery in *Urban Encounters* (2003) – a book that combines street level images with a celebration of the delights and dangers of the (American) urban experience.

Postmodernism/discourse theory/ the postmodern landscape

For absolute newcomers to the study of postmodernism, a useful starting point is Jim Powell's *Postmodernism for Beginners* (1998). Charles Jencks (ed.) (1992)

contains a selection of key postmodernist texts by authors including Lyotard, Baudrillard, Habermas and Harvey. Explicitly urban-oriented postmodern studies include Ellin (1996), Dear (2000), Scott and Soja (1996), Soja (2000b). The Open University (Wetherell *et al.*, 2001) has produced a very useful reader on discourse analysis and theory that assesses the contributions of Saussure, Goffman and Foucault and others to the development of narrativity and structural analysis in language and communication. The volume also provides applied analyses of discourse theory in social interaction. Mike Featherstone, 'City Cultures and Post-modern Lifestyles' in Amin (ed.) (1994) refers to postmodern culture and leisure pursuits as Simmel-type 'bridges' and 'doors' that can unite discrete cultures and life worlds, and provides a useful complement to Zukin (1991, 1995). On postmodern architecture and landscape, once again Ellin (1996) is essential reading, as is Michael Sorkin's *Variations on a Theme Park* (1992), *The Language of Post-Modern Architecture* by Charles Jencks ([1977] 1991) remains a classic of the genre, while

his *Heteropolis: Los Angeles, the riots and the strange beauty of hetero-architecture* (1993) gives the analysis of the violent events of 1992 a postmodern twist.

The Cinematic City

David Clarke (ed.) *The Cinematic City* (1997) draws on the urban theory of Benjamin, Baudrillard, Foucault and Lacan to provide analyses of Ridley Scott's *Blade Runner* among several other classic urban movies. Shiel and Fitzmaurice (eds) (2001) discuss the interactions between power, space and representation through the medium of cinema in cities around the world including North America, Europe, Africa, Asia and the Pacific. Shonfield analyses the close relationship between cinema and architectural form in *Walls Have Feelings – Architecture, Film and the City* (2000). Kennedy (2000) provides an interesting treatment of the cinematic representation of the city as a 'paranoid spatiality'. Chapter 9 of Michael Dear's *The Postmodern Urban Condition* (2000) considers film, architecture and urban space from a postmodern perspective.

9

PUTTING THE CITY IN ITS PLACE

Urban futures and the future of urban theory

'I have also thought of a model city from which I deduce all the others', Marco answered. 'It is a city made only of exceptions, exclusions, incongruities, contradictions. If such a city is the most improbable, by reducing the number of elements we increase the probability that the city really exists. So I have only to subtract exceptions from my model, and in whatever direction I proceed, I will arrive at one of the cities which, always as an exception, exist. But I cannot force my operation beyond a certain limit: I would achieve cities too probable to be real.'

Italo Calvino, *Invisible Cities*

- In 1970 there were as many metropolitan cities in the developing world as in the industrialised world.
- In the year 2000, 117 of the world's million plus cities were in the developing world.

INTRODUCTION

The aim of this book has been to fit together the pieces of the urban complex in order to demonstrate as fully as possible the theoretical foundations and empirical concerns of those writers who have seen, and continue to see, the city as the cornerstone of modern civilisation. The picture presented is inevitably partial and selective, but I hope it is broadly representative of the issues and concerns that have animated urban studies over the past hundred and fifty years or so. In this last chapter I want to consider how the traditional approaches to the study of the city are being challenged and reinvigorated by a wave of new scholarship that has been stimulated at least in part by the transformation of the city in a variety of important directions.

First, the scale of the larger world cities has increased to such an extent that thirteen 'megacities' now have populations of over 10 million. But only four of them (New York, Los Angeles, Tokyo and Osaka) are to be found in the advanced industrialised nations, while the remainder are to be found in South America and Asia. Today, more than half the world's population live in an urban environment of one form or another. With the rise of the megalopolis we have also witnessed the emergence of 'edge cities' and urban corridors running for hundreds of miles along interstate highways that fundamentally challenge accepted notions of the city propounded by urban sociologists such as Burgess, while appearing to confirm the near-future speculations of imaginative writers such as H.G. Wells and William Gibson (Sudjic, 1995). As the European model becomes just one among several varieties of metropolitan form, urban theory has begun to liberate itself from the evolutionary approach to urban development that Max Weber and Lewis Mumford did so much to propagate.

Second, as Amin and Thrift argue, there is no longer a meaningful divide between the country and the city in the developed world and this is increasingly true for the less developed world also (Amin and Thrift, 2002). In their account, the country is permeated by the urban to such an extent that it is increasingly meaningless to talk about the city as a specific place. Rather, the city is a spatially open phenomenon 'cross-cut by many different kinds of mobilities, from flows of people to commodities and information' (ibid.: 3). Whether it is Manuel Castells' 'space of flows' (Castells, 1989 and 1999) or Michael Peter Smith's notion of 'transnational urbanism' (Smith, 2001), or Graham and Marvin's 'splintering urbanism' (Graham and Marvin, 2001) the new urban research emphasises transitivity and morphology in the urban system and increasingly rejects the static functionalism of most twentieth-century urbanism.

Third, there is a welcome interaction between empirical analysis and more theoretical speculations that has done much to breach the gap we have identified in the previous chapters between continental European and Anglo-American urbanism. The rise of the so-called 'L.A. School' of urban geography including figures such as Davis, Dear, Flusty, Sorkin and Soja reveals a concern with the lived experience of the most brutally excluded Los Angelinos while at the same time trying to make sense of the postmodern landscape of this most unconventional city through the lens of post-structuralism, post-colonial theory and discourse analysis (Dear, 2002). A concern with issues of power, inequality, racism and the changing nature of contemporary capitalism also informs the 'new urban sociology' of writers such as Joe Feagin and Mark Gottdiener (Feagin, 1998).

In making sense of the changing urban world we can identify three major research agendas that, to quote Reyner Banham, are looming ever larger in the rear view mirror of urban theory (Banham, [1971]

2001). The first concerns the fate of what we might call the majority urban world – conventionally but often misleadingly referred to as 'the Third World City'. As we noted in Chapter 5, the vogue for urban ethnography in the developing world almost killed off urban sociology as a discrete subject in the Western Academy as scholars rushed to see 'cities in the making', believing that the urban condition had been so diffused in western societies as to make its study redundant. Infused by a renascent Marxism in the 1970s, the western city re-assumed its pre-eminence within critical urbanism, while 'developing cities' became the preserve of regional specialists in human geography, anthropology and development economics. There is though, a growing recognition that the majority urban world is a hugely diverse and highly articulate urban complex that can be contained neither within the paradigmatic straitjacket of western urban ecologies nor within a *tiers mondiste* approach that sees the cities of the global South as superconcentrated examples of global uneven development.

Second, the transition from the machine age to the information age has entailed major reconfigurations of the urban environment and changes in the organisation of capital, labour and space. Developments in the means of communication have always involved new urban concentrations. In the nineteenth century we saw the advance of the railway town, in the twentieth century the motor cities predominated. In the twenty-first century new hyper nodes are emerging that combine air, sea and land connections with broadband and satellite based communications technologies allowing both distribution and command and control functions to take place within the same urban space. Highly capitalised, ultra-networked, information rich 'spaces of flows' are much less place dependent than they were in the past, thanks in large measure to the creation of a global service intelligentsia chiefly centred on the continent of Asia but with major diasporas in North America, Oceania and Europe. But can a hard-wired, ungoverned, non-space global urban realm preserve those Enlightenment values of civility, culture and democracy for which the Renaissance city proved such a compelling model?

Third, one of the most striking features of the processes we have just noted is how the synergy between urbanisation and the rise of the informational city has produced dramatic benefits for capital while degrading and polluting not just the urban environment but 'the global commons' on which the life of cities is so precariously dependent. Sustainability, however, is not just about the physical environment; it is about preserving and, indeed, enhancing the quality of life for future generations. Urban researchers who urge governments to build sustainability into their policies for the regeneration of cities are echoing the sentiments of the Victorian social reformers but with a greater awareness that technology can never offer a quick fix for problems that, as Nietzsche reminds us, are 'human, all too human'.

EAST OF HELSINKI, SOUTH OF SAN DIEGO: THE MAJORITY URBAN WORLD

It is hard to escape the conclusion that even though very few cities in the advanced industrial nations compare in size and extent to the megalopolises of the developing world, we persist in seeing 'the Third World' city through western eyes. In particular, it is common to find analyses of the non-western metropolis that such cities are either 'pre-modern' settlements at an earlier stage of evolution towards the western capitalist model, or simply densely populated sites of 'under-development' *sui generis*. The question that future urban theories have to address is, does this provide an adequate perspective on a diversity of urban experiences whose only conceptual linkage is geographic ('non-western') and economic (predominantly poor)?

How for example does the experience of living in Manila compare with that of Calcutta? Can we even speak of a 'universal urban experience'? Given that 'western' urban theorists now accept that we can only talk in a very limited sense of a common urban experience, should we not be focusing our attention on discrete segments of the urban population and

urban workforce? Also, to what extent do the colonial designs of Asian, Latin American and African cities still resonate, and how far have new urban conglomerations offered an alternative organisation and layout to the European city? Is the 'informal city' taking over the 'rational city' to such an extent that new urban forms, routines and urban experiences are being created – as in the case of Brasilia? Does the informalisation of the 'western city', described by Mike Davis in its Latino aspect as 'magical urbanism' (Davis, 2000a) herald a new future for the dense metropolises of the advanced capitalist world?

To these series of questions I can only provide signposts to possible answers because a comprehensive response is beyond the scope of this volume. The first point to make is that certain differences exist in all cities – thus if we take Lefebvre's notion of the right to the city, it is clear that the urban context and the identity of the subject have a direct relationship to the extent of one's involvement in the urban experience. In some contexts gender will heavily circumscribe access to the city, in others sexuality, religion or ethnicity will be significant. But we must be careful not to assume too readily that 'subalternity' to use a Gramscian phrase reworked by Gayatri Spivak (1987) produces a convenient universal set of winners and losers. For example, in acknowledging the difficulties many women experience in walking the streets without fear of harassment or intimidation, we also need to remember that young males are more likely to be the victims of violent assault or murder than any other population group.

In order to make any such generalisations meaningful there will continue to be a need for comparative sociological and ethnographic studies of urban populations using perhaps different methods than pioneers such as Mayhew, Booth and Riis, but with the same focus on what makes the urban experience different for different categories of city dwellers and workers. One major advantage contemporary investigators have over their predecessors is the range and geographical scope of statistical data that offers the possibility of producing detailed comparisons of urban populations around the world. As a new generation of scholars begins to emerge from the majority

urban world – and already many university programmes in Latin America, Africa and Asia focus on
the problems of the city – we can expect some
exciting and challenging alternative readings of the
urban experience that, as with postcolonial narratives
in the field of cultural studies, will reinvigorate urban
theory wherever it is taught and studied.

Were we ever in any doubt about the importance
of the scale and pace of urbanisation and development in the 'majority world' it is worth engaging in
a little thought experiment. Imagine for a moment
that Friedrich Engels became the first passenger in
H.G. Wells' Time Machine, but instead of landing
in Salford, England in the 1840s – the epicentre
of the 'workshop of the world' in the nineteenth
century – he instead programmed the computer to
find him a city of equivalent importance to Victorian
Manchester somewhere in the world in 2010. Where
might the machine have taken our time-travelling
historical materialist?

New York, Los Angeles, Tokyo, or London all
spring to mind as possible contenders, but if one
wanted to see the birth of a new urban world happening before one's eyes, it would be to the Asian cities
of the Pacific Rim that one would journey, such
as Singapore, the Hong Kong–Guangzhou–Macao
region of the Pearl River delta and Shanghai where
the world's most rapid urbanisation and economic
growth is expected to take place over the next three
decades. Just as the Victorian cities of the European
industrial revolution saw a doubling of their population over the space of fifty years, so the heartland
cities of Asia-Pacific have emerged as major metropolitan centres with astonishing speed and success in
the latter half of the twentieth century.

In 1949 Shanghai's population was 5.2 million,
but by the end of 2000 it had risen to 13.2 million
official residents, while the figure including the floating population was some 16.7 million with urban
area population densities of 2,897 per square kilo-
metre.[59] From 1992 to 2000 the city registered
double-digit annual growth for nine consecutive
years, and between 1995 and 1999 the annual growth
rate averaged 11.4 per cent. In 1997 the five cities
of the Pearl River Delta (PRD) had a combined

population of twelve million. By 2020 the population of the PRD is expected to rise to thirty-six
million (Koolhaas/Harvard Project on the City, 2001:
281).

The speed and efficiency of this infrastructural
development would be unattainable in any western
state due entirely to the peculiarities of a socialist
market economy where land and skilled labour
(including that of architects and designers) can be
deployed at ultra low cost and at the whim of powerful Communist Party functionaries who are incon-
venienced by none of the burdensome planning
restrictions, Jane Jacobs-style conservation movements, or temperamental labour unions associated
with capitalist democracies. As Le Corbusier re-
marked, 'a city for the machine age could never
emerge from discussion and compromise: that was the
path to chaos' (Fishman, 1987: 239 in Mabin, 2000:
558). Indeed Le Corbusier's dream of building on
'clear sites' and the replacement of the artisan mason
by the rationalist theodolite of the structural engineer
has finally been realised in the New China where the
totalising vision of the modernist planner can project
destinationless 40-kilometre long bridges and can
make International Airports spring out of the paddy
fields with more flourish and audacity than any Baron
von Haussmann or Robert Moses.

East of Helsinki, in the Russian Federation, a
rather different picture of the post-socialist city
confronts us. About 107 million Russians (73.1 per
cent) live in urban areas, including 1,091 cities and
1,922 urban-type settlements. A little over 45 per
cent of the total population live in large cities with
a population of 100,000, and 17 per cent of the population lives in 13 cities with populations over a
million. Since the end of communist dominance in
1991, the Russian urban economy has faced a number
of difficulties, most of which have been borne by
urban residents themselves. Overcrowding is a major
problem, with living space averaging less than 20
square metres, while households contain an average
of 2.85 people. As the Russian Federation's own
report to the United Nations admits, it is practically
impossible 'to acquire social housing' in contemporary Russia, while, '[t]he majority of the Russian

population cannot afford to buy or build housing at their own expense', chiefly because incomes are low, and interest rates put home loans outside the reach of ordinary workers (Russian Federation, 2000: 4). At the same time, Russia has increasingly become a nation of property owners by virtue of 'right to buy legislation', which offered tenants the possibility of purchasing their apartments at a discount. Home ownership rose to one third of the population in 1990 and by the end of 1999 stood at 55 per cent – a figure higher than some member countries of the European Union. Municipal governments own most of the remaining housing stock (Guzanova, 1997).

Life expectancy rates have fallen and mortality rates have increased due to what the Russian Federation coyly refers to as 'adaption syndrome' – a euphemism for the stress and insecurity resulting from the rapid shift from a planned welfarist society to a market economy with minimal social guarantees manifested in high rates of hypertension and heart disease among all adult cohorts. The spread of major diseases is reaching epidemic proportions in Russia's cities with TB growth running at 65.7 per cent in the 1990s, syphilis increasing three-fold and HIV infection rates spreading even faster (Russian Federation, 2000: 10). Across a key range of indicators one could, therefore, describe the post-Soviet city as assuming a trajectory far closer to that of South Asia and Africa than the western-growth model of the China-Pacific metropolitan regions, even though there are some regional enclaves (such as the oil fields of the Caucasus) where the global economy is helping to stimulate some development and modernisation.

South of San Diego, the Maquiladora corridor just inside Mexico's border with the US provides the regional counter-example to the global uneven development that has seen much of Russia relegated to a peripheral status. The conclusion of the North American Free Trade Agreement (NAFTA) and the dismantling of tariffs and restrictions on foreign direct investments has encouraged the phenomenal growth in bonded assemblage plants (maquila or maquiladora) on the Mexican side of the Rio Bravo/ Grande – more than half of which are subsidiaries of

US firms (Kamel and Hoffman, 1999).[60] These special economic zones employ Mexican labour, much of it un-unionised and un-regulated at a fraction of the equivalent cost in the US. Once assembled, the goods are then usually exported to processing and distribution facilities on the American side of the border. Some idea of the value of this industrial zone to California businesses can be appreciated by the fact that $15 billion dollars worth of goods and components were exported to the maquiladora region in 1999, while a total value of $20 billion was created in finished goods imported back into California.[61]

As a result of this jobs bonanza, the population in the Mexican border region has risen to 12 million, but little provision has been made in terms of housing, sanitation and schooling for this floating population. Typical of the shantytowns that have grown up in the shadow of the high-tech maquila business parks is Cardboard City in the province of Acuña where 'almost all the houses are made of cardboard', but where unemployment is zero thanks to the post-1980 maquila boom which has brought 60 plants to the area. Still more workers are needed but 'the city has no capacity to provide the most basic services to its new inhabitants. In just five years, from 1990 to 1995, the population of Acuña grew by 49 per cent' (Cano, 1999: 9–13). Meanwhile, the environmental impact of this sudden urbanisation 'is verging on the catastrophic' (UN Human Development Report, 1999: 584).

The 'informal cities' of the global South, as writers such as Naomi Klein have shown, are fast becoming the new sweatshops of the world. But in these special economic zones the floating population of poor migrants is kept in constant circulation in order to prevent the emergence of community or workplace solidarities that might challenge the hegemony of the transnational corporations and their client governments in the minority urban world (Sassen, 1999; Klein, 2001). The example of the Comité Fronterizo de Obreras (Committee of Workers on the Frontier) who are working to improve wages and conditions for the maquiladora workers, nevertheless, demonstrates the potential for collective action among a poor and desperate workforce that has been denied even the protection of its own government (Kamel and Hoffman, 1999: 1–4).

If Broadacre City was the blue-print for the western 'functional city', Lagos in the African state of Nigeria must qualify as one of the least functional of the world's major metropolises. Lagos 'inverts every essential characteristic of the so-called modern city', but yet, 'it is still . . . a city; and one that works' (Koolhaas *et al.*, 2001: 652). The Harvard Project on the City's reading of Lagos rejects developmentalist and dependency theory models that would see the Nigerian megalopolis at an earlier evolutionary stage of the western capitalist city in favour of it representing 'a developed, extreme, paradigmatic case-study of a city at the forefront of globalizing modernity' (ibid.: 653). This is a very important insight because it highlights the ways in which the complexes of consumption are not just questions for the sociologists of affluence, but that the poor inhabitants of some of the world's most overcrowded cities have a capacity for shifting family resources, goods and capital around global spaces that would leave the well-healed Cosmocrats of Battery Park or London's Docklands open-mouthed with wonder (see also Smith, 2001).

Typical of many African cities is the expropriation of 'public infrastructure' such as the Oshodi road-rail intersection in Lagos which has become a large informal market, where 'all along its length, the roadsides have been annexed and overrun with trading activities' (ibid.: 693). 'Oshodi's traders and transport businesses have literally annexed the transport infrastructure . . . and have even taken measures to construct new roads and new right-of-ways' (ibid.: 694). This is Lefebvrian transduction in action, the remaking and refashioning of urban space to meet the exigencies of the users in a manner that calls to mind the 'spontaneous market' that Benjamin encountered around a dug-up street in Paris. This hyper-informalised city is not so much the untamed past of modernity's future as the untamed future of modernity's present. As the functional city begins to atrophy and shed its skin, new paths and channels are being cut through the subsoil as the anarchy of the DIY satellite dish trumpets the new urban order of cyberville.

THE INFORMATIONAL CITY: LINKING THE VIRTUAL AND MATERIAL URBAN WORLDS

[The Net] . . . will play as crucial a role in twenty-first-century urbanity as the centrally located, spatially bounded, architecturally celebrated agora did (according to Aristotle's *Politics*) in the life of the Greek polis and in the prototypical urban diagrams like that so lucidly traced by the Milesians on their Ionian rock.

W.J. Mitchell, *City of Bits*

the simplistic and parallel reification of cities, neoliberal 'market forces' and 'new technologies' often serves to obfuscate the broader power relations, political economies and practices bound up with the reconstruction of (parts of) cities as premium network spaces.

Graham and Marvin, *Splintering Urbanism*: 410

The discussion around the impact of new information technologies on 'the urban' can be broken down into three broad areas. The first is the impact such technologies have on 'real' or 'actually existing' cities – are they making traditional cities more or less viable? What impact are they having on work and leisure patterns and on the urban economy as a whole? The second concerns the impact of new technologies such as computer aided design (CAD) and geographical information systems on urban design and planning. While a third strand deals with the nature of the urban within cyberspace itself – how does the organisation and distribution of space in the real world carry over into the virtual world, how do real world urban geographies compare to those of hyperspace? How are our interactions with the virtual city feeding back into the way we design and experience 'real spaces'?

What the information revolution facilitates, above all, is networking, and because one of the essential functions of the city is to bring buyers and sellers together it is not surprising that all the key technological innovations involving the use of networks first began by connecting the largest urban business centres to one another, such as the first transatlantic telegraph cable that linked London with New York. With the advent of the steam locomotive new lines of communication ran from pre-existing city to pre-existing city, but along the way new settlements were formed at the junction of these communication nodes. The Internet 'backbone' is being constructed along similar principles with a super concentration of network infrastructure and traffic between the principal business cities of the US (especially New York to Los Angeles) so that the Internet cyber-map reflects, with few variations, the commercial supremacy of First World, and particularly American, enterprise.[62]

Simon Marvin, reporting on research conducted by New York University in the late 1990s, observes that the top 15 metropolitan cores in the US (the centre cities only) are home to some 4.5 per cent of the national population but contain 20 per cent of all Internet domain names (Parker, 2000a: 124). Castells, drawing on the work of Matthew Zook, points out that in the US (controlling for income) urban residents were more than twice as likely to have Internet access as rural dwellers. In other parts of the world the spatial concentration of Internet access is even starker. In Russia, for example, 75 per cent of all Internet users live in the three cities of Moscow, St Petersburg and Yekaterinburg (Castells, 2000a: 377–82). This suggests that far from creating a counter-revolution away from the industrialised conurbation to remote 'silicon villages', the information revolution is concentrating ever greater stores of value in traditional dense cities as well as in new agglomeration economies such as those found in Santa Clara County (Silicon Valley) in California (Castells, 2000a: 62).

Increasingly, the availability of specialised knowledge centres is the key factor in business location. For example Silicon Valley has a decentralised industrial system that is organised around regional networks. Like firms in Japan, and parts of Germany and Italy, Silicon Valley companies tend to draw on local knowledge and relationships to create new markets, products and applications. These specialist firms compete intensely while at the same time learning from one another about changing markets and technologies. The region's dense social networks and open labour markets encourage experimentation and entrepreneurship. The boundaries within firms are porous, as are those between firms themselves and between firms and local institutions such as trade associations and universities (Saxenian, 1994: 44 in De Landa, 1997: 95).

Place marketers set increasing store by a city's information technology infrastructure in order to lure investors to their region as do loft conversion developers keen to attract the monied IT professionals who require quick and reliable access to the Internet trunk routes using the fastest possible broadband technology. But the privileging of these forms of network connectivity can mean that the information aristocracy attracts a disproportionate amount of state funding and subsidy at the expense of public transportation or welfare services, further exacerbating the polarities of the dual city (Graham, 2002).

That the new information technologies have had a significant impact on human geography there can be no denying, but does this mean we have to agree with William Mitchell that the city as understood by Plato, Aristotle, Mumford and Jacobs is 'finally flatlining' (Mitchell, 2000: 3)? Mitchell has seen the e-future and it is a world of live/work dwellings, twenty-four-hour neighbourhoods, loose-knit, far-flung configurations of electronically mediated meeting places, flexible, decentralised production, marketing and distribution systems, and electronically summoned and delivered services (ibid.: 7).

The silicon-based utopia that many high-tech oriented growth coalitions believe is about to arrive has many parallels with the steam-age liberation from useless toil that the pre-proletarian working class were promised nearly two centuries ago.[63] However, if for Mitchell there is no alternative but to build e-topias 'that encompass virtual places as well as physical ones' (ibid.: 8), Kevin Robins takes an altogether more sceptical view of cyber boosterism, arguing that

> global cities are places of cultural encounter and confrontation. They are spaces of disorder. And for those whose imagination (with its Enlightenment roots) is centred around the values of coherence and order, this kind of urbanism must be deeply problematical. In this context, we may see the ideal of the virtual city as a defensive and protective response: in one respect it is about the denial or disavowal by technological means,

of this chaotic and difficult reality; and in another, it involves the attempt to sustain or restore the values (coherence, order, community ...) of the older (European) ideal of urbanism that now seems in crisis. The new technological urbanism is in fact a conservative urbanism.

(Robins, 1999: 35)

Even Mitchell concedes that the digital revolution has its down side in terms of promoting 'the right to the city' noting that:

Urban areas could well continue to congeal into introverted, affluent, gated communities intermixed with 'black holes' of disinvestments, neglect, and poverty – particularly if, as the unrestrained logic of the market seems to suggest, low-income communities turn out to be the last to get digital infrastructure and the skills to use it effectively. As Manuel Castells has vividly warned, we could end up with *dual cities* – urban systems that are 'spatially and socially polarized between high value-making groups and functions on the one hand and devalued social groups and downgraded spaces on the other'.

(Mitchell, 2000: 81)

This is a concern that Castells articulates in another context when he writes, 'new information technologies are not in themselves the source of the organizational logic that is transforming the social meaning of space: they are, however, the fundamental instrument that allows this logic to embody itself in historical actuality' (Castells, 1989: 348). The replacement of 'the space of places' by what Castells calls 'the space of flows' poses an enormous challenge to local and national policy-makers and those groups and individuals who are affected by the globalising logic of the informational city. In order to avoid a digitally assisted 'race to the bottom' as cities compete to direct flows of investment, business location and qualified labour in their direction – as Saskia Sassen has proposed – we need to establish something akin to a UN for cities, a United Cities Organisation that will act as an advocate for the currently voiceless global spaces on which the prosperity of the information age depends.[64] Urban governments might then be in a position to substitute the short-term 'clicks and bricks' ideology of the electronic frontier for a sustainable long-term future in which the existence

of cohesive communities is regarded by decision-makers as a higher priority than their collective consumption potential.

Stephen Graham writes that

it is now clear that cyberspace is largely an urban phenomenon. It is developing out of the old cities, and is associated with new degrees of complexity within cities and urban systems, as urban areas across the world become combined into a single, globally interconnected, planetary metropolitan system.

(Graham, 1999: 10)

Since the early 1990s when Internet technology, and especially the World Wide Web, began to reach a mass audience, there have been a proliferation of 'multiple user domains' (MUDs) that have their own 'virtual space'.

Some, such as Activeworlds.com's Alphaworld,[65] are based on imaginary cities, which users colonise by occupying spare plots – in reality a grid coordinate that identifies the 'address' of the inhabitant. Some of Alphaworld's talented inhabitants even engage in virtual civil engineering – for example there is an impressive monorail system, an airport, and even a perfect replica of the Titanic!

One of the fascinating aspects of these virtual worlds is how closely the layout and settlement patterns of the virtual neighbourhoods correspond to those of 'real cities'. In the virtual, as in the real world, 'location, location, location' is everything, with the most 'fashionable' coordinates (such as the furthest edge of the compass points) being the first to be occupied. What is also revealing is that new users tend to locate where there is already an existing cluster of users, even though there is no special advantage in doing so in the virtual metropolis. Such contiguity is encouraged by peer-to-peer interaction by the use of 'avatars' (three-dimensional characters familiar to the world of computer gaming) that allow one to talk and move in the same domain as other users.

Another feature of virtual cities is what Graham calls 'grounded cities' such as *De Digital Stad*, which is directed at the Amsterdam on-line community. The site uses interlocking city cells (virtual city squares) providing access to information, chat rooms and links to other sites to a 50,000-strong user base that has even elected its own 'electronic' mayor (Aurigi and Graham, 2000: 490–2). Others such as Bologna City Council's Iperbole site use three-dimensional graphic icons of local landmarks such as the town hall, the police station, the airport and the railway station as a link in to relevant information sources that are specific to these facilities. The city also provides free dialup access to its network and an increasing number of council services can now be accessed or requested on-line. Such initiatives are welcomed by proponents of 'local' cyberspace as an alternative virtual future to 'the "Utopian school" of future "cyber-lifestyles", which sees cities becoming depopulated, "instant electronic democracy" replacing the need for governmental structures and services and a dominant "ruralist" lifestyle emerging' (Carter 1997: 139 in ibid.: 494).

'Ungrounded' MUD-type virtual cities and 'grounded' holistic virtual cities are expressions of the dichotomy between what we might broadly term a postmodern non-place urban imaginary – a place of irony, play and transgression, and a modernist conceptualisation of cyberspace as serving the ends of the synecdochal community – a replica of the 'real world' in microcosm, the aerial city as seen from Le Corbusier's marvellous flying machines. Both systems of representation have intended and unintended material consequences. City 'portals' have a civic function, but they are also a means by which 'the virtual city growth machine' tries to reach potential tourists and investors in the local economy. As we saw in Chapter 5, the world Garreau describes of scattered edge-city networks is heavily reliant on telematics and real-time communication systems but, at the same time, the human need for face to face interaction appears to be stimulated rather than diminished by the presence of information technologies. Far from making cities redundant, 'the global rush to urbanization and rising, although highly uneven, physical mobility is happening *at the same time* as the pervasive . . . growth of electronic communications' (Aurigi and Graham, 2000: 490). In other words, as we shall see in the following section, the rapid growth of the virtual world is paralleled by a growth in city

users and city use that many commentators warn is already at an unsustainable level with significant negative consequences for the health and well-being of urban populations around the world.

THE NEW ECOLOGY OF THE CITY: SUSTAINABILITY AND THE URBAN FUTURE

Three-fourths of those joining the world's population during the next century will live in Third World cities. Unless these cities are able to provide decent livelihoods for ordinary people and become ecologically sustainable, the future is bleak. The politics of livelihood and sustainability in these cities has become the archetypal challenge of twenty-first century governance.
(Peter Evans, 2002: I)

one fourth of the world's urban population is living below the poverty line. In many cities, which are confronted with rapid growth, environmental problems and the slow pace of economic development, it has not been possible to meet the challenges of generating sufficient employment, providing adequate housing and meeting the basic needs of citizens.
(United Nations General Assembly, June 2001)

As Rodney White notes, '[s]ustainable development became an issue of widespread concern with the publication of "Our Common Future", the summary report of the World Commission on Environment and Development (the Brundtland Commission)' (WCED, 1987; White, 2001). The aim of the report was to promote continuing economic and social development especially of poorer communities alongside the preservation of the natural and physical environment. Richardson defines this sustainable development in the urban context as, 'the development of a city's physical structure and systems and its economic base in such a way as to enable it to provide a satisfactory human environment with minimal demands on resources and minimal adverse effects on the natural environment' (Richardson, 1992: 148 in Chowdhury and Furedy, 1994: 6).

Sustainability is clearly, therefore, not just an issue of environmental concern. A recent conference volume dealt with nine themes in urban sustainability including strategy and development; environmental management and pollution; land-use and management; transport, environment and integration, cultural heritage and architectural issues; planning, development and management; restructuring and renewal; the community and the city; and public safety and security (Mattrisch, 2001). However, critics point out that if by 'development' we mean more consumption growth, increased production and higher carbon emissions, then this goal is certainly not compatible with ecological sustainability.

Urban growth in the developing world is increasingly associated with the growth in slums around the fringes of established cities. The United Nations estimates that as many as 30 per cent of the world's urban population (or 712 million people) were living in slums in 1993. By 2001 the number of slum-dwellers had reached 837 million worldwide. The UN is committed to improving the lives of some 100 million slum-dwellers by 2020 by which time the total number of slum-dwellers is likely to have reached 1 billion or one seventh of the predicted world's total population in 2010 (UN Global Urban Observatory). Welcome though these interventions will be if words are turned into deeds, such an aspiration reveals that even the most optimistic plans fail to scratch the surface of a global urban crisis that is reaching dangerous levels of violence and discontent.[66]

The degradation of the urban environment also manifests itself through the atrophication of its communication systems. While many western local governments have agreed to sign up to traffic containment, waste reduction and environmental protection programmes agreed at the Earth Summit in Rio de Janeiro in 1991 (known as Local Agenda 21), most urban authorities in the cities of the South lack even the limited political and economic freedoms available in the west to pursue such objectives. Governments in the global South are often faced with the Hobson's choice of accepting lower environmental standards than transnational corporations would be able to operate in the west or face the prospect of disinvest-ments and job losses in often desperately poor communities. However, against all odds, it is encouraging to see that many politicians and NGOs in the developing world are working to improve the quality of their urban environment and are active supporters of the urban sustainability agenda of the Habitat II Summit.

There is also an argument that 'sustainability' for all its unwelcome consequences for individual city users is a form of spatial regulation that is essential for the long-term survival of capitalism and its workforce. World Health Authority data collected in Austria, France and Switzerland found that exposure to pollution caused an estimated 21,000 deaths a year, and found that car fumes were responsible for 300,000 extra cases of bronchitis in children and 15,000 extra hospital admissions for heart disease problems exacerbated by pollution.[67] In the more heavily polluted industrial cities of Russia, children are 1.3 times less healthy than they were in the 1980s, and 0–7-year-old children fall ill 1.5 times more frequently (Russian Federation, 2000). In economic terms, road congestion (chiefly urban, but also increasingly suburban) has been estimated to cost $100 billion a year in the US alone.[68]

The industrial legislation that was introduced to prevent the cities of Victorian England from descending into squalor, dirt and disease were violently resisted by the business interests who confidently predicted that such regulations would strangle enterprise in red tape and make its activities unprofitable. When, in the middle of the nineteenth century the urban middle classes also began to die of water-borne diseases such as cholera, the 'laissez-faire' city soon gave way to an efficient local state armed with all the utilities and resources it needed to guarantee clean air and water, pleasant parks and efficient public transportation. However, 150 years later the succession of the successful to privatopias and gated communities is leaving our cities bereft of the influential patrons who once made London, Paris and New York the envy of the world.

The hollowing out of the state has been accompanied by the hollowing out of the city and as the urban experience has been transformed we need to rethink many of the assumptions on which urban theory has been based. In particular, we have to ask, in light of the contemporary crisis of modernity and the 'retreat from reason' in many aspects of social and cultural life, not 'What is a city?', but the more troubling interrogative, 'Wither the city?'.

EXPERIENCE TEACHING THEORY: FROM CHICAGO TO LA TO GROUND ZERO

What I hope to have demonstrated in the previous chapters is that by encountering the city, ideas about its nature, purpose and routines emerge and develop. Whether we bracket them as 'empiricists' or 'theorists', almost without exception, the urban witnesses we have met have walked the city's streets, mingled with its crowds, and breathed its heavy air. 'The city', in whatever form it assumes, stimulates thought and creativity by its mere existence. It was no accident that the early development of urban sociology happened in the environs of one of America's fastest growing and brashest 'can do cities'. The fact that Engels could see the misery and squalor of the English industrial revolution from the windows of his own Salford mills meant that he had an immediate example of the depredations of modern capitalism on which to base his essay. It was also significant that Simmel and Benjamin were able to experience with their own eyes the transition of Berlin and Paris from the static world of the *belle époque* to the modern metropolis of the S-Bahn and the Metro. It is also apparent that those associated with the new 'LA School' of urban theory share a common interest in the city and environs of Los Angeles (and to an extent the wider Southern California region) as characteristic of an urban form that is beginning to replace that typically associated with East Coast and Mid-West cities. It is a bitter irony therefore that New York City, a city increasingly associated with America's urban past should be so brutally cast as the emblem of its beleaguered urban future.

The city after September 11 . . .

Urban analysts have been struggling to come to terms with the terrorist attacks on New York City's World Trade Center in 2001 and what this event means both for the future of New York and for the metropolis as a viable urban form (Sorkin and Zukin, 2002).[69] Those who have been persistent critics of featureless high-rise blocks took the opportunity to point out that the architect of the WTC, Minoru Yamasaki was also the designer of the Pruitt Igoe apartment blocks in St Louis, Missouri which were deliberately demolished back in 1972 – in an event that was meant to herald the death of mega blocks as a solution to the city's housing needs in the late twentieth century. But since September 11, high-rise blocks continue to spring up in the downtown areas as well as in the increasingly commercialised suburbs and the new edge cities. Far from heralding the death of the skyscraper, 9/11 has, if anything, helped to promote a popular revival in very tall 'statement' buildings as a celebration of capitalist modernity and a hymn to western liberty and progress.

All of the nine finalists in the competition for the rebuilding of the WTC site included very tall towers. One proposed building breaks the symbolic 2,000 feet barrier, while Daniel Libeskind's winning design rises to precisely 1776 feet – the year of the birth of American Liberty – in case we were in any doubt about the potency of these architectural statements in terms of shaping the urban discourse of the post-apocalyptic city. Each scheme emphasised its 'green' credentials by the use of natural light, or wind turbines to power the buildings' elevators. Several designs incorporated cultural themes such as the THINK consortium, which proposed a $1 billion World Cultural Centre and a 16-acre rooftop public park. But it was Libeskind's evocative use of the well of the ex-trade tower as a memorial space that finally swung opinion in his favour. Nevertheless, controversy is likely to continue if the owners of the site insist on a higher volume of office space than that currently planned in order to make the building commercially viable, despite the fact that there is already a glut of vacant office space in Lower Manhattan.

Unable to escape the fatal dichotomies of global trade and world peace (the two epitaphs of Yamasaki's Twin Towers), all the new imagineers of Ground Zero can do is to 'collaborate on the production of images of security, comfort, and memory' (Wigley, 2002: 84). Although the Lower Manhattan Development Corporation appear not to have seized it, as Michael Sorkin argues,

> [t]he radical act of the terrorists opens a space for us to think radically as well, to examine alternatives for the future of New York City. It is no coincidence that we have created a skyline in the image of a bar graph. This is not simply an abstraction but a multiplication, an utterly simple means of multiplying wealth: where land is scarce, make more. Lots more. There is a fantasy of Manhattan as driven simply by a pure and perpetual increase in density. But while our dynamism is surely a product of critical mass, not all arguments for concentration are the same. Viewed from the perspective of the city as a whole, the hyperconcentration of the World Trade Center was not necessarily optimal by any standard other than profit, and even that proved elusive.
>
> (Sorkin, 2002: 202)

Mayor Giuliani urged New Yorkers and visitors to help restore the city's wounded economy by going shopping, but it was not enough to prevent the loss of up to 100,000 jobs in the city alone after September 11 (Harvey, 2002: 62). In truth, New York's economy had been having a hard time of it before 9/11 and, as Harvey argues, 'September 11 appeared more and more as a wonderful excuse for companies and industries to do what they were preparing to do anyway (including moving out of a highly congested and very much overpriced Manhattan) (ibid.).[70] The questions many journalists began to ask were – 'Do heavily built up, densely populated cities such as New York have a future?' 'Isn't life so much safer in the corporate park, the gated community, and the privately policed atria of the mall?' 'Who needs downtowns anyway? There's nowhere to park, the sidewalks are noisy, crowded and "unsafe", and it takes me an hour and a half just to get there'.

All this may be true, but would it not be a pity if the Woody Allens of the future, instead of producing movies such as *Manhattan* or *Broadway Danny Rose*

were obliged to call their films '287 and 78' or 'Bullets Over the Bridgewater Mall Area'? How different would our image of the urban modern world be without the Manhattan skyline, even allowing for the disappearance of the Twin Towers? *Delirious New York* (Koolhaas, 1978) has long been the fantastical toy box of urban design makers, filmmakers, comic strip artists, novelists, composers, digital artists, painters and poets. If Çatal Hürük or Ur are the fore-runners of the ancient city (Soja, 2000a), then Manhattan must be the *ur*city, the originary city of modernity.

But is the celebrated Gotham that filled every urban imaginary from Lang to Scorsese destined to become a mere movie set? It may be that 'the money' is leaving Manhattan in droves for the business parks of New Jersey or Long Island, but the city still has a thriving textile and fashion trade, some of the most prestigious American universities, and a high concentration of young entrepreneurs at the cutting edge of information technology, digital media and e-commerce who all have a stake and belief in the coffee house contiguities that make for a successful urban economy (Pratt, 2000). In other words, as we saw in Chapter 6, agglomeration economies require dense cities, and if we are to cope with the environmental consequences of global urbanisation in the next hundred years we have to find new ways to make such cities livable (Rogers, 1997).

A MANIFESTO FOR THE CITY

Nonetheless, large, dense cities have very high support needs – they require adequately funded public administrations at local, regional and national level, and political leaders that are engaging in more than mere rhetoric when they talk of regeneration and social inclusion. They need responsible and committed businesses that are prepared to contribute some of their profits to local communities rather than demanding tax breaks and labour deregulation. Such cities will also only thrive if the richest quarter of society has an income no greater than 2.5 times that of the poorest quarter – Sweden has proved that it is possible while still being home to some of the most successful companies in the world such as Ericsson, Ikea, Volvo and Saab. Currently, American cities have some of the worst income disparities in the 'First World' and this must mean that with the inflationary impact on rental costs, utilities, food and other essentials, the rights to the city are being denied to all but the most affluent.

The Dutch who co-founded Manhattan still have much to teach Americans and, indeed, the rest of the world about what makes a livable city (van Kempen and Priemus, 1999; Fainstein, 1999). While New York City is losing population and jobs, Amsterdam is gaining them – and, indeed, there is a nine-year wait to rent a publicly owned apartment in the city centre. What accounts for this success?

We can summarise the reasons as follows: (i) the city government actually owns the real estate on which Amsterdam is built (Fainstein, 2000a: 96). This means that the municipality is much better able to control development and speculation in sought-after sites (see Figure 9.1). Profits from commercial rents and leases can also be ploughed back into the social rented housing sector in order to keep rents economical for lower income groups.[71] (ii) The motorcar is discouraged as a form of transport and bicycle tracks, tramways and pedestrian zones take priority in the city's traffic system. This makes the city noticeably less polluted than other capital cities, and it encourages families with children to use the downtown area as they would any other part of the city. (iii) There are no 'hyperghettos' in Amsterdam. It is true that certain housing projects have large populations from the former Dutch colonies and more recently Africa and Turkey, but

Figure 9.1 Public housing development, Amsterdam, the Netherlands.
Copyright David Smith, courtesy of the Architectural Association Photo Library

there are also many traditional neighbourhoods with mixed populations.[72] Of course, racism and discrimination still exist (and, indeed, the assassination of the right-wing Dutch politician Pim Fortuyn in 2002 proved that anti-immigration is a very live political issue) but this has not resulted in institutionalised segregation, and the city still retains an international reputation for tolerance of different lifestyles.[73]

The Amsterdam experience shows that to adapt the slogan of the World Social Forum 'another urban world is possible'. We might summarise such a practical utopia in the following terms:

• Successful cities must be public – the city government should call the shots in terms of what business and the market can do, not the other way round.[74]

• Cities must guarantee the full citizenship of all users and residents – no one should be denied the right to the city because of their income, family status, sex, religion, sexuality or ethnicity. National, regional or local governments should not be able to override these fundamental rights, which should be protected in the form of a city charter (such as in the case of Toronto).[75]

• While recognising that the management of cities is too important to be left to city managers, it is important not to lose sight of the need for purposive civic leadership. This leadership must be accountable, democratic, transparent and representative of the diversity of its urban communities. It should also be equipped with the political, legal and financial resources necessary to fulfil its mandate.

• Cities should be prized and utilised by national governments as engines of sustainable economic growth, cultural diversity and advanced social policy, but they should not become reservations for certain groups of populations.

• Finally, governments should strive to maintain a distinction between the urban and the rural in order to maintain the character of the city, to avoid the problems of sprawl, and to preserve the earth's rapidly diminishing natural resources and green space.

The Dutch Golden Age cities of Amsterdam and Rotterdam were at the forefront of the global economy in the seventeenth century, and they succeeded in establishing a pride and investment in public space and public buildings that still survives to this day (Schama, 1987). But the contemporary renaissance of the city-state is likely to be stifled by the moral failure of the modern equivalent of the Dutch East India Company – the transnational corporation – because the death of the Keynesian Welfare State has not been accompanied by the return of corporate philanthropy and the civic mission. The de-spacing and de-personalisation of capitalism means that we are unlikely ever to see entrepreneurs erecting model towns such as Bourneville, Saltaire and New Earswick that offered a utopian alternative to the dark satanic mills of the English industrial revolution, or the New Deal federal government owned Garden Cities that briefly offered a glimpse of how European-style community planning could work in the US.

At the same time, in the majority urban world, globalisation is producing an increasingly market oriented urban hierarchy but it is also generating movements that are demanding a greater stake in their urban futures. In endorsing Weber's view that the ideal city *is* the democratic city, we do not have to accept that the function of democracy should be to serve the logic of capitalist accumulation. Democracy is also in the 'magical urbanism' of the barrio, the Islamic charitable organisations that do more than state governments to feed and clothe the urban poor, and the transnational networks of migrants who organise and defend their communities against terrible odds (Davis, 2000a; Smith, 2001; Lubeck and Britts, 2002). Whether one is a medieval German peasant or a Dinka tribesman, city air may not make you free, but it offers the glimmer of a job prospect, education, shelter and food, which most westerners now take for granted. Third World urbanisation will not cease because UN agencies, western governments and NGOs say it is a bad thing. Therefore, policymakers and academic urbanists should be thinking about how we can make 80–100 million strong urban agglomerations such as exist in northern Indonesia at least semi-functional and tolerable for their residents.

This is where theory may have something to teach those living the urban experience, and it is to the future directions of urban theory that I now want to turn.

CONCLUSION: THE URBAN EXPERIENCE AND THE FUTURE OF URBAN THEORY

What are the prospects for a new spatial dialectic that is capable of paying equal respect to the material and imaginary urban world while not subordinating one to the other? And is, what Ed Soja calls, this new 'trialectics of space' (Soja, 2000b) capable of holding a meaningful conversation between the world of urban theory and that of the urban experience? Perhaps, less than ten years ago, the answer to this question would have been an unqualified 'no'. Now I think there is sufficient evidence of boundary crossing to answer with a qualified 'yes'. My reasons for optimism are partly related to what we might call exogenous, or external factors to the world of urban scholarship, and partly the result of endogenous, or internal developments within the broad domain of the social sciences and humanities that have had an important impact on cross-disciplinary research.

Let us begin with the exogenous factors first. The most obvious scalar shift in the geographical imagination of late modernity has undoubtedly been globalisation. The 'G' word has been a recurring motif in this volume, and its impact can be seen in the debates and experiences that have taken place across the spectrum of the urban complex. The second, and related but not coterminous theme is de-industrialisation, or post-industrialisation, perhaps best summed up in the term post-Fordism. Of course, Fordist style manufacturing has not disappeared, far from it, in fact there is rather more manufacture going on than there was 50 years ago, but it is happening in different places, especially in and around the larger urban settlements of the global south, and less and less in the established cities of the global north. This means the city that was once organised around the work rhythms of the industrial factory and the consumption and leisure habits of a contiguous work force no longer needs to be organised in such a fashion, and is becoming increasingly 'disorganised' into a patchwork quilt of habitats, business parks, shopping malls, expressways, entertainment villages, and so on. Or, at least some urban agglomerations are beginning to exhibit these features – predominantly in the 'sunshine belt' of the US. Outside of the US there are hints of such scattered 'exopolises' (Soja, 2000b) along the St Petersburg to Moscow expressway, in Guangdong province and also Shanghai in China, but the context – especially the social and political context is very different in each locale making generalisations based on the appearance of urban form problematic.

This brings us to our third exogenous factor that points to a 'new' urban order, the so-called 'crisis of the state'. Commentators have been talking about a crisis of the state ever since it emerged as a meaningful political concept in the time of Plato and Aristotle, but what we are concerned with from an urban theory perspective is the demise of the nation-state (i) as the apex of a hierarchical territorial 'command and control' system in which the city is held subordinate, (ii) as the 'rain maker' of the national economy – capable of controlling key economic indices, slowing or accelerating economic growth, and directing economic activity at a sub-national level according to national priorities, and (iii) as 'a sovereign among sovereigns' in the international system with the power to opt in and opt out of international organisations, treaties, protocols, alliances and trading systems according to the 'national interest'. If – and it is a big 'if' – this changing status of the nation-state is verifiable then its implications for city government are enormous – it has even led some commentators to talk of the return of 'city-states' (Castells, 1994; McNeill, 1999).

Can we, therefore, synthesise all these related processes as symptomatic of a new phase of advanced capitalism – what Castells calls 'the information age' – that is leading to fundamental changes in the form and function of cities? The difficulty is that capitalism is not changing at the same pace, in the same places at the same time, and even if it were, the fact that the nature of capitalist development was already very uneven before this latest phase upped the stakes means that the experience of change will be different for each (urban) locale.

Turning now to endogenous factors, the biggest obstacle to an integrated approach not just to urban studies, but to the study of social phenomena in general has been the continuing intellectual division of labour. There are, I think, two main causes for this trend, though doubtless other factors also have an impact. First, the continuing 'professionalisation' of higher education requires those who wish to succeed in career terms to become ever more 'expert' in their subject. But because the sheer volume of books and research papers produced every year is so vast, scholars are encouraged to become masters of a very parochial universe in which if one goes far enough to the right of the decimal place in the Dewey classification it is just possible to have 'read everything'. This is fine, even necessary for acquiring a doctorate, but the existence of specialist journals, scholarly associations and conferences further reinforces 'inflexible specialisation', despite the increasingly plaintive appeals from journal editors and conference organisers for inter-disciplinary submissions.

A second reason for the general absence of integral treatments of urban studies is that the faculty and departmental structure of many universities tends to prioritise the discipline over the subject area, and this is only partially offset by the existence of dedicated research centres specialising in urban research, because in order to attract vital external funding 'blue skies' urban research (particularly in the area of theory) is hard to pursue. Even the several excellent journals dedicated explicitly to the study of urban phenomena (one thinks in particular of *Urban Studies*, the *International Journal of Urban and Regional Research*, the *Journal of Urban Affairs*, the *Urban Affairs Review, Cities, City*) are portmanteaux of specialist papers by disciplinary experts, with the occasional conspicuous exception.[76]

The third reason is that as the 'heroic era' in the social sciences came to an end in the late 1960s, a comprehensive, holistic account of the city was seen as a theological rather than a scientific exercise and therefore unworthy of serious academic endeavour. Such attitudes encouraged many researchers to take refuge in their adoptive branch of the intellectual division of labour rather than develop what, to paraphrase C. Wright Mills we might call 'the urban imagination'. As a result, urban theory became increasingly separated from empirical urban research just as social theory and empirical social science research continued on their separate paths and viewed each other's work with unconcealed scepticism.

Are urban narratives beginning to converge?

However, in the last twenty years I believe the trends towards convergence of these two modes of enquiry or narratives have become manifest in the emergence of what we might call 'the empirical theorist'. All of the most prominent figures in urban theory have cut their teeth on 'real world' investigations of urban problems, urban movements and/or urban phenomena. If one looks at the résumés of leading urbanists such as Michael Peter Smith, Bob Beauregard, Saskia Sassen, Janet Abu-Lughood, Norman and Susan Fainstein, Doreen Massey, Mark Gottdiener, Sharon Zukin and company, each has a direct experience of what Robert Park called 'dirty hands' research, and their theoretical writings are, therefore, informed by a strong sense of how the processes we observe at an abstract or universal level are experienced at the grass roots.

Another great impulse towards convergence has come from the demonstration effect of urban theory's most prolific researchers, and here I am particularly thinking of David Harvey and Manuel Castells. Having begun as essentially orthodox Marxist urban geographers/urban sociologists, their intellectual trajectories have spanned out to include theories of philosophy, culture, art, architecture, planning, gender, sexuality, race and ethnicity, social movements, criminology, identity and post-colonialism, along with many other areas of intellectual concern that have not traditionally been bracketed with urban studies. New spaces have thus opened up for cross-fertilisations of critical approaches and methods that have found, in the study of the city, a particularly rich soil for theoretical hybridisation. At the same time, one only has to note the amount of space given over to tables, figures and diagrams in the recent work of major urbanists such as Sassen and Castells to recognise that a close empirical study of the urban experience is informing urban theory to an even greater extent than in the era of Engels, Booth and Burgess.

Beyond the urban fragments

> the city of late capitalism is too complex, and too fragmented in its physical and ideological formations, to ever permit a unitary comprehension.
>
> (Borden *et al.*, 2000)

Classical and contemporary urban theorists and investigators have tried to describe the urban experience using the moral, aesthetic and philosophical language available to them at the time. But few, with the notable exception of Benjamin and Lefebvre, I would argue, have avoided historicising the city as a way of distinguishing previous urban experiences from those we see today. A significant obstacle to a heterochronic/heterotopic account of the city is the widespread recourse to historical context as an explanatory cause for some general phenomenon. Yet, the 'Victorian city' is no more meaningful a description of the urban experience than Rudyard Kipling's 'City of dreadful night'.

However, rather than abandon metaphor to the misappropriations of historical cause and effect we need to rescue this tropological space (White, 1987) for a more compelling urban critique. Here, Foucault's distinction between metaphor, metonymy, synecdoche and irony that we discussed in Chapter 8 is important because it offers the prospect for a non-historicist reading of urban form, thus providing a meta-language for urban discourse that can simultaneously accommodate contingent narratives of social change such as 'globalisation', 'postmodernism', 'splintered urbanism', 'neo-traditionalism', and so forth. In order for these powerful and suggestive narratives of the changing urban condition to acquire

meaning as 'urban experience' we need to ground them in genuine rather than speculative investigations while not conflating the particular and the universal. One way of achieving this objective is to recognise the ways in which identities and social relations bind with locales to produce structurations of urban power.

Urban structurations can be visualised as what Raymond Williams described as 'structures of feeling' (Williams, 1961) which are denser and more articulated wherever the urban life-world is at its most compact and reflexive. Hence, for example, the East London Jews who fled the Russian and Baltic pogroms in the late nineteenth and early twentieth centuries, forged a structure of feeling – a unique blend of radical socialism, messianic Judaism and Zionism – in the dense network of streets in and around London's Whitechapel. Economic and geographical mobility has almost entirely dispersed this and other established East End communities, but the cultural economy of migrant space continues to define the settlement patterns and labour of new arrivals from Bangladesh and the Horn of Africa as 'communities of resistance' (Parker, 2001a, 2001b). Thus, the idea of the city can survive its physical disembodiment by the corrosive tides of the space of flows, as Benjamin intuited, through the reclamation of its salvaged past in a critical affirmation of a dynamic urban present.

In recognising that the *nomos* 'city' or 'urban' is every bit as contestable as other value-terms such as 'community' or 'society', as many urban analysts do, it is important to remember that the social construction of the city is, itself, an expression of the material conditions of its physical development. In other words, it is possible to identify a common 'urban language', while at the same time appreciating that many urban discourses result from the contested interpretations of place, identity and rights.

The four C's of the city that I mentioned in the introduction – consumption, culture, community and conflict – each structure, reproduce and mutate the discourse of urbanity, just as the urban complex shapes and stretches the modes of modernity on a global scale. The urban experience will continue to set the terms in which the city is conceived and imagined, but unless we continue to develop our theoretical vocabulary, the language that we need in order to perceive and understand it will be unequal to the challenge of our common urban futures.

FURTHER READING

On *alternative urban futures*, the Partners for Livable Communities' publication, *The Livable City. Revitalizing Urban Communities* (2000) shows that another urban world is also possible in the US. The way to achieve it, argue the authors, is by getting communities to work with each other. Although the emphasis is on self-help, the volume acknowledges that government and businesses also need to do more to make America's cities livable. Amin *et al.*'s *Cities for the Many not the Few* (2001) offers an alternative urban vision from the perspective of British radical geography, and argues for a practical realisation of Lefebvre's 'right to the city' by making the democratic city a reality. Susan Fainstein also endorses Amsterdam as an alternative urban future, though with some qualifications (Fainstein, 2000). Manuel Castells' lecture on Sustainability and the Information City published in *City* in April 2000 (Castells, 2000b) is an inspiring call to make cities better places for children and, indeed, for all of us. The UNESCO Growing Up in Cities publication *Creating Better Cities with Children and Youth* has practical suggestions on how children and young people can be involved in determining their own urban environment (Driskell, 2002).

The majority urban world

Jeremy Seabrook's *In the Cities of the South* (1996) is a powerful account of the very different types of urban experience found outside the west. Michael Peter Smith's *Transnational Urbanism* (2001) takes the study of 'third world urbanism' on to a new plane by radically questioning traditional assumptions about the geography of uneven development. For a good general discussion of the city and uneven development see Savage and Ward ([1993] 2003), also excerpted in Le Gates and Stout (various editions). J. John Palen's *The Urban World* (2001) is a student-friendly introduction to urban sociology and urbanisation around the world. David Drakakis-Smith's *Third World Cities* (2000) provides analyses of developing cities from around the world including Bangkok, Delhi, Manila, Mexico City, Singapore and Zimbabwe and looks at various issues affecting such cities including population growth, environmental problems, human rights, and planning and urban management.

A book that stimulates the imagination as well as informs, is the Harvard Design School's *Great Leap Forward* (Harvard Design School, 2001). This is a beautifully illustrated volume full of fascinating accounts of the Pearl River Delta's spectacular transformation – though the copyrighting of commonplace terms and slogans throughout the volume can be off-putting. Although not exclusively concerned with urban life in the developing world, Graham and Marvin's *Splintering Urbanism* (2001) is a path breaking approach to the study of the twenty-first century city.

Virtual cities

A longer version of Aurigi and Graham (2000) on urbanisation and cybercities can be found in *City* 7, (1997) 18–39. On the private world of virtual cities, Bruce Damer's *Avatars! Exploring and Building Virtual Worlds on the Internet* (1998) is a good reference source. 'Cyberbooster' accounts of virtual cities include Negroponte (1995) and Mitchell (1995, 2000). The definitive sci-fi accounts of cyberspace are William Gibson's *Neuromancer* ([1984] 2001) and Philip K. Dick's *Do Androids Dream of Electric Sheep?* ([1968] 1999), the novel that inspired Ridley Scott's movie, *Blade Runner*.

GLOSSARY

CID Common interest development. A private housing scheme in which the land and infrastructure is held in common by the owners of each housing unit. Management of CIDs can be elected representatives of the owners or can be delegated to professional managers or lawyers. Such schemes are often weakly regulated by outside authorities although a number of US states have introduced laws tightening the rules regarding their operation.

commodified space The transformation of land, territories, waterways, etc. into marketable assets (exchange value). A practice predominantly, though not exclusively, associated with capitalist type economies.

dialectic A process for arriving at truth by the statement of a thesis, which is then counter-posed by its antithesis, and finally resolved as a coherent synthesis. This is also the causative process of historical change in Hegel's philosophy, as well as that of Marx, except that the latter sees dialectical oppositions as being generated by antagonistic class forces.

embourgeoised The conversion of an environment, a belief, or a social group to a middle-class or bourgeois character.

epistemology The theory of knowledge and systems of thought.

flânerie The activities of a *flâneur*.

flâneur A term popularised by Walter Benjamin to describe someone who engages in the drama of city life, a seeker after the hidden truths, mysteries and pleasures of the urban experience.

Fordist/Fordism A system of industrial production associated with the founder of the Ford Motor Company, Henry Ford, in which mass production, within a single company, on a large site is typical, and where the workforce normally resides within a short distance of the factory gates. The concentration of specific types of manufacture in particular towns or cities is another feature associated with Fordism because of the need to concentrate specialist suppliers in the same district and to exploit the benefits of economies of scale. See also **post-Fordism**.

Frankfurt School Also known formally as the Institute for Social Research founded by Max Horkheimer in Frankfurt-am-Main, Germany in 1923. Later, the Institute transferred to the US and included figures such as Theodor Adorno, Herbert Marcuse and Ernst Bloch. Its members often took different positions on important philosophical and aesthetic questions, but 'the common language' of the group derived from their concern to develop a 'critical theory' from the dialectical materialism of Marxism and its antecedents in the work of Hegel and Kant. The 'later Frankfurt School' is particularly associated with the figure of Jürgen Habermas whose work has focused on universal themes such as legitimacy, authority and communicative action.

historical materialism Marxist theory of social and economic relations which sees the material exploitation of one class by another as the animating principle of productive relations and historical change. The long term effects of the unequal distribution of labour value (surplus value) will lead eventually to the crisis, breakdown and eventual replacement of exploitative modes of production until eventually the classless society of pure communism is achieved (the intermittent stages being primitive communism, slavery, feudalism, capitalism and socialism). In some formulations, historical materialism is an equivalent term to dialectical materialism. See also **dialectic**.

historicist   An explanation of social phenomena that sees historical conditions and context as determining structure and agency.

immanent critique   A method of social analysis associated with Marxism and, in particular, the Frankfurt School of Critical Theory (q.v.), which seeks to reveal the radical possibilities of social transformation through a critical and dialectical (q.v.) engagement with actually existing systems of power and domination (the theological etymology meaning 'of this world'). Less systematic variations of this critical method have been adopted by a broad range of radical philosophies including feminism, deconstruction, post-structuralism and post-colonialism.

new urbanism   See **TND**

oeuvre   A term devised by Henri Lefebvre to describe the city in all its aspects – real and imaginary, concrete and symbolic.

ontological   Relating to the nature of being.

post-Fordism   A type of industrial production associated with a transition from 'vertical' to 'horizontal' manufacturing processes whereby larger firms typically subcontract part or all of the manufacturing process to specialist (mostly) smaller firms. Costs are kept low by keeping stock levels to a minimum through 'just in time' delivery processes and by prioritising low labour costs over the proximity of labour. Due to lower transport and communication costs, not only manufacturing industry but also the service-based economy has found it possible to improve profitability by shifting labour-intensive activities to the newly industrialising countries where wage levels are generally lower than in the western industrialised nations.

smart growth   A planning movement that originated in the US which advocates tighter planning guidelines, a shift to re-development of traditional cities and suburbs rather than arable and forest land, a diversion of subsidies away from private to public transportation and less federal inducements for sprawl (such as highway subsidies and tax breaks for those raising mortgages on newly built housing). Advocates of smart growth include the new urbanist (q.v.) architects Duany Plater-Zyberg.

sociation   Associated with the social theory of Georg Simmel, it is the process in which individuals grow together into a unity and within which their interests are collectively realised.

TND   Traditional Neighbourhood Design (also known as Traditional Neighbourhood Development) – an architectural design and town planning movement associated with the 'new urbanist' school of architects such as Duany Plater-Zyberg and Leon Krier that celebrates historic building styles and village style arrangements of housing, retail and services. Typical examples include Seaside in Florida and Poundbury in Dorset, England.

transduction   A term used by Henri Lefebvre to describe the process of constructing a theoretical object (such as a new city) on the basis not only of information relating to current urban reality, but by the problematic posed by other possible urban realities. As new theoretical imaginings are given concrete form, the feedback mechanism ensures that subsequent projections/projects are informed by this newly altered reality in an endless loop of speculation-investigation-critique-implementation.

trope   Figurative use of language or systems of meaning – i.e. 'metaphorical', 'ironical', etc.

NOTES

2 THE FOUNDATIONS OF URBAN THEORY

1 Weber was not alone in failing to develop a distinct 'urban theory' as Peter Saunders points out, Durkheim and Marx were also more interested in the changing basis of social relations under capitalism than assessing the uniqueness and distinctiveness of the urban experience per se (Saunders, 1981: 12).

2 Prebendary derives from the Latin term for a living or a pension and refers to the income paid to individual clergy from the income of church property. Here, Weber is using the term in relation to the form of support available to Muslim clerics or Hindu priests. The point being that eastern religious elites were dependent on fixed land-holdings, whereas by the early Middle Ages, European feudal elites were increasingly able to liquidise their capital assets.

3 Benjamin was certainly aware of Simmel's work and refers approvingly to a passage in *The Philosophy of Money* where Simmel makes a 'distinction between the concept of culture and the spheres of autonomy in classical Idealism' when he writes: 'It is essential that the independent values of aesthetic, scientific, ethical,. . . and even religious achievements be transcended, so that they can all be integrated as elements in the development of human nature beyond its natural state' (Georg Simmel (1900) *Philosophie des Geldes*, Leipzig: 476–7 in Walter Benjamin, *The Arcades Project*, Konvolut N, 'On the Theory of Knowledge, Theory of Progress' (Benjamin, 1999b: 480)).

4 Benjamin met Lacis, a revolutionary theatre director from Riga, during a holiday in Capri in 1924. The two were to become lovers and Benjamin intended to marry her once his unhappy union to his German wife had been dissolved, but Lacis returned to the USSR and she subsequently spent ten years in a labour camp as a result of Stalin's purges.

5 Here, however, Benjamin is probably also referring to the Italian towns associated with the seven deadly sins where Milan is to greed what Naples is to indolence or Genoa to pride. Benjamin, 'Naples' in *One-Way Street* (Benjamin, 1977: 173).

6 In all, there were thirty-six files compiled on different themes relating to city life. Some referred to significant writers (such as Baudelaire, Saint-Simon, and Jung), socio-psychological themes (such as the figure of the *flâneur*, 'Social Movement', 'idleness', boredom), institutions (the Stock Exchange, the École Polytechnique) or physical features (the Seine, Railroads, the Streets of Paris) and perception (Mirrors, Panorama, Modes of Lighting).

7 Marcel Proust (1939) *Du Côté de chez Swann*, Vol. 1, Paris, 256 in Benjamin, *The Arcades Project*, 420.

8 Although thoroughly acquainted with Bloch and Adorno's work, Benjamin only began his reading of *Das Kapital* in the last few years of his life according to Susan Sontag ('Introduction' in Benjamin, 1977: 18). Benjamin's admiration for, and critical appreciation of, Marx is nevertheless attested by his dedication of an entire section to Marx's writings in the *Passagen-Werk*.

9 One thinks here of Philip Glass's powerful score for Godfrey Regio's 1982 movie *Koyaanisqatsi*.

3 THE CITY DESCRIBED

10 While the authorship of the pamphlet is frequently attributed to Mearns alone, it is now accepted that *The Bitter Cry* was, in fact, written by W.C. Preston (who had been a newspaper editor in Lancashire before moving to London), although Mearns is believed to have conducted the fieldwork, along with the Reverend James Munro (Chaloner in Hill, 1970: 2).

11 Riis notes that in Crosby Street and Mulberry Street it was not uncommon to find seventy or eighty children

living in a four to six storey tenement building with a corresponding number of adults.

12　But Riis' prejudices are often not lacking in sympathy. Writing on the plight of the Polish and Russian Jewish immigrants fleeing persecution from Europe, Riis observes: 'The curse of bigotry and ignorance reaches half-way across the world, to sow its bitter seed in fertile soil in the East Side tenements. If the Jew himself was to blame for the resentment he aroused over there, he is amply punished. He gathers the first-fruits of the harvest here' (Riis, 1891: 123).

13　Indeed, the later discovery of Charles Booth's unpublished notebooks proved to be 'far more engaging than the published text', and were, 'full of interesting detail . . . display[ing] a sensitivity to the language and sentiments of working people' (Englander in Englander and O'Day, 1995: 132).

14　This is not to say that the general view has gone unchallenged. For example, L.H. Lofland studied the titles of doctoral and masters dissertations between 1915 and 1935 in the Department of Sociology and Anthropology at the University of Chicago and concluded that 'the heritage of Chicago . . . is the virtual absence of a specifically urban sociology' (Lofland, 1983: 105 in Harvey, 1987: 115). But, as Harvey points out, this hypothesis relies on a narrow interpretation of urban sociology and no detailed investigation of the content of the research. But even if her argument did stand closer scrutiny, it is surely the esteem and the frequency of reference to Chicago sociologists by later practising urban sociologists that is the true test of the contribution of 'the Chicago School' to the sub-discipline.

15　A separate Sociology Department was founded in 1929. See Wax, 2000: 70.

16　However, this is not 'philosophy as praxis' favoured by Karl Marx, but the 'philosophy of pragmatism' formulated by the Chicago philosopher George Herbert Mead and contemporaries such as William James and John Dewey.

4　VISIONS OF UTOPIA

17　'[Y]et I freely confess that in the Utopian commonwealth there are very many features that in our own societies I would wish rather than expect to see' (More, 1995).

18　The term Euclidian is a reference not to the celebrated Greek geometer but to a decision of the US Supreme Court in *Village of Euclid, Ohio vs. Amber Realty Co.*, 272 US 365 (1926). The judgment gave sanction under the US Constitution to the restriction of development according to zones of activity and building type. See Mitchell-Weaver, 2000: 853.

19　The 'Thirteen Points' are available at http://www. newurbanist.com/newurban.htm (accessed July 2002).

20　Named after the hand pulled truck (charrette) that toured the studios of apprentice painters collecting material for the final year exhibition at the École des Beaux-Arts in Paris. If your work was not in the charrette it did not get exhibited – hence, the frantic productive activity of the students in the week before its arrival.

21　Robert Steuteville, NewUrbanNews, online edition (n.d.) http/www.newurbannnews.com/AboutNew Urbanism.html

22　See Chapter 7, 'public choice and urban politics' for a fuller discussion of the implications of utility theory for residential segregation and municipal fragmentation.

23　Nearly two-thirds of British 18 to 24 year olds are attracted to the idea of living in a village style community surrounded by a security perimeter according to the Royal Institution of Chartered Surveyors (Martin Wainwright, *The Guardian*, 28 November 2002).

24　The movie begins with the discovery of a detached human ear smothered by ants. This macabre symbol links the fate of the characters in this typical American small town to the horrific criminality of the nearby city.

5　BETWEEN THE SUBURB AND THE GHETTO

25　See note 18, chapter four.

26　Mumford's prescription for regional development makes this explicit when he calls for the re-grouping and nucleation of population *and* the decentralisation of industry.

27　William H. 'Holly' Whyte was a colleague of Jane Jacobs on *Architecture Forum* and was instrumental in securing a publisher for *The Death and Life of Great American Cities* (see Chapter 4), a book that developed from her original contribution 'Downtown is for People' to *The Exploding Metropolis*.

28　Garreau sets five conditions for an area to be deemed a genuine edge city – (i) It must have five million square feet or more of leasable office space, (ii) it must have 600,000 square feet or more of leasable retail space, (iii) it must have more jobs than bedrooms – i.e. it must be a working destination, (iv) it must be perceived by its population as one place, and (v) it must have been nothing like a city thirty years previously – i.e. it should not be the result of add on sprawl (Garreau, 1991: 6–7).

29　The Internet – a medium essential to the communication strategies of edge city dwellers and cosmocrats –

is also stimulating the demand for these facilities as some virtual relationships will eventually result in face to face encounters in real places such as convention centres and airport hotels (Mitchell, 2000: 91).

30 Anna Minton, 'Utopia Street', *The Guardian*, 27 March 2002.

31 Though Smith particularly contests Hamnett's distinction between his and David Ley's approach to gentrification, particularly in their more recent work (Smith, 1992: 111).

32 Segregation is calculated using the Index of Dissimilarity on a range of 0 to 1. A score of zero would indicate that groups are distributed exactly in proportion to their general population in a given area. A score of 1 would indicate that the entire population was located in one area and not at all in any others. Scores of 0.2 and under are considered relatively low and indicate a broad population dispersal, scores of 0.6 and above would be considered high and indicate high levels of population concentration or segregation (See Logan, 2000: 185, note 1, also Peach, 1996: 218). In the case of African Americans and Afro Caribbeans in the New York Metropolitan Area census for 1990, scores of segregation were high for every ethnic group other than between 'white' ethnic populations (English, German, Irish, Italian and Russian), with 'black/white' segregation the highest at 0.83 and 0.84 degrees of dissimilarity (ibid.: 178).

33 A gated community is defined by Blakely and Snyder as a place where 'some citizens secede from public contact' (1997: 3). According to the authors as many as half a million Californians were living in walled enclaves in the early 1990s.

34 Although it would be misleading to suggest that the highest murder rates are always to be found in the largest cities. For example, the New York Metropolitan Survey Area with a total population of 8 million plus registered 7.2 homicides per 100,000 population in 2001 compared to Memphis, Tennessee, which registered a homicide rate of 15.3 with a population of less than 700,000. Source: Federal Bureau of Investigations Uniform Crime Reports, 2001.

35 US Census Bureau, Mapping Census 2000: The Geography of US Diversity Data Tables, Washington, DC, 7 December 2001.

6 URBAN FORTUNES

36 It was not until 1892 that the English edition appeared (Short, 1996: 25).

37 W.O. Henderson and W.H. Chaloner (eds) 'Introduction' to Friedrich Engels, *The Condition of the Working Class in England*, Oxford, Basil Blackwell, 1971, xiv–xv.

38 David MacLellan, 'Introduction' to Friedrich Engels, *The Condition of the Working Class in England*, Oxford University Press, Oxford, 1993, xix.

39 Interestingly, Walter Benjamin cites the same passage from Engels in Konvolut M (The Flâneur) in the Arcades Project where the latter laments the 'narrow self-seeking' individualism that the crowding of the great city engenders (W. Benjamin, (1999b): 427–8). Here, the parallels with Simmel's conception of the 'blasé attitude' and Durkheim's notion of 'anomie' are remarkably clear.

40 Cited in Frisby, 1988: 23.

41 Although it is more accurate to describe such places not as 'open spaces' but 'commodified public space' since ownership resides, in nearly every case, with the local or national state. Marxist urbanists such as Lojkine (1976: 122) are wrong, in my view, to argue that collectively provided services are not commodities. This error is a result of seeing collectively provided services as somehow 'market free', whereas a cursory glance at the privatisation policies of almost any western government in the last twenty years shows how profitably collectively provided 'public goods' such as railways, leisure services or prisons can be commodified.

42 The regulation approach to political economy is most associated with French economists such as Michel Aglietta (1976), Alain Lipietz and Robert Boyer, all of whom have worked at one time at the University of Paris. Their interpretation of the role of state institutions in economic management owes much to the structuralist Marxism of Louis Althusser and Nicos Poulantzas, and the belief that the most important function of the state at all levels is to rescue the capitalist economy from its periodic productions crises – especially following the decline of the Fordist manufacturing system. Regulationists now include a large number of scholars in Europe and America working across a range of disciplines with an interest in how different institutional, temporal and territorial contexts combine to produce different types of capitalist systems. For a good survey of the regulation approach see Jessop (1997).

43 See in particular Habermas, 1976.

44 Althusser was probably more important as a reference point for systematic Marxist analyses of society, but it would be wrong to suggest, as Szelenyi does that Maxist urbanists were uncritical endorsers of his philosophy (Hill, 1984: 125).

45 Castells, M. *City, Class & Power* (1978). See especially Chapter One, 'Urban Crisis, Political Process & Urban Theory' in which he argues, 'the urban crisis is a particular form of the more general crisis linked to the contradiction between productive forces and relations of production which are at the basis of the ecological stake' (Castells, 1978: 5).

46 Ruth Glass, 'Verbal Pollution', *New Society*, 29 September 1977, 667–8, cited in Katznelson, 1992: 103.

47 For a fascinating account of the shrinking of time-space horizons and its effect on globalisation since the fifteenth century, see Thrift (1996).

48 Africa's population in 1997 was 600 million, the per capita GDP was $US 315 per annum. Source, Habitat, 2001: 17.

49 It is of course true that communist societies experienced a degree of cosmopolitanism – for example in the education of students from Marxist African and Asian states in the universities of the former Soviet Union, or in the internal migration of non-Russian minorities to Moscow. However, such movements were tightly controlled by the state and did little to vary the cultural and political monoculture of the typical communist city.

7 THE CONTESTED CITY

50 Other notable urban regime theorists include Shefter (1985), and DiGaetano and Klemanski (1993) as well as Mollenkopf in his later work (1992).

51 'tall buildings can be a very efficient way of using land and can make an important contribution to creating an exemplary sustainable world city. In Central London they can offer a supply of premises suited to the needs of global firms – especially those in the finance and business services sector. More generally, they can support the strategy of creating the highest levels of activity at locations with the greatest transport capacity. Well-designed tall buildings can also be landmarks identifying their locations, and can contribute to regeneration' (Mayor of London, *Draft London Plan*, 2002: 249).

8 FROM PILLAR TO POST

52 A 'field' is 'a structured system of social positions' occupied by individuals or institutions (Jenkins, 2002: 85). Bourdieu also describes fields as configurations of forces that are defined in terms of their access to the four varieties of capital – economic, social, symbolic and cultural. The more of each aspect of capital an individual or institution has, the more it or s/he is able to organise and manage the field's borders and relations with other fields. 'Stronger' fields such as the field of power (or politics) are able to 'over-determine' other fields such as 'the field of philosophy'.

53 Stonewall was a bar in Greenwich Village, New York City, that had been frequently raided by the police as part of the city authorities' policy of suppressing well-known gay haunts. During one such raid in 1969, the clientele and their supporters decided to fight back and the ensuing disturbances lasted for three days. The anniversary is celebrated by the gay community around the world as the moment when gays and lesbians took a stand against the denial of their civil and human rights because of their sexual orientation.

54 The Los Angeles rebellion (Keil, 1998a) followed the acquittal of four white Los Angeles Police Department officers on 29 April 1992. The officers were accused of assaulting King after his arrest following a high-speed chase. Videotape shot by a local resident, George Holliday, appeared to directly implicate the arresting officers in the beating of King. The officers claimed that King's previous arrest record, and his suspected use of drugs after two discharges of a 50,000-volt stun gun failed to contain him, justified the use of extraordinary force. In the riots that followed 53 people died, 10,000 people were arrested, 2,300 people were injured, 1,000 buildings were burned down, and several thousand jobs were lost at a total cost to the city of over $1 billion. It was these events that prompted King's questioning appeal, 'Can we all just get along?'. The phrase has inspired a number of initiatives aimed at improving race relations in the US and, indeed, has passed into the common lexicon of mainstream media culture – usually in the revised negative interrogative form 'Can't we all just get along?'.

55 Richard Rodriguez, 'A Letter from the Future – Fifty Years After the L.A. Riots', in *Pacific News Service-JINN Magazine*, Issue No. 3.09, April–May, 1997.

56 The reference is taken from the 'Who We Are' section of the Pacific News Service website. http://www.pacificnews.org

57 If measured in terms of published items dealing with the argument or application of post-structuralist ideas. But that is not to say, as post-structuralism's many detractors argue, that ubiquity is any necessary indication of the validity of an argument.

58 Postmodernism is a generic term that is applied (often erroneously) to any cultural expression or critique that attempts to disrupt or challenge the claims of 'the Enlightenment project'. This could be crudely summarised as the belief in philosophical reason as the foundation of all thought and enquiry, that moral and logical truth must exist independently of the subject and, finally, that modernism should be celebrated for applying rationalism to our everyday life and for repudiating the conservatism of romanticism and traditionalism. Insofar as post-structuralism contests the idea of absolute, non-contingent truth it can be said to be 'postmodernist', but this is just one element in the kaleidoscope of milieus associated with the postmodern. Post-structuralism is, thus, better described

as radical hermeneutics (or critical interpretation), whereas postmodernism is also associated with heterodox styles in architecture and cultural production (from pastiche and collage in design and painting to sampling in music and 'magical realism' in literature). Postmodernism is also a sociological assertion that we live in a qualitatively different era to modernism where the old political and moral certainties and identities no longer shape our value systems.

9 PUTTING THE CITY IN ITS PLACE

59 All figures are from the official Shanghai government website www.shanghai.gov.cn

60 Although foreign firms are not able to directly own maquiladora within a 100-kilometre strip of the border, a foreign owned maquiladora may become a beneficiary of a trust established by a Mexican bank allowing it to lease the buildings and facilities and to repatriate any profits earned from the enterprise.

61 However, the added value for American companies derives almost entirely from low labour costs and customs exemptions since up to half of the components supplied to the maquiladora are actually sourced in Asia. Source: San Diego East County Economic Development Council.

62 One of the pioneers of Internet 'weather maps' is John Quarterman, founder of the company Matrix Systems (www.mids.org), which provides cartographic visualisations of the density of net traffic and server concentrations. A wonderful visualisation of the hyperconcentration of Internet networks into and out of the world's major 'information rich' cities has been developed by the Internet cartographers Tamara Munzner and K. Claffy using 3D Mbone maps. Arching tunnels of different thicknesses and colours are used to show the density and speed of traffic between different principal cities. See Martin Dodge, 'Internet Arcs Around the Globe', *Mappa Mundi Magazine*, 1996, http://mappa.mundi.net/maps/maps_003.

63 William Gibson and Bruce Sterling bring these parallels to light in their steam-age allegory on the new electronic frontier in their novel *The Difference Engine* (1990).

64 The United Nations Human Settlement Programme (UN-Habitat) represents an attempt to provide such a coalition, but its focus is on issues such as urban governance, the provision of adequate housing, and the discouragement of unplanned urbanisation (all themes that emerged from the UN Habitat II Declaration in Istanbul in June 1996) rather than addressing the profound structural changes that cities are experiencing as a consequence of capitalist globalisation. If anything

the UN Declaration appears to see globalisation as an opportunity for 'forging public–private partnerships and strengthening small enterprises and microenterprises'. The declaration goes on to state that: 'Cities and towns hold the potential to maximize the benefits and offset the negative consequences of globalization'. This is not a view shared by many of the leading theorists of urban globalisation. See UN General Assembly, 'Declaration on Cities and Other Human Settlements in the New Millennium', 9 June 2001.

65 Its creators, Greg Roelofs and Pieter van der Meulen claim that Activeworld is the biggest and fasted growing virtual world on the net. See the Alphaworld FAQs page at http://mapper.activeworlds.com/aw/faq.htm#satellite. The Activeworlds website claims that as of 26 October 2001 Alphaworld users had placed more than 100 million building blocks on its virtual real estate using more than 3,000 objects and textures. Alphaworld is only one of the over 1,000 worlds hosted by Activeworlds.com, which is only one of many providers offering virtual world hosting. See Damer (1998).

66 Michael Safier pointed to no fewer than 30 cities around the world that had recently experienced or were experiencing collective cultural conflicts in the 1990s (Safier, 1996).

67 UN-Habitat/UNEP Press Release, 8 February 2000.

68 Victoria Transport Planning Institute, 2003. Based on reports by the Texas Transportation Institute which showed that in the 68 major urban regions in the US in 1999, congestion costs were $78 billion – equivalent to 4.8 billion hours of delay and 6.8 billion gallons of excess fuel consumed. The figure of $100 billion is arrived at by estimating the congestion costs for the areas not covered by the study. See David Schrank and Tim Lomax, *Urban Mobility Study*, Texas Transportation Institute, 2001.

69 See also *City* 5, 3, November 2001 and the special issue of the *International Journal of Urban and Regional Research* 26, 3, September 2002.

70 Sorkin and Zukin note that 95,000 jobs were estimated to have been lost as a result of the Twin Towers attack, but 75,000 jobs had been lost in the previous year. As a result 'city finances are teetering on the brink of fiscal meltdown' (Sorkin and Zukin, 2002: viii).

71 I am grateful to the Dutch urbanist Pieter Terhorst for providing this information. Another important element in Dutch urban policy is the fact that the central government contributes around 90 per cent of municipal expenditure and has a national policy of welfare equity that makes its cities far less subservient to footloose capital than American cities (see Chapter 7 of this volume) (Fainstein, 2000: 106).

72 The most well-known example is the Le Corbusier-style public housing project of Bijlmermeer located on the southeastern outskirts of the city. Here, about 40 per cent of the 53,000 residents are of Surinamian or Antilles origin, with Africans making up 17 per cent of the population. However, the Dutch government is concerned enough about the risk of Bijlmermeer taking on the characteristics of an American-style housing project that it is demolishing the most dilapidated tower blocks and replacing them with high-quality low-rise accommodation with the aim of producing a better income and ethnic balance in the area (Fainstein, 2000: 105).

73 This is not to say that life in Amsterdam is without problems. The tolerance of prostitution and soft drug use has attracted organised traffickers to the city, with the result that violent crime has increased in recent years. Amsterdam now has one of the highest homicide rates in Europe, though still half that of New York City. See Table 5.1.

74 In fact, America's foremost urbanist, Lewis Mumford, accepted this argument when he observed: 'The accretion of the debt structure in the great metropolises, the toppling pyramid of land values, make economic life precarious and effective social planning an impossibility. Hence, the real need is to deflate this burdensome structure as a deliberate public policy and to set up a responsible public body capable of directing the flow of investment into social channels and to liquidate with the least possible hiatus the present speculative structure' (Mumford, 1938: 391). Mumford was, of course, merely echoing the prophetic words of Henry George in *Progress and Poverty* (see Chapter 4).

75 For background on how the Toronto Charter came about see Keil and Young (2001).

76 Indeed, the former editors of the *International Journal of Urban and Regional Research* confess that their aim when re-launching the journal in 1977 as an inter-disciplinary project did not produce 'as ... much integrated interdisciplinary research as we had hoped'. Harloe, M., *et al.* 'IJURR: looking back twenty-one years later', *International Journal of Urban and Regional Research*, 22, 1, 1998, i–iv (ii).

BIBLIOGRAPHY

Abbott, A. (1999) *Department and Discipline. Chicago Sociology at One Hundred*, Chicago: University of Chicago Press.

Abu-Lughod, J. (ed.) (1994) *From Urban Village to East Village: The Battle for New York's Lower East Side*, Oxford: Blackwell.

Ackerman, D. (1990) *A Natural History of the Senses*, New York: Random.

Addams, J. (1895) *Hull-House Maps and Papers. A Presentation of Nationalities and Wages in a Congested District of Chicago*, New York: Library of Economics and Politics.

—— (1898) 'Why the Ward Boss Rules', *The Outlook* 58, 14 (April): 879–82.

—— (1909) *The Spirit of Youth and the City Streets*, New York: Macmillan.

—— (1930) *The Second Twenty Years at Hull-House*, New York: Macmillan.

—— (1967) *Twenty Years at Hull-House with Autobiographical Notes*, New York: Macmillan.

Aglietta, M. (1976) *Régulation et crise du capitalisme: l'expérience des Etats-Unis*, Paris: Calmann-Levy.

Alland, A. (1975) *Jacob A. Riis, Photographer and Citizen*, London: Gordon Frazer.

Allen, M. (1997) *Ideas That Matter. The Worlds of Jane Jacobs*, Owen Sound, Ontario: Ginger Press.

Almond, G.A. (1990) *A Discipline Divided: Schools and Sects in Political Science*, 1st edn, Newbury Park, CA: Sage.

American Planning Association (2002) *Planning for Smart Growth 2002. State of the States*, Washington, DC: APA.

Amin, A. (ed.) (1994) *Post-Fordism. A Reader*, Oxford: Blackwell.

—— and Thrift, N. (1992) 'Neo-Marshallian Nodes on Global Networks', *International Journal of Urban and Regional Research* 16, 4: 571–87.

—— and —— (2002) *Cities. Reimagining the Urban*, Cambridge: Polity Press.

—— , Massey, D. and Thrift, N. (2001) *Cities for the Many not the Few*, Bristol: Policy Press.

Anderson, E. (1993) 'Sex Codes and Family Life among Poor Inner-City Youths', in Wilson, W.J. (ed.) *The Ghetto Underclass. Social Science Perspectives*, Newbury Park, CA: Sage, 76–95.

Anderson, N. (1967) *The Hobo. The Sociology of the Homeless*, Chicago: University of Chicago Press.

Anderson, P. (1998) *The Origins of Postmodernism*, London: Verso.

Arrow, K.J. (1951) *Social Choice and Individual Values*, London: Chapman and Hall.

Atkinson, B. (1997) 'Jane Jacobs Author of Book on Cities, Makes the Most of Living in One', in Allen, M. (ed.) *Ideas that Matter. The Worlds of Jane Jacobs*, Owen Sound, Ontario: The Ginger Press, 52–3.

Atkinson, R. (2000) 'The Hidden Costs of Gentrification: Displacement in Central London', *Journal of Housing and the Built Environment* 15, 4: 307–26.

Augé, M. (1995) *Non-Places. Introduction to an Anthropology of Supermodernity*, London: Verso.

Aurigi, A. and Graham, S. (2000) 'Cyberspace and the City: The "Virtual City" in Europe', in Bridge, G. and Watson, S. (eds) *A Companion to the City*, Oxford: Blackwell, 489–502.

Bachrach, P. and Baratz, M.S. (1962) 'Two Faces of Power', *American Political Science Review* 56: 947–52.

Bailey, R.W. (1999) *Gay Politics, Urban Politics: Identity and Economics in the Urban Setting*, New York: Columbia University Press.

Banfield, E.C. (1961) *Political Influence*, Glencoe, IL: Free Press.

Banham, R. (2001) *Los Angeles. The Architecture of Four Ecologies*, Berkeley, CA: University of California.

Barclay, G. and Tavares, C. (2002) *International Comparisons of Criminal Justice Statistics 2000*, London: UK Home Office.

Barth, G. (1980) *City People: The Rise of Modern City Culture in Nineteenth-Century America*, New York and London: Oxford University Press.

Baruzi, J. (1988) 'L'Esthetique de Georg Simmel' (Georg Simmel's Aesthetic), *Sociétés* 19, Sept.: 3–5.

Baudrillard, J. (1997) 'The Beaubourg-Effect: Implosion and Deterrence', in Leach, N. (ed.) *Rethinking Architecture. A Reader in Cultural Theory*, London: Routledge, 210–24.

Beames, T. (1970) *The Rookeries of London*, London: Frank Cass.

Beauregard, R.A. (1985) 'Politics, Ideology and Theories of Gentrification', *Journal of Urban Affairs* 7, 4: 51–62.

—— (1986) 'The Chaos and Complexity of Gentrification', in Smith, N. and Williams, P. (eds) *Gentrification of the City*, Winchester, MA, and London: Allen and Unwin, 35–55.

—— (1995) 'Edge Cities: Peripheralizing the Center', *Urban Geography*, 16, 8: 708–21.

—— and Body-Gendrot, S. (eds) (1999) *The Urban Moment. Cosmopolitan Essays on the Late-20th Century City*, Urban Affairs Annual Reviews ed., Thousand Oaks, CA: Sage.

Bech, H. (1998) 'Citysex: Representing Lust in Public', *Theory, Culture and Society* 15, 3–4: 215–41.

Beevers, R. (1988) *The Garden City Utopia. A Critical Biography of Ebenezer Howard*, London: Macmillan.

Bell, C. and Newby, H. (1971) *Community Studies. An Introduction to the Sociology of the Local Community*, London: Allen and Unwin Ltd.

Bellamy, E. (1996) [1888] *Looking Backward, 2000–1887*, London and New York: Dover.

Bendix, R. (1998) *Max Weber: An Intellectual Portrait*, London: Routledge.

Benevolo, L. (1977) *La progettazione della città moderna*, Rome: Laterza.

Benfield, F.K., Terris, J. and Glendening, P. (2001) *Solving Sprawl: Models of Smart Growth in Communities Across America*, New York: Natural Resources Defence Council.

Benjamin, W. (1986) *Moscow Diary*, Cambridge, MA: Harvard University Press.

—— (1997) *One-Way Street*, London: Verso.

—— (1999a) *Illuminations*, London: Pimlico.

—— (1999b) *The Arcades Project*, Cambridge, MA, and London: Belknap Press/Harvard University Press.

Berman, M. (1982) *All that is Solid Melts into Air*, London: Verso.

Berry, B.J.L. (1985) 'Islands of Renewal in Seas of Decay', in Petersen, P. (ed.) *The New Urban Reality*, Washington, DC: The Brookings Institution, 69–96.

——, Portney, K. E. and Thomson, K. (1993) *The Rebirth of Urban Democracy*, Washington, DC: Brookings Institution.

Bevers, A.M. (1982) *Geometrie van de Samenleving: Filosofie en Sociologie in het Werk van Georg Simmel*, Deventer: Van Loghum Slaterus.

Bhabha, H. (1994) *The Location of Culture*, London: Routledge.

Bish, R.L. and Ostrom, V. (1973) *Understanding Urban Government. Metropolitan Reform Reconsidered*, Washington, DC: American Enterprise Institute for Public Policy.

Blakely, E.J. and Snyder, M.G. (1997) *Fortress America. Gated Communities in the United States*, Washington, DC: Brookings Institution Press.

Bondi, L. (1991) 'Gender Divisions and Gentrification. A Critique', *Transactions of the Institute of British Geographers* 16, 2: 190–8.

Bonney, N. (1996) 'The Black Ghetto and Mainstream America', *Ethnic and Racial Studies* 19, 1: 193–200.

Booth, C. (1902) *Life and Labour of the People in London*, London: Macmillan.

Borden, I., Rendell, J., Kerr, J. and Pivaro, A. (eds) (2000) *The Unknown City*, Cambridge, MA: MIT Press.

Boudreau, J.A. and Keil, R. (2001) 'Seceding from Responsibility? Secession Movements in Los Angeles', *Urban Studies* 38, 10: 1701–32.

Bourdieu, P. (1986) *Distinction*, London and New York: Routledge.

—— (2000a) *Habitus*, Aldershot: Ashgate.

—— (2000b) 'Social Space and Symbolic Space', in Robbins, D. (ed.) *Pierre Bourdieu*, Vol. 6, London: Sage, 3–16.

Boyer, C. (1983) *Dreaming the Rational City. The Myth of American City Planning*, Cambridge, MA: MIT Press.

Boyle, T.C. (1996) *The Tortilla Curtain*, New York: Penguin USA.

Brenner, N. (1997) 'Global, Fragmented, Hierarchical: Henri Lefebvre's Geographies of Globalization', *Public Culture* 10, 1: 135–67.

—— (1998a) 'Globalisation and Reterritorialisation: The Re-scaling of Urban Governance in the European Union', *Urban Studies* 26, 3: 431–51.

—— (1998b) 'Global Cities, Glocal States: Global City Formation and State Territorial Restructuring in Contemporary Europe', *Review of International Political Economy* 5, 1: 1–37.

—— (2000) 'The Urban Question as a Scale Question: Reflections on Henri Lefebvre, Urban Theory and the Politics of Scale', *International Journal of Urban and Regional Research* 24, 2: 361–79.

—— (2002) 'Decoding the Newest "Metropolitan Regionalism" in the USA: A Critical Overview', *Cities* 19, 1: 3–21.

Bressi, T. (2002) *The Seaside Debates*, New York: Rizzoli.

Bridge, G. and Watson, S. (eds) (2000) *A Companion to the City*, Oxford: Blackwell.

—— and —— (2002) *The Blackwell City Reader*, Oxford: Blackwell.

Briggs, A. (1982) *Cities and Countrysides. British and American Experience*, Leicester: Leicester University Press.

—— (2001) *Michael Young. Social Entrepreneur*, Basingstoke: Palgrave.

—— and Macartney, A. (1984) *Toynbee Hall. The First Hundred Years*, London: Routledge and Kegan Paul.

Briggs, X. de Souza (1997) 'Social Capital and the Cities: Advice to Change Agents', *National Civic Review* 86, 2 (Summer): 111–18.

Brody, K.M. (1982) 'Simmel as a Critic of Metropolitan Culture', *The Wisconsin Sociologist* 19: 75–83.

Brotchie, J.F. *et al.* (eds) (1995) *Cities in Competition. Productive and Sustainable Cities for the 21st Century*, Melbourne: Longman Australia.

Brown, K. (1981) 'Race, Class and Culture: Towards a Theorization of the "Choice/Constraint" Concept', in Jackson, P. and Smith, S.J. (eds) *Social Interaction and Ethnic Segregation*, London and New York: Academic Press, 185–203.

Brownill, S. (2000) 'Regen(d)eration: Woman and Urban Policy in Britain', in Darke, J., Ledwith, S. and Woods, R. (eds) *Woman and the City: Visibility and Voice in Urban Space*, Basingstoke: Palgrave, 114–29.

Bryan, M.L. and Davis, A.F. (1990) *100 Years at Hull-House*, Bloomington, IN: Indiana University Press.

Buchanan, J.M. (1957) 'Social Choice, Democracy, and Free Markets', *Journal of Political Economy* 62, 2: 114–23.

—— (1965) 'An Economic Theory of Clubs', *Economica* 32, 125: 1–14.

Buck-Morss, S. (1989) *Dialectics of Seeing: Walter Benjamin and the Arcades Project*, Cambridge, MA: MIT Press.

Bulmer, M. (1984) *The Chicago School of Sociology. Institutionalization, Diversity, and the Rise of Sociological Research*, Chicago and London: University of Chicago Press.

Burgess, E.W. (1925) 'The Growth of the City: An Introduction to a Research Project', in Park, R.E., Burgess, E.W. and McKenzie, R.D. *The City*, Chicago: University of Chicago Press, 47–62.

—— (1938) 'Personality Traits of the Brothers', in Shaw, C. *et al.*, *Brothers in Crime*, Chicago: University of Chicago Press, 326–35.

Butler, J. (1999) *Gender Trouble. Feminism and the Subversion of Identity*, London and New York: Routledge.

Butler, O. (1993) *Parable of the Sower*, New York: Four Walls Eight Windows.

Butler, T. and Robson, G. (2001) 'Social Capital, Gentrification and Neighbourhood Change in London: A Comparison of Three South London Neighbourhoods', *Urban Studies* 38, 12: 2145–62.

Byrne, D. (2001) *Understanding the Urban*, Basingstoke and New York: Palgrave.

Caldeira, T.P. (1996) 'Fortified Enclaves: The New Urban Segregation', *Public Culture* 8, 2: 303–28.

Calthorpe, P. (1993) *The Next American Metropolis*, New York: Princeton Architectural Press.

Cano, A. (1999) 'The Cardboard Door', in Kamel, R. and Hoffman, A. (eds) *The Maquiladora Reader*, Philadelphia, PA: American Friends Service Commission, 9–13.

Carey, J.T. (1975) *Sociology and Public Affairs. The Chicago School*, Beverly Hills, CA, and London: Sage.

Carr, J.B. and Feiock, R.C. (1999) 'Metropolitan Government and Economic Development', *Urban Affairs Review* 34, 3: 476–88.

Carr, J.H. and Megbolugbe, I.F. (1993) *HMDA Data and Mortgage Market Research*, Washington, DC: Fannie Mae Research Roundtable Series.

Carson, M. (1990) *Settlement Folk. Social Thought and the American Settlement Movement, 1885–1930*, Chicago: University of Chicago Press.

Carter, D. (1997) 'Digital Democracy or Information Aristocracy?', in Loader, B. (ed.) *The Governance of Cyberspace*, London: Routledge, 136–52.

Castells, M. (1968) 'Y a-t-il une sociologie urbaine?', *Sociologie du Travail* 1: 72–91.

—— (1972) *La question urbaine*, Paris: Maspero.

—— (1974) *Monopolville. Analyse des rapports entre l'entreprise, l'Etat et l'urbain à partir d'une enquête sur la croissance industrielle et urbaine de la région de Dunkerque*, Paris: Mouton.

—— (1977) *The Urban Question*, London: Edward Arnold.

—— (1978) *City, Class and Power*, London: Macmillan.

—— (1983) *The City and the Grassroots*, London: Edward Arnold.

—— (1989) *The Informational City. Information Technology, Economic Restructuring, and the Urban Regional Process*, Oxford: Basil Blackwell.

—— (1994) 'European Cities, the Informational Society, and the Global Economy', *New Left Review* 204 (March–April): 18–32.

—— (1996) *The Rise of the Network Society*, 1st edn, Oxford: Blackwell.

—— (1997) *The Power of Identity*, Vol. 2 of *The Information Age*, Oxford: Blackwell.

—— (1999) *End of Millennium*, Vol. 3 of *The Information Age*, Oxford: Blackwell.

—— (2000a) *The Rise of the Network Society*, 2nd edn, Vol. 1 of *The Information Age*, Oxford: Blackwell.

—— (2000b) 'Urban Sustainability in the Information Age', *City* 4, 1: 118–22.

—— and Borja, J. (1997) *Local and Global. Management of Cities in the Information Age*, London: Earthscan.

—— and Hall, P. (1994) *Technopoles of the World. The Making of 21st Century Industrial Complexes*, London: Routledge.

Cavan, R. (1983) 'The Chicago School of Sociology, 1918–1933', *Urban Life* 11, 4: 407–20.

Caygill, H. (1998) *Walter Benjamin. The Colour of Experience*, London: Routledge.

——, Coles A. and Klimowski, A. (with Appignanesi, R.) (1998) *Walter Benjamin for Beginners*, Duxford: Icon.

Celebration Company (2001) 'Innovative Town Focuses Spotlight on Community Building', 30 October. Available on-line at http://www.celebrationfl.com/press_room/011030.html (accessed 27 March 2003).

Chauncey, G. (1994) *Gay New York. Gender, Urban Culture and the Making of the Gay Male World 1890–1940*, New York: Basic Books.

Cherry, G. (1996) *Town Planning in Britain since 1900: The Rise and Fall of the Planning Ideal*, Oxford: Blackwell.

Chowdhury, T. and Furedy, C. (1994) *Urban Sustainability in the Third World: A Review of the Literature. Issues in Urban Sustainability* 5, Winnipeg: Institute of Urban Studies, University of Winnipeg.

Ciucci, G., Dal Co, F., Maniera-Elia, M. and Tafuri, M. (1979) *The American City*, Cambridge, MA: MIT Press.

Claes, T. (1994) 'Op wandel in de grootstad. Een schets van de moderne samenleving in Simmeliaans perspectief' (Living in a Metropolis. A Sketch of Modern Society from a Simmelian Point of View), *Tijdschrift voor Sociale Wetenschappen* 39, 4: 357–83.

Clapson, M. (1998) *Invincible Green Suburbs, Brave New Towns. Social Change and Urban Dispersal in Post-War England*, Manchester: Manchester University Press.

Clark, T.J. (1999) *Farewell to an Idea. Episodes from a History of Modernism*, New Haven, CT, and London: Yale University Press.

Clarke, D.B. (ed.) (1997) *The Cinematic City*, London: Routledge.

Clarke, S.E. and Gaile, G.L. (1998) *The Work of Cities*, Minneapolis, MN: University of Minnesota Press.

Clavel, P. and Kleniewski, N. (1990) 'Space for Progressive Local Policy: Examples from the United States and the United Kingdom', in Logan, J.R. and Swanstrom, T. (eds) *Beyond the City Limits. Urban Policy and Economic Restructuring in Comparative Perspective*, Philadelphia, PA: Temple University Press, 199–234.

Cockburn, C. (1977) *The Local State*, London: Pluto Press.

Cohen, R.B. (1981) 'The New International Division of Labour, Multinational Corporations and Urban Hierarchy', in Dear, M. and Scott, A. (eds) *Urbanization and Urban Planning in Capitalist Society*, London: Methuen, 287–315.

Congress of the New Urbanism (2003) 'Charter of the New Urbanism'. Available on-line at http://www.cnu.org/cnu_reports/Charter.pdf (accessed March 2003).

Copjec, J. (1999) 'The Tomb of Perseverance: On Antigone', in Sorkin, M. and Copjec, J. (eds) *Giving Ground. The Politics of Propinquity*, London: Verso, 233–66.

Cox, K.R. (ed.) (1997) *Spaces of Globalization. Reasserting the Power of the Local*, New York and London: Guilford Press.

Crawford, M. and Cenzatti, M. (1998): '"The Right to the City": Commodification and Decommodification of Public Space in Los Angeles', International Sociological Association Paper.

Crenson, M. (1971) *The Un-Politics of Air Pollution: A Study of Non-Decisionmaking in the Cities*, Baltimore, MD: Johns Hopkins University Press.

Cressey, P.G. (1932) *The Taxi-Dance Hall. A Sociological Study in Commercialized Recreation and City Life*, Chicago: University of Chicago Press.

Cutler, D.M., Glaeser, E.L. and Vigdor, J.L. (1997) *The Rise and Decline of the American Ghetto*, Cambridge, MA: National Bureau of Economic Research.

Dahl, R. (1961) *Who Governs?* New Haven, CT: Yale University Press.

Dahrendorf, R. (1995) *LSE. A History of the London School of Economics and Political Science*, Oxford: Oxford University Press.

Damer, B. (1998) *Avatars! Exploring and Building Virtual Worlds on the Internet*, Berkeley, CA: Beachpit Press.

Davidson, A. (1992) 'Henri Lefebvre', *Thesis Eleven* 33: 152–5.

Davis, A.F. (1967) *Spearhead for Reform: The Social Settlements and the Progressive Movement*, Oxford: Oxford University Press.

—— (1973) *American Heroine. The Life and Legend of Jane Addams*, New York: Oxford University Press.

Davis, J.S., Nelson, A.C. and Dueker, K.J. (1994) 'The new 'burbs: the exurbs and their implications for planning policy', *Journal of the American Planning Association* 60, 1: 45–59.

Davis, M. (1992) *City of Quartz. Excavating the Future of Los Angeles*, London: Vintage.

—— (2000a) *Magical Urbanism. Latinos Reinvent the US City*, London and New York: Verso.

—— (2000b) *Ecology of Fear. Los Angeles and the Imagination of Disaster*, London: Picador.

Dear, M. (ed.) (2000) *The Postmodern Urban Condition*, Oxford and Malden, MA: Blackwell.

—— (2002) *From Chicago to L.A. Making Sense of Urban Theory*, Thousand Oaks, CA, and London: Sage.

de Certeau, M. (1984) *The Practice of Everyday Life*, Berkeley: University of California Press.

Deegan, M.J. (1988) *Jane Addams and the Men of the Chicago School, 1892–1918*, New Brunswick, NJ: Transaction Books.

de Landa, M. (1997) *A Thousand Years of Non-Linear History*, New York: Zone Books.

Della Porta, D. and Andretta, M. (2002) 'Social Movements and Public Administrations: Spontaneous Citizens' Committees in Florence', *International Journal of Urban and Regional Research* 26, 2: 244–65.

Derrida, J. (1972) *Marges de la philosophie*, Paris: Les Editions de Minuit.

de Saussure, F. (1974) *Course in General Linguistics*, London: Fontana.

Dick, P.K. (1999) *Do Androids Dream of Electric Sheep?*, London: Millennium.

Dickens, P. (1990) *Urban Sociology. Society, Locality and Human Nature*, London: Harvester Wheatsheaf.

Dieleman, F.M. (1994) 'Social Rented Housing: Valuable Asset or Unsustainable Burden', *Urban Studies* 31, 3: 447–63.

DiGaetano, A. and Klemanski, J.S. (1993) 'Urban Regimes in Comparative Perspective: The Politics of Urban Development in Britain', *Urban Affairs Quarterly* 29, 1: 54–83.

Dililberto, G. (1999) *A Useful Woman. The Early Life of Jane Addams*, New York: Scribner/A Lisa Drew Book.

Doherty, R.J. (1981) *The Complete Photographic Work of Jacob A. Riis*, New York: Macmillan.

Domhoff, G. (1970) *The Higher Circles: The Governing Class in America*, New York: Random House.

—— (1978) *Who Really Rules? New Haven Community Power Re-Examined*, Santa Monica, CA: Goodyear.

Donald, J. (1992) 'Metropolis: The City as Text', in Bocock, R. and Thompson, K. (eds) *Social and Cultural Forms of Modernity*, Cambridge: Polity Press/Open University, 417–75.

Dowding, K. (1996) *Power*, Buckingham: Open University Press.

Downs, A. (1957) *An Economic Theory of Democracy*, New York: Harper and Row.

—— (1968) 'Alternative Futures for the American Ghetto', *Daedalus* XCVII, 4 (Fall): 1331–78.

Drakakis-Smith, D. (2000) *Third World Cities*, London: Routledge.

Dreier, P., Mollenkopf, J. and Swanstrom, T. (2001) *Place Matters. Metropolitics for the Twenty-First Century*, Lawrence, KA: University Press of Kansas.

Driskell, D.C. (2002) *Creating Better Cities for Children and Youth*, London: Earthscan.

Duany, A., Plater-Zyberg, E. and Speck, J. (2001) *Suburban Nation: The Rise of Sprawl and the Decline of the American Dream*, New York: North Point Press.

——, —— and —— (2003) *The Smart Growth Manual*, New York: McGraw-Hill Professional.

Duncan, N. (1996) 'Renegotiating Gender and Sexuality in Public and Private Spaces', in Duncan, N. (ed.) *BodySpace: Destabilising Geographies of Gender and Sexuality*, London and New York: Routledge, 127–45.

Duncan, S. and Goodwin, M. (eds) (1988) *The Local State and Uneven Development. Behind the Local Government Crisis*, Cambridge: Polity Press.

Dunleavy, P. (1980) *Urban Political Analysis. The Politics of Collective Consumption*, Basingstoke: Macmillan.

Durkheim, E. (1947) *The Division of Labor in Society*, Glencoe, IL: The Free Press.

Dutton, J.A. (2000) *New American Urbanism. Re-Forming the Suburban Metropolis*, Milan and London: Skira.

Eisenman, P., Krauss, R. and Tafuri, M. (1987) *House of Cards*, New York: Oxford University Press.

Elkin, S. (1987) *City and Regime in the American Republic*, Chicago: University of Chicago Press.

Ellin, N. (1996) *Postmodern Urbanism*, Oxford and Cambridge, MA: Blackwell.

Elliott, B. and McCrone, D. (1982) *The City: Patterns of Domination and Conflict*, London: Macmillan.

Elshtain, J.B. (ed.) (2002) *The Jane Addams Reader*, New York: Basic Books.

Engels, F. (1971) *The Condition of the Working Class in England*, 2nd edn, Oxford: Blackwell.

—— (1993) *The Condition of the Working Class in England*, Oxford: Oxford University Press.

Englander, D. (1995) 'Comparisons and Contrasts: Henry Mayhew and Charles Booth as social investigators', in Englander D. and O'Day R. *Retrieved Riches: Social Investigation in Britain 1840–1914*, Aldershot: Scholar Press, 105–42.

—— and O'Day, R. (1993) *Retrieved Riches: Social Investigation in Britain 1840–1914*, Aldershot: Scholar Press.

Evans, P. (2002) 'Introduction. Looking for Agents of Urban Livability in a Globalized Political Economy', in Evans, P. (ed.) *Livable Cities? Urban Struggles for Livelihood and Sustainability*, Berkeley, CA: University of California Press, 1–30.

Faderman, L. (1991) *Odd Girls and Twilight Lovers. A History of Lesbian Life in Twentieth-Century America*, New York: Columbia University Press.

Fainstein, N. and Fainstein, S. (1974) *Urban Political Movements. The Search for Power by Minority Groups in American Cities*, Urban Affairs Annual Reviews, Englewood Cliffs, NJ: Prentice-Hall.

—— and —— (eds) (1982) *Urban Policy Under Capitalism*, Vol. 22, Beverly Hills, CA, and London: Sage.

Fainstein, S. (1993) *The City Builders*, Oxford: Blackwell.

—— (1999) 'Can We Make the Cities We Want?', in Beauregard, R. and Body-Gendrot, S. (eds) (1999) *The Urban Moment. Cosmopolitan Essays on the Late-20th Century City*, Urban Affairs Annual Reviews, Thousand Oaks, CA: Sage, 249–72.

—— (2000) 'The Egalitarian City. Images of Amsterdam', in Deben, L., Heinemeijer, W. and van der Vaart, D. (eds) *Understanding Amsterdam. Essays on Economic Vitality, City Life and Urban Form*, Amsterdam: Het Spinhuis Publishers, 93–115.

—— and Campbell, S. (eds) (1996) *Readings in Urban Theory*, Oxford: Blackwell.

——, Gordon, I. and Harloe, M. (1992) *Divided Cities: New York and London in the Contemporary World*, Oxford: Blackwell.

Faris, R.E.L. and Dunham, H.W. (1939) *Mental Disorders and Urban Areas*, Chicago: University of Chicago Press.

Farrell, J.T. (1993) *Studs Lonigan. A Trilogy Comprising Young Lonigan, The Young Manhood of Studs Lonigan, and Judgement Day*, Urbana, IL, and Chicago: University of Illinois Press.

Feagin, J.R. (1998) *The New Urban Paradigm. Critical Perspectives on the City*, Lanham, MD: Rowman and Littlefield.

—— and Smith, M.P. (eds) (1987) *The Capitalist City. Global Restructuring and Community Politics*, Oxford and Cambridge, MA: Blackwell.

Ferguson, P.P. (1994) 'The Flâneur on and off the Streets of Paris', in Test, K. (ed.) *The Flâneur*, London and New York: Routledge, 22–42.

Ferrell, J. (2003) *Tearing Down the Streets: Adventures in Crime and Anarchy*, Basingstoke: Palgrave.

Fine, G.A. (1995) *A Second Chicago School? The Development of a Postwar American Sociology*, Chicago: University of Chicago Press.

Fisher, I.D. (1986) *Frederick Law Olmsted and the City Planning Movement in the United States*, Ann Arbor, MI: UMI Research Press.

Fishman, R. (1977) *Urban Utopias in the Twentieth Century. Ebenezer Howard, Frank Lloyd Wright, and Le Corbusier*, New York: Basic Books.

—— (1987) *Bourgeois Utopias. The Rise and Fall of Suburbia*, New York: Basic Books.

—— (2000) 'The American Planning Tradition: An Introduction and Interpretation', in Fishman, R. (ed.) *The American Planning Tradition*, Washington, DC: The Woodrow Wilson Center Press, 1–29.

Foucault, M. (1997a) *The Order of Things*, London: Routledge.

—— (1997b) 'Of Other Spaces: Utopia and Heterotopias', in Leach, N. (ed.) *Rethinking Architecture. A Reader in Cultural Theory*, London: Routledge, 350–6.

—— (2001) *Madness and Civilization*, London: Routledge.

—— (2002) *The Archaeology of Knowledge*, London and New York: Routledge.

Frantz, D. and Collins, C. (2000) *Celebration, USA. Living in Disney's Brave New Town*, New York: Henry Holt.

Fried, A. and Elman, R.M. (1969) *Charles Booth's London*, London: Hutchinson.

Friedmann, J. (1986) 'The World City Hypothesis', *Development and Change* 17, 10: 69–83.

—— (1995) 'Where We Stand. A Decade of World City Research', in Knox, P.L. and Taylor, P. (eds) *World Cities in a World System*, Cambridge: Cambridge University Press, 21–47.

—— and Wolff, G. (1982) 'World City Formation: An Agenda for Research and Action', *International Journal of Urban and Regional Research* 6, 3: 309–43.

Frisby, D. (1984) *Georg Simmel*, Chichester: Horwood.

—— (1985) *Fragments of Modernity. Theories of Modernity in the Work of Simmel, Kracauer, and Benjamin*, 1st edn, Cambridge: Polity Press.

—— (1987) 'The Ambiguity of Modernity: Georg Simmel and Max Weber', in Mommsen, W. (ed.) *Max Weber and His Contemporaries*, London: Allen and Unwin, 422–33.

—— (1988) *Fragments of Modernity. Theories of Modernity in the Work of Simmel, Kracauer, and Benjamin*, 2nd edn, Cambridge, MA: MIT.

—— (1992a) *Sociological Impressionism. A Reassessment of Georg Simmel's Social Theory*, 2nd edn, London: Routledge.

—— (1992b) *Simmel and Since. Essays on Georg Simmel's Social Theory*, London and New York: Routledge.

—— (1994) *Georg Simmel. Critical Assessments*, Vol. 3, London and New York: Routledge.

Froebel, V., Heinrichs, J. and Kreye, O. (1980) *The New International Division of Labour. Structural Unemployment in Industrial Countries and Industrialisation in Developing Countries*, Cambridge: Cambridge University Press.

Fujita, M., Krugman, P. and Venables, A.J. (1999) *The Spatial Economy. Cities, Regions, and International Trade*, Cambridge, MA: MIT Press.

Fulton, W. (1996) *The New Urbanism: Hope or Hype for American Communities?*, Cambridge, MA: Lincoln Institute of Land Policy.

Gambacorta, C. (1989) 'Experiences of Daily Life', *Current Sociology* 37, 1: 121–40.

Gandal, K. (1997) *The Virtues of the Vicious. Jacob Riis, Stephen Crane, and the Spectacle of the Slum*, Oxford: Oxford University Press.

Gans, H.J. (1962) *The Urban Villagers. Group and Class in the Life of Italian-Americans*, 1st edn, Glencoe, IL: The Free Press.

—— (1967) *The Levittowners. Ways of Life and Politics in a New Suburban Community*, London: Allen Lane/The Penguin Press.

—— (1982) *The Urban Villagers. Group and Class in the Life of Italian-Americans*, 2nd edn, London: The Free Press/Collier Macmillan Publishers.

—— (ed.) (1991) *People, Plans, and Policies: Essays on Poverty, Racism, and Other National Urban Problems*, New York: Columbia University Press.

Gärling, T. (1995) 'Introduction: How Do Urban Residents Acquire, Mentally Represent, and Use Knowledge of Spatial Layout?' in Gärling, T. (ed.) *Urban Cognition*, London: Academic Press.

Garreau, J. (1992) *Edge City: Life on the New Frontier*, New York: Anchor Books.

Garside, P. and Hebbert, M. (eds) (1989) *British Regionalism 1900–2000*, London: Mansell.

Gbah Bear, L. (1994) 'Miscegenations of Modernity: Constructing European Respectability and Race in the Indian Railway Colony, 1857–1931', *Women's History Review* 3, 4: 531–48.

Gee, M. (1997) 'Deciphering the Great Code', in Allen, M. (ed.) *Ideas that Matter. The Worlds of Jane Jacobs*, Owen Sound, Ontario: The Ginger Press, 159–60.

Genocchio, B. (1995) 'Discourse, Discontinuity, Difference: The Question of "Other" Spaces', in Watson, S. and Gibson, K. (eds) *Postmodern Cities and Spaces*, Oxford: Blackwell, 35–46.

George, H. (1976) [1880] *Progress and Poverty*, London: Dent.

Gephart, W. (1992) 'L'image de la modernité chez Georg Simmel' (The Image of Modernity in Georg Simmel), *Sociétés* 37: 267–78.

Gerth, H.H. and Mills, C.W. (eds) (1970) *From Max Weber. Essays in Sociology*, London: Routledge & Kegan Paul.

Gibson, W. (2001) *Neuromancer*, London: Voyager.

—— and Sterling, B. (1990) *The Difference Engine*, London: Gollancz.

Giddens, A. (1972) *Politics and Sociology in the Thought of Max Weber*, London: Macmillan.

—— (1979) *Central Problems in Social Theory. Action, Structure and Contradiction in Social Analysis*, Basingstoke: Macmillan.

—— (1984) *The Constitution of Society. Outline of the Theory of Structuration*, 1st edn, Cambridge: Polity Press.

—— (1994) 'Living in a Post-Traditional Society', in Beck, U., Giddens, A. and Lash, S. (eds) *Reflexive Modernization. Politics, Tradition and Aesthetics in the Modern Social Order*, Cambridge: Polity Press, 56–109.

Gilloch, G. (1996) *Myth and Metropolis. Walter Benjamin and the City*, Cambridge: Polity Press.

Glass, R. (1955) 'Urban Sociology in Great Britain', *Current Sociology* IV, 4: 5–76.

—— (1960) *Newcomers. The West Indians in London*, London: Centre for Urban Studies/Allen and Unwin.

—— (1966) 'Conflict in Cities', in Glass, R. (ed.) *Conflict in Society*, London: Churchill.

—— *et al.* (eds) (1964) *London: Aspects of Change*, London: Macgibbon and Kee.

Goldsmith, W.W. and Blakely, E.J. (1992) *Separate Societies. Poverty and Inequality in U.S. Cities*, Philadelphia, PA: Temple University Press.

Goodman, R. (1971) *After the Planners*, Harmondsworth: Penguin.

Gordon, D. (1977) 'Class Struggle and the Stages of Urban Development', in Watkins, A. and Derry, R. (eds) *The Rise of the Sunbelt Cities*, Beverly Hills, CA: Sage, 55–82.

—— (1984) 'Capitalist Development and the History of American Cities', in Tabb, W. and Sawers, L. (eds) *Marxism and the Metropolis*, 2nd edn, New York: OUP, 21–53.

Gottdiener, M. (1987) 'Crisis Theory and Socio-Spatial Restructuring: the US Case', in Gottdiener, M. and Komninos, N. (eds) *Capitalist Development and Crisis Theory: Accumulation, Regulation and Spatial Restructuring*, Basingstoke: Macmillan, 365–90.

—— (1992) 'Critique of Everyday Life', Vol. 1, *Sociology* 26, 3: 530–2.

—— (1993) 'A Marx for Our Time: Henri Lefebvre and The Production of Space', *Sociological Theory* 11, 1: 129–34.

—— (1994) *The Social Production of Urban Space*, 2nd edn, Austin, TX: University of Texas Press.

—— (1996) 'Alienation, Everyday Life, and Postmodernism as Critical Theory', in Geyer, F. (ed.) *Alienation, Ethnicity, and Postmodernism*, Westport, CT: Greenwood, 139–48.

—— (2000) 'Lefebvre and the Bias of Academic Urbanism. What Can We Learn from the "New" Urban Analysis?', *City* 4, 1: 93–100.

—— Collins, C.C. and Dickens, D. (1999) *Las Vegas: The Social Production of an All-American City*, Oxford: Blackwell.

Graham, S. (1999) 'Towards Urban Cyberspace Planning: Grounding the Global through Urban Telematics Policy and Planning', in Downey, J. and McGuigan, J. (eds) *Technocities*, London: Sage, 9–33.

—— (2002) 'Digital Space Meets Urban Place. Socio-technologies of Urban Restructuring in Downtown San Francisco', *City* 6, 3: 369–82.

—— and Marvin, S. (1996) *Telecommunications and the City. Electronic Spaces, Urban Places*, London and New York: Routledge.

—— and —— (2001) *Splintering Urbanism: Technology, Globalization and the Networked Metropolis*, London and New York: Routledge.

Gramsci, A. (1971) *Selections from the Prison Notebooks*, Hoare, Q. and Nowell-Smith, G. (eds) London: Lawrence & Wishart.

Green, S.F. (1997) *Urban Amazons. Lesbian Feminism and Beyond in the Gender, Sexuality, and Identity Battles of London*, Basingstoke: Macmillan.

Greer, S. (1962) *The Emerging City. Myth and Reality*, Glencoe, IL: Free Press.

Gregory, D. and Urry, J. (eds) (1985) *Social Relations and Spatial Structures*, Cambridge: Cambridge University Press.

Gregory, S. (1999) *Black Corona: Race and the Politics of Place in Urban Community*, Princeton, NJ: Princeton University Press.

Grosz, E. (1995) *Space, Time and Perversion: Essays on the Politics of Bodies*, London: Routledge.

Guerrero, C. (1997) 'Regional Development Strategies of a New Regional Government: the Junta de Andalucia, 1984–1992', *Progress in Planning* 48, 2: 67–160.

Guest, A.M. (1997) 'Robert Park and the Natural Area: A Sentimental Review', in Plummer, K. (ed.) *The Chicago School. Critical Assessments*, London: Routledge, 5–27.

Gurr, T.R. and King, D.S. (1987) *The State and the City*, Basingstoke: Macmillan.

Guzanova, A.K. (1997) *The Housing Market in the Russian Federation: Privatization and its Implications for Market Development*, Washington, DC: World Bank.

Habermas, J. (1976) *Legitimation Crisis*, London: Heinemann Educational Books.

—— (1989) *The Structural Transformation of the Public Sphere*, Cambridge, MA: MIT Press.

Habitat/United Nations Centre for Human Settlements (2001) *Cities in a Globalizing World. Global Report on Human Settlements*, London: Earthscan.

Hall, P. (1966) *The World Cities*, New York: McGraw-Hill.

—— (1985) *Urban and Regional Planning*, London: Allen and Unwin.

—— (1988) *Cities of Tomorrow: An Intellectual History of Urban Planning and Design in the Twentieth Century*, Oxford and Cambridge, MA: Blackwell.

—— (1998) *Cities in Civilisation. Culture, Innovation and Urban Order*, London: Weidenfeld and Nicolson.

—— and Ward, C. (1998) *Sociable Cities. The Legacy of Ebenezer Howard*, Chichester: J. Wiley.

Halpern, R. (1997) 'Respatializing Marxism and Remapping Urban Space', *Journal of Urban History* 23, 2: 221–30.

Hamel, P., Mayer, M. and Lustiger-Thaler, H. (eds) (2000) *Urban Movements in a Globalising World*, London: Routledge.

Hamnett, C. (1991) 'The Blind Men and the Elephant: The Explanation of Gentrification', *Transactions of the Institute of British Geographers* 16, 2: 173–89.

—— (1992) 'Gentrifiers or Lemmings? A Response to Neil Smith', *Transactions of the Institute of British Geographers* 17, 1: 116–19.

Hannigan, J. (1998) *Fantasy City: Pleasure and Profit in the Postmodern Metropolis*, London: Routledge.

Hansen, M. (1987) 'Benjamin, Cinema and Experience: The Blue Flower in the Land of Technology', *New German Critique* 40: 179–224.

Haraway, D.J. (1997) *Modest-Witness@Second-Millennium. FemaleMan-Meets-OncoMouse^{TM} Feminism and Technoscience*, New York and London: Routledge.

Harding, A. (1994) 'Urban Regimes and Growth Machines: Towards a Cross-national Research Agenda', *Urban Affairs Quarterly* 29, 1: 356–82.

—— (1997) 'Urban Regimes in a Europe of the Cities?', *European Urban and Regional Studies* 4, 4 (Oct.): 291–314.

—— (1999) 'North American Urban Political Economy, Urban Theory and British Research', *British Journal of Political Science* 29, 4: 673–98.

Hartman, C. (1979) 'Comment on "Neighborhood Revitalization and Displacement: A Review of the Evidence"', *Journal of the American Planning Association* 45, 4: 488–91.

Harvard Design School (2001) *Great Leap Forward*, Cologne: Taschen.

Harvey, D. (1973) *Social Justice and the City*, London: Edward Arnold.

—— (1975) 'The Geography of Capitalist Accumulation: A Reconstruction of Marxian Theory', *Antipode* 7, 2: 9–21.

—— (1982) *The Limits to Capital*, Oxford: Blackwell.

—— (1985) *Consciousness and the Urban Experience*, Oxford: Basil Blackwell.

—— (1989) *The Condition of Postmodernity*, Oxford: Basil Blackwell.

—— (2000) *Spaces of Hope*, Edinburgh: Edinburgh University Press.

—— (2002) 'Cracks in the Edifice of the Empire State', in Sorkin, M. and Zukin, S. (eds) *After the World Trade Center. Rethinking New York City*, New York: Routledge, 57–67.

Harvey, Lee (1987) *Myths of the Chicago School of Sociology*, Aldershot: Avebury.

Hausner, V.A. *et al.* (1987) *Urban Economic Change. Five City Studies*, Oxford: Clarendon Press.

Haussermann, H. (1995) 'Die Stadt und die Stadtsoziologie. Urbane Lebensweise und die Integration des Fremden' (City and the Sociology of Cities. Urban Lifestyles and the Integration of the Foreign), *Berliner Journal für Soziologie* 5, 1: 89–98.

Hayden, D. (2000) *Model Houses for the Millions. The Making of the American Suburban Landscape, 1820–2000*, Cambridge, MA: Lincoln Institute of Land Policy.

Hetherington, K. (1997) *The Badlands of Modernity: Heterotopia and Social Ordering*, London and New York: Routledge.

Hill, O. (1970) *Homes of the London Poor* (featuring a new impression of Mearns, A. *The Bitter Cry of Outcast London: An Inquiry into the Condition of the Abject Poor*, with a note on the authors by W.H. Chaloner), London: Frank Cass and Co.

Hill, R.C. (1984) 'Urban Political Economy: Emergence, Consolidation and Development', in Smith, M.P. (ed.) *Cities in Transformation: Class, Capital and the State*, London and Thousand Oaks, CA: Sage, 123–37.

Hommann, M. (1993) *City Planning in America. Between Promise and Despair*, Westport, CT: Praeger.

Hooghe, L. (1995) 'Subnational Mobilisation in the European Union', *West European Politics* 18, 3: 175–98.

—— and Keating, M. (1994) 'The Politics of European Union Regional Policy', *Journal of European Public Policy* 1, 3: 53–78.

Hook, D. and Vrdoljak, M. (2002) 'Gated Communities, Heterotopia and a "Rights" of Privilege: A Heterotopology of the South African Security-Park', *Geoforum* 33, 2: 195–219.

Howard, E. (1985) [1902] *Garden Cities of To-morrow*, Eastbourne: Attic Books.

HRH the Prince of Wales (1989) *A Vision of Britain. A Personal View of Architecture*, London: Doubleday.

—— (1998) 'Why I'm Modern but not a Modernist', *The Spectator*, 8 August.

—— (2001): 'Tall Buildings', speech given to the Invensys Conference, QE2 Centre, London, 11 December.

Humpherys, A. (1977) *Travels in a Poor Man's Country. The Work of Henry Mayhew*, Athens, GA: Georgia University Press.

Hunter, F. (1953) *Community Power Structure. A Study of Decision Makers*, Chapel Hill: University of North Carolina Press.

—— (1959) *Top Leadership, U.S.A.*, Chapel Hill, NC: University of North Carolina Press.

Hymer, S. (1972) 'The Multinational Corporation and the Law of Uneven Development', in Bhagwati, J. (ed.) *Economics and World Order from the 1970s to the 1990s*, London: Macmillan, 31–46.

Jackson, F. (1985) *Sir Raymond Unwin. Architect, Planner and Visionary*, London: Zwemmer.

Jackson, K.B. (1965) *Dark Ghetto. Dilemmas of Social Power*, New York: Harper and Row.

Jackson, K.T. (1985) *Crabgrass Frontier. The Suburbanization of the United States*, Oxford: Oxford University Press.

Jacobs, J. (1958) 'Downtown is for People', in Whyte, W.H. (ed.) *The Exploding Metropolis*, New York: Doubleday, 140–68.

—— (1986) *Cities and the Wealth of Nations*, Harmondsworth: Penguin.

—— (1992) *The Death and Life of Great American Cities*, New York: Vintage Books.

Jameson, F. (1998) *The Cultural Turn. Selected Writings on the Postmodern 1983–1998*, London: Verso.

—— (1999) 'The Theoretical Hesitation: Benjamin's Sociological Predecessor', *Critical Inquiry* 25, 2: 267–88.

Jaworski, G.D. (1997) *Georg Simmel and the American Prospect*, Albany, NY: State University of New York Press.

Jay, M. (1973) *The Dialectical Imagination. A History of the Frankfurt School and the Institute of Social Research 1923–1950*, Boston: Little Brown & Co.

Jencks, C. (1980) *Late-Modern Architecture and Other Essays*, London: Academy Editions.

—— (1987) *Le Corbusier and the Tragic View of Architecture*, Harmondsworth: Penguin.

—— (1988) *The Prince, the Architects, and the New Wave Monarchy*, London: Wiley-Academy.

—— (1991) *The Language of Post-Modern Architecture*, New York: Rizzoli.

—— (ed.) (1992) *The Post-Modern Reader*, London: Academy Editions.

—— (1993) *Heteropolis: Los Angeles, the Riots and the Strange Beauty of Hetero-architecture*, London: Academy Editions.

Jenkins, R. (2002) *Pierre Bourdieu*, revised edn, London: Routledge.

Jessop, B. (1997) 'Survey Article: The Regulation Approach', *Journal of Political Philosophy* 5, 3: 287–326.

—— (2000) 'Globalisation, Entrepreneurial Cities and the Social Economy', in Hamel, P., Lustiger-Thaler, H. and Mayer, M. (eds) *Urban Movements in a Globalising World*, London and New York: Routledge, 81–100.

—— and Sum, N.-L. (2000) 'An Entrepreneurial City in Action: Hong Kong's Emerging Strategies in and for (Inter)Urban Competition', *Urban Studies* 37, 12: 2287–313.

——, Peck, J. and Tickell, A. (1999) 'Retooling the Machine: Economic Crisis, State Restructuring, and Urban Politics', in Jonas, A.E.G. and Wilson, D. (eds) *The Urban Growth Machine. Critical Perspectives Two Decades Later*, Albany, NY: State University of New York Press, 141–59.

Joas, H. (1991) 'Berlin lieu et objet de la recherche sociologique' (Berlin as the Location and the Object of Sociological Research), *Critique* 47: 531–2, 643–54.

Jonas, S. (1991) 'La Metropolisation de la société dans l'oeuvre de Ferdinand Tönnies' (The Metropolitanization of Society in the Work of Ferdinand Tönnies), *Sociétés* 31: 79–86.

—— (1995) 'La Groszstadt-Metropole européene dans la sociologie des pères fondateurs allemands' (The European Metropolis in the Sociology of the German Founding Fathers), in Rémy, J. (ed.) *Georg Simmel: Ville et Modernité*, Paris: Editions L'Harmattan, 19–35.

Judd, D.R. and Kantor, P.B. (eds) (2001) *The Politics of Urban America*, New York: Longman.

Judge, D., Stoker, G. and Wolman, H. (1995) *Theories of Urban Politics*, London: Sage.

Kaesler, D. (1988) *Max Weber: An Introduction to His Life and Work*, Cambridge: Polity Press.

Kajaj, K. (1998) 'Stadtluft macht frei: ou la ville comme lieu de liberté et d'épanouissement personnel' (City Air Makes Free, or, the City as a Place of Liberty and Personal Expansion), *Revue des Sciences Sociales de la France de l'Est* 25: 28–34.

Kamel, R. and Hoffman, A. (1999) *The Maquiladora Reader. Cross-Border Organizing since NAFTA*, Philadelphia, PA: American Friends Service Committee.

Kantor, P. and Savitch, H.V. (2002) 'Can Politicians Bargain with Business? A Theoretical and Comparative Perspective on Urban Development', in Judd, D.R. and Kantor, P. (eds) *The Politics of Urban America. A Reader*, New York: Longman, 276–95.

Katznelson, I. (1981) *City Trenches. Urban Politics and the Patterning of Class in the United States*, Chicago and London: University of Chicago Press.

—— (1992) *Marxism and the City*, Oxford: Oxford University Press.

Keating, M. (1995) 'Size, Efficiency and Democracy: Consolidation, Fragmentation and Public Choice', in Judge, D., Stoker, G. and Wolman, H. (eds) *Theories of Urban Politics*, London: Sage, 117–34.

Keil, R. (1998a) *Los Angeles. Globalization, Urbanization and Social Struggles*, Chichester: J. Wiley and Sons.

—— (1998b) 'Globalization Makes States: Perspectives of Local Governance in the Age of the World City', *Review of International Political Economy* 5, 4: 616–46.

—— and Ronneberger, K. (2000) 'The Globalization of Frankfurt am Main: Core, Periphery and Social Conflict', in Marcuse, P. and van Kempen, R. (eds) *Globalizing Cities. A New Spatial Order?*, Oxford: Blackwell, 228–48.

—— and Young, D. (2001) *A Charter for the People? The Debate on Municipal Autonomy in Toronto*, Toronto: York University.

Kelley, R.D.G. (1998) *Yo' Mama's Disfunktional. Fighting the Culture Wars in Urban America*, Boston: Beacon Press.

Kennedy, L. (2000) 'Paranoid Spatiality: Postmodern Urbanism and American Cinema', in Balshaw, M. and Kennedy, L. (eds) *Urban Space and Representation*, London: Pluto Press, 116–28.

King, D. and Stoker, G. (eds) (1996) *Rethinking Local Democracy*, Basingstoke: Macmillan/ESRC Local Governance Programme.

Kipfer, S. (1998): 'On the Possibilities of the Urban: Rereading Henri Lefebvre's Open and Integral Marxism', International Sociological Association Paper.

Kish Sklar, K. (1985) 'Hull-House in the 1890s: A Community of Women Reformers', *Signs* 10 (Summer): 657–77.

Klein, Naomi (2001) *No Logo*, London: Flamingo.

Kling, R. and Poster, M. (eds) (1995) *Beyond the Edge: The Dynamism of Posturban Regions*, Berkeley, CA: University of California Press.

Knox, P.L. (1995) 'World Cities and the Organisation of Global Space', in Johnston, R.J., Taylor, P.J. and Watts, M.J. (eds) *Geographies of Global Change*, Oxford: Blackwell, 232–48.

—— and Taylor, P.J. (1995) *World Cities in a World System*, Cambridge: Cambridge University Press.

Kofman, E. and Lebas, E. (eds) (1996) *Henri Lefebvre: Writings on Cities*, Oxford: Blackwell.

Koolhaas, R. (1978) *Delirious New York. A Retroactive Manifesto for Manhattan*, London: Thames and Hudson.

—— and Harvard Project on the City (2001) *Mutations*, Bordeaux: Arc en Rêve Centre d'Architecture.

Korllos, T.S. (1988) 'The Physical City: A Reinterpretation of Wirth's Theory', North Central Sociological Association Paper.

Krier, L. (1988) 'God Save the Prince', *Modern Painters* 1, 2: 23–5.

—— (1998) *Architecture: Choice or Fate*, Windsor: Andreas Papadakis Publisher.

Krugman, P. (1991) *Geography and Trade*, Leuven: Leuven University Press.

Kunstler, J.H. (1993) *Geography of Nowhere*, New York: Simon and Schuster.

—— (2001) Interview with Jane Jacobs, *Metropolis Magazine*, March. Available on-line at http://www.kunstler.com/mags_jacobs1.htm (accessed September 2002).

Kurtz, L.R. (1984) *Evaluating Chicago Sociology. A Guide to the Literature, with an Annotated Bibliography*, Chicago and London: The University of Chicago Press.

LaFarge, A. (ed.) (2000) *The Essential William H. Whyte*, New York: Fordham University Press.

Lal, B.B. (1990) *The Romance of Culture in an Urban Civilization: Robert E. Park on Race and Ethnic Relations in Cities*, London: Routledge.

Landesco, J. (1968) *Organized Crime in Chicago*, Chicago: University of Chicago Press.

Landman, K. (2000) *Gated Communities: An International Review*, CSIR Building and Construction Technology, BP 449, BOU/I 186.

Landry, B. (1987) *The New Black Middle Class*, Berkeley, CA: University of California Press.

Lang, M.H. (1999) *Designing Utopia. John Ruskin's Urban Vision for Britain and America*, Montreal and London: Black Rose.

Lave Johnston, J. and Johnston Dodds, K. (2002) *Common Interest Developments: Housing at Risk?*, Sacramento, CA: California State Library Research Bureau.

Lebas, E. (1982) 'Urban and Regional Sociology in Advanced Industrial Societies', *Current Sociology* 30, 1: 1–264.

Leccese, M. and McCormick, K. (eds) (1999) *Charter of the New Urbanism*, New York: McGraw-Hill Professional.

Le Corbusier (1986) *Towards a New Architecture*, London and New York: Dover.

—— (1996) 'A Contemporary City' from 'The City of To-morrow and Its Planning', reprinted in Le Gates R.T. and Stout F. (eds) *The City Reader*, London: Routledge 368–75.

Lefebvre, H. (1968) *Le droit à la ville*, Paris: Anthropos.

—— (1972) *La pensée marxiste et la ville*, Paris: Castermann.

—— (1976) *The Survival of Capitalism. Reproduction of the Relations of Production*, London: Allison and Busby.

—— (1984) *Critique of Everyday Life*, London: Verso.

—— (1991) *The Production of Space*, Oxford: Basil Blackwell.

—— (2002) *Critique of Everyday Life*, Vol. 2, London: Verso.

—— (2003) *Critique of Everyday Life*, Vol. 3, London: Verso.

—— and Regulier, C. (1986) 'Essai de rythmanalyse des villes mediterranéennes' (An Essay on a Rhythm Analysis of Mediterranean Cities), *Peuples Mediterranéens* (Mediterranean Peoples) 37, Oct.–Dec.: 5–16.

Le Galès, P. (1999) 'Is Political Economy Still Relevant to Study the Culturalization of Cities?', *European Urban and Regional Studies* 6, 4: 293–302.

Le Gates, R.T. and Stout, F. (eds) (1996) *The City Reader*, 1st edn, London: Routledge.

—— and —— (eds) (1999) *The City Reader*, 2nd edn, London: Routledge.

—— and —— (eds) (2003) *The City Reader*, 3rd edn, London: Routledge.

Leslie, E. (1999) 'Space and West End Girls: Walter Benjamin versus Cultural Studies', *New Formations* 38: 110–24.

Levine, D.N., Carter, E.B. and Miller Gorman, E. (1976) 'Simmel's Influence on American Sociology', *American Journal of Sociology* 81: 813–45.

Ley, D. (1987) 'Reply: The Rent Gap Revisited', *Annals of the Association of American Geographers* 77, 3: 465–8.

Library of Congress, 'Frank Lloyd Wright: Designs for an American Landscape, 1922–1932', Madison Gallery, Library of Congress, 14 November 1996 to 15 February 1997. Available on-line at http://lcweb.loc.gov/exhibits/flw/flw.html (accessed September 2002).

Liggett, H. (2003) *Urban Encounters*, Minneapolis, MA: University of Minnesota Press.

—— and Perry, D.C. (1995) *Spatial Practices. Critical Explorations in Social/Spatial Theory*, Thousand Oaks, CA: Sage.

Lindner, B. (1986) 'The Passagen-Werk, the Berliner Kindheit, and the Archaelogy of the "Recent Past"', *New German Critique* 39: 25–46.

Lindner, R. (1996) *The Reportage of Urban Culture: Robert Park and the Chicago School*, Cambridge: Cambridge University Press.

Linn, J.W. (1935) *Jane Addams. A Biography*, New York and London: Appleton-Century Co.

Livingstone, K. (1987) *If Voting Changed Anything, They'd Abolish It*, London: Fontana/Collins.

Llewelyn-Davies Planning (1996) *Four World Cities. A Comparative Study of London, Paris, New York and Tokyo*, London: Comedia.

Lofland, L.H. (1971) *Analyzing Social Settings*, Belmont, CA: Wadsworth.

Logan, J.R. (2000) 'Still a Global City: The Racial and Ethnic Segmentation of New York', in Marcuse, P. and Van Kempen, R. (eds) *Globalizing Cities. A New Spatial Order?*, Oxford and Malden, MA: Blackwell, 158–85.

—— and Molotch, H. (1987) *Urban Fortunes: The Political Economy of Place*, Berkeley, CA: University of California Press.

—— and —— (1996) 'The City as a Growth Machine', in Fainstein, S. and Campbell, S. (eds) (1996) *Readings in Urban Theory*, Oxford: Blackwell, 291–337.

—— and Swanstrom, J. (eds) (1990) *Beyond the City Limits. Urban Policy and Economic Restructuring in Comparative Perspective*, Philadelphia, PA: Temple University Press.

Lojkine, J. (1976) 'A Marxist Theory of Capitalist Urbanisation', in Pickvance, C.G. (ed.) *Urban Sociology. Critical Essays*, London: Tavistock, 119–46.

—— (1977) *Le Marxisme, L'Etat et La Question Urbaine*, Paris: Presses Universitaires de France.

Lubeck, P.M. and Britts, B. (2002) 'Muslim Civil Society in Urban Public Spaces: Globalization, Discursive Shifts, and Social Movements', in Eade, J. and Mele, C. (eds) *Understanding the City. Contemporary and Future Perspectives*, Oxford: Blackwell, 305–35.

Lynch, K. (1960) *The Image of the City*, Cambridge, MA: Technology Press and Harvard University Press.

Lynd, R.S. and Lynd, H.M. (1929) *Middletown*, New York: Harcourt Brace.

—— and —— (1937) *Middletown in Transition*, New York: Harcourt Brace.

Lyotard, J.-F. (1979) *La condition postmoderne: rapport sur le savoir*, Paris: Les Editions de Minuit.

—— (1984) *The Postmodern Condition. A Report on Knowledge*, Manchester: Manchester University Press.

Mabin, A. (2000) 'Varied Legacies of Modernism in Urban Planning', in Bridge, G. and Watson, S. (eds) *A Companion to the City*, Oxford: Blackwell, 555–66.

MacCannell, D. (1999) '"New Urbanism" and its Discontents', in Copjec, J. and Sorkin, M. (eds) *Giving Ground. The Politics of Propinquity*, London: Verso, 106–28.

McDowell, L. (1999) *Gender, Identity and Place. Understanding Feminist Geographies*, Cambridge: Polity Press.

Mackenzie, E. (1994) *Privatopia: Homeowner Associations and the Rise of Residential Private Government*, New Haven, CT, and London: Yale University Press.

McNeill, D. (1999) 'Globalization and the European City', *Cities* 16, 3: 143–8.

Magnusson, W. (1996) *The Search for Political Space. Globalization, Social Movements and the Urban Political Experience*, Toronto and London: University of Toronto Press.

Marcuse, P. (1995) 'Not Chaos, but Walls: Postmodernism and the Partitioned City', in Watson, S. and Gibson, K. (eds) *Postmodern Cities and Spaces*, Oxford: Blackwell, 243–53.

—— and van Kempen, R. (eds) (2000) *Globalizing Cities. A New Spatial Order?*, Oxford: Blackwell.

Marshall, A. (2000) *How Cities Work: Suburbs, Sprawl, and the Roads not Taken*, Austin, TX: University of Texas Press.

Martínez-Herrera, E. (2002) 'From Nation-Building to Building Identification with Political Communities: Consequences of Political Decentralisation in Spain, the Basque Country, Catalonia and Galicia, 1978–2001', *European Journal of Political Research* 41: 421–53.

Marx, K. (1964) *Pre-Capitalist Economic Formations*, New York: International Publishers.

—— (1973) *Grundrisse*, Harmondsworth: Penguin.

—— (1974) [1846] *The German Ideology*, London: Lawrence and Wishart.

—— (1976) *Capital*, Volume 1, Harmondsworth: Penguin.

Massey, D. (1995) *Spatial Divisions of Labour. Social Structures and the Geography of Production*, 2nd edn, Basingstoke: Macmillan.

Massey, D.A. and Denton, N.A. (1993) *American Apartheid. Segregation and the Making of the Underclass*, Cambridge, MA: Harvard University Press.

Matthews, F.H. (1977) *Quest for an American Sociology. Robert E. Park and the Chicago School*, Montreal and London: McGill-Queen's University Press.

Mattrisch, G. (2001) 'Sustainable Cities: Future Challenges and Research Issues', in Brebbia, C.A., Ferrante, A., Rodriquez, M. and Terra, B. (eds) *The Sustainable City. Urban Regeneration and Sustainability*, Southampton: WIT Press, 3–11.

Mayhew, H. (1968) *London Labour and the London Poor*, London: Constable.

—— (1980) *Morning Chronicle Survey of Labour and the Poor*, 6 volumes, Firle: Caliban Books.

—— (1985) *London Labour and the London Poor*, Harmondsworth: Penguin.

Mayor of London (2002) *The Draft London Plan. Draft Spatial Development Strategy for Greater London*, London: Greater London Authority.

Meacham, S. (1987) *Toynbee Hall and Social Reform 1880–1914. The Search for Community*, New Haven, CT, and London: Yale University Press.

Micklethwait, J. and Wooldridge, A. (2000) *A Future Perfect. The Challenge and Hidden Promise of Globalization*, New York: Times Books.

Mila, N. (1998) 'Imaginaries of the Modern City', *Simmel Newsletter* 8, 2: 107–13.

Miller, M. (1992) *Raymond Unwin: Garden Cities and Town Planning*, Leicester: Leicester University Press.

Mills, C.W. (1956) *The Power Elite*, New York: Oxford University Press.

Miner, H. (1952) 'The Folk-Urban Continuum', *American Sociological Review* 17, 5: 529–37.

Mingione, E. (1981) *Social Conflict and the City*, Oxford: Blackwell.

Mitchell, W.J. (1995) *City of Bits*, Cambridge, MA: MIT Press.

—— (2000) *E-topia. 'Urban Life, Jim – but not as we know it'*, Cambridge, MA: MIT Press.

Mitchell-Weaver, C., Miller, D. and Deal, R.J. (2000) 'Multilevel Governance and Metropolitan Regionalism in the USA', *Urban Studies* 37, 5–6: 851–76.

Mollenkopf, J. (1992) *A Phoenix in the Ashes: The Rise and Fall of the Koch Coalition in New York City Politics*, Princeton, NJ: Princeton University Press.

—— and Castells, M. (eds) (1991) *Dual City. Restructuring New York*, New York: Russell Sage.

Molotch, H. (1976) 'The City as a Growth Machine. Toward a Political Economy of Place', *American Journal of Sociology* 82, 2: 309–32.

—— (1993) 'The Space of Lefebvre', *Theory & Society* 22, 6: 887–95.

More, Sir Thomas (1995) *Utopia*, Latin text and English translation, Logan, G.M., Adams, R.M. and Miller, C.H. (eds), Cambridge: Cambridge University Press.

Mort, F. (2000) 'The Sexual Geography of the City', in Bridge, G. and Watson, S. (eds) *A Companion to the City*, Oxford: Blackwell, 307–15.

Moynihan, D.P. (1965) *The Negro Family: The Case for National Action*, Washington, DC: Office for Policy Planning and Research, US Department of Labor.

Muller, P. and Surel, Y. (1998) *L'analyse des politiques publiques*, Paris: Montchrestien.

Mumford, E. (2000) *The CIAM Discourse on Urbanism, 1928–1960*, Cambridge, MA: MIT Press.

Mumford, L. (1924) *Sticks and Stones. A Study of American Architecture and Civilization*, New York: W.W. Norton.

—— (1938) *The Culture of Cities*, London: Secker and Warburg.

—— (1961) *The City in History*, New York: Harcourt Brace and World.

Munt, S.R. (2001) 'The Lesbian Flâneur', in Borden, I., Kerr, J., Rendell, J. and Pivaro, A. (eds) *The Unknown City. Contesting Architecture and Social Space*, Cambridge, MA: MIT Press, 246–60.

Murray, C. (1984) *Losing Ground. American Social Policy, 1950–1980*, New York: Basic Books.

Negroponte, N. (1995) *Being Digital*, London: Hodder and Stoughton.

Newman, O. (1972) *Defensible Space: Crime Prevention Through Urban Design*, London: Macmillan.

Nippel, W. (1995) 'Max Weber und die okzidentale Stadt' (Max Weber and the Western City), *Berliner Journal für Soziologie* 5, 3: 359–66.

North, D.C. (1990) *Institutions, Institutional Change and Economic Performance*, Cambridge: Cambridge University Press.

Obraniak, P. (1997) 'Zmiany w spoleczno-przestrzennej strukturze Lodzi w drugiej polowie XX wieku' (Changes in the Social and Spatial Structure of the City of Lodz in the Second Half of the 20th Century), *Przeglad Socjologiczny* 46: 29–51.

O'Day, R. and Englander, D. (1993) *Mr Charles Booth's Inquiry: Life and Labour of the People in London Reconsidered*, Rio Grande, OH, and London: Hambledon Press.

Olson, M. (1965) *The Logic of Collective Action: Public Goods and the Theory of Groups*, Cambridge, MA: Harvard University Press.

Orlans, H. (1971) *Stevenage. A Sociological Study of a New Town*, London: Greenwood Press.

Osborne, D. and Gaebler, T. (1992) *Reinventing Government: How the Entrepreneurial Spirit is Transforming the Public Sector*, Reading: Addison-Wesley.

Pahl, R.E. (1970) *Whose City? And Other Essays on Sociology and Planning*, London: Longman.

Palen, J.J. (2001) *The Urban World*, New York: McGraw-Hill Education.

Paquin, J. (2001) 'World City Theory: The Case of Seoul', *Research in Urban Sociology* 6: 337–56.

Park, R.E. (1915) 'The City: Suggestions for the Investigation of Human Behavior in the City Environment', *American Journal of Sociology* 20, 5 (March): 577–612.

—— (1925) 'The City: Suggestions for the Investigation of Human Behavior in the Urban Environment', in Park, R.E., Burgess, E.W. and McKenzie, R.D. *The City*, Chicago: University of Chicago Press, 1–46.

—— (1929) 'The City as a Social Laboratory', in Smith, T.V. and White, L.D. (eds) *Chicago: An Experiment in Social Science Research*, Chicago: University of Chicago Press.

—— (1967) *On Social Control and Collective Behaviour. Selected Papers*. Selected and with an introduction by Ralph H. Turner, Chicago: University of Chicago Press.

——, Burgess, E.W. and McKenzie, R.D. (1925) *The City*, Chicago: University of Chicago Press.

Parker, S. (1999) 'From the Slums to the Suburbs. Labour, the LCC and the Woodberry Down Estate, Stoke Newington (1934–1963)', *London Journal* 24, 2: 51–69.

—— (2000a) 'Sustainability and the Information City', *City* 4, 1: 123–35.

—— (2000b) 'Tales of the City. Situating Urban Discourse in Place and Time', *City* 4, 2: 233–46.

—— (2001a) 'Community, Social Identity and the Structuration of Power in the Contemporary European City. Part One: Towards a Theory of Urban Structuration', *City* 5, 2: 189–202.

—— (2001b) 'Community, Social Identity and the Structuration of Power in the Contemporary European City. Part Two: Power and Identity in the Urban Community: A Comparative Analysis', *City* 5, 3: 281–309.

Parkin, F. (1982) *Max Weber*, Chichester: Ellis Horwood.

Parks, R. and Oakerson, R. (1989) 'Metropolitan Organization and Governance: A Local Public Economy Approach', *Urban Affairs Quarterly* 25, 1: 18–29.

Parsons, T. (1937) *The Structure of Social Action. A Study in Social Theory with Special Reference to a Group of Recent European Writers (Alfred Marshall, Vilfredo Pareto, Emile Durkheim, Max Weber)*, New York and London: McGraw-Hill.

Partners for Livable Communities (2000) *The Livable City. Revitalizing Urban Communities*, Washington, DC: McGraw-Hill.

Peach, C. (1996) 'Does Britain Have Ghettos?', *Transactions of the Institute of British Geographers* 21, 1: 216–35.

Peterson, P.E. (1981) *City Limits*, Chicago: University of Chicago Press.

Pfautz, H.W. (1967) *Charles Booth On the City: Physical Pattern and Social Structure*, Chicago and London: The University of Chicago Press.

Phillips, W.R.F. (1994): *Urban Sociology and City Planning at the Turn-of-the-Century: Reevaluating Significant Links in Britain, Germany, and the United States*, International Sociological Association Paper.

Philpott, T.L. (1978) *The Slum and the Ghetto. Neighbourhood Deterioration and Middle Class Reform, Chicago, 1880–1930*, New York: Oxford University Press.

Pickvance, C.G. (1985) 'The Rise and Fall of Urban Movements and the Role of Comparative Analysis', *Environment and Planning D: Society and Space* 3: 31–53.

Pierre, J. (1999) 'Models of Urban Governance: The Institutional Dimension of Urban Politics', *Urban Affairs Review* 34, 3: 372–96.

Pile, S. and Thrift, N. (eds) (2000) *City A-Z*, London: Routledge.

Plant, S. (1998) *Zeros + Ones. Digital Women + the New Technoculture*, London: Fourth Estate.

Platt, H.L. (2000) 'Jane Addams and the Ward Boss Revisited: Class, Politics, and Public Health in Chicago, 1890–1930', *Environmental History* 5, 2 (April): 194–222.

Plummer, K. (ed.) (1997) *The Chicago School. Critical Assessments*, London: Routledge.

Polikoff, B.G. (1999) *With One Bold Act. The Story of Jane Addams*, Chicago: Boswell Books.

Polsby, N.W. (1963) *Community Power and Political Theory*, New Haven, CT: Yale University Press.

Portes, A. and Landolt, P. (1996) 'Unsolved Mysteries: The Tocqueville Files II', *The American Prospect* 7, 26, 1 May–1 June. Available on-line through www.prospect.org.

Powell, J. (1998) *Postmodernism for Beginners*, London: Writers and Readers.

Pratt, A.C. (2000) 'New Media, the New Economy and New Spaces', *Geoforum* 31, 4: 425–36.

Putnam, R.D., with Leonardi, R. and Nanetti, R.Y. (1993a) *Making Democracy Work. Civic Traditions in Modern Italy*, Princeton, NJ: Princeton University Press.

—— (1993b) 'The Prosperous Community: Social Capital and Public Life', *The American Prospect* 13 (Spring): 35–42.

—— (1996) 'The Strange Disappearance of Civic America', *The American Prospect* 24 (Winter): 34–48.

—— (2000) *Bowling Alone. The Collapse and Revival of American Community*, New York: Simon and Schuster.

Rabinow, P. (ed.) (1984) *The Foucault Reader*, New York: Pantheon Books.

Reckless, W. (1933) *Vice in Chicago*, Chicago: University of Chicago Press.

Redfield, R. (1947) 'The Folk Society', *American Journal of Sociology* 3, 4: 293–308.

Reeder, D.A. (1984) *Charles Booth's Descriptive Map of London Poverty 1889*, London: Topographical Society.

Rex, J. (1973) *Race, Colonialism and the City*, London: Routledge and Kegan Paul.

—— (1988) *The Ghetto and the Underclass. Essays on Race and Social Policy*, Aldershot: Avebury.

Riesman, D.N. with Glazer, N. and Denney, R. (1950) *The Lonely Crowd*, New Haven, CT: Yale University Press.

Riis, J. (1891) *How the Other Half Lives. Studies among the Poor*, London and New York: Sampson, Low, Marson, Searle & Rivington. [Also sub-titled 'Studies among the Tenements of New York'].

Riker, W.H. (1992) 'The Entry of Game Theory into Political Science', in Weintraub, E.R. (ed.) *Toward a History of Game Theory*, Durham, NC, and London: Duke University Press, 207–3.

Robins, K. (1999) 'Foreclosing on the City? The Bad Idea of Virtual Urbanism', in Downey, J. and McGuigan, J. (eds) *Technocities*, London: Sage, 34–59.

Rodaway, P. (1994) *Sensuous Geographies: Body, Sense and Place*, London: Routledge.

Rogers, R. (1997) *Cities for a Small Planet*, London: Faber and Faber.

Rosenlund, L. (2000) 'Cultural Changes in a Norwegian Urban Community: Applying Pierre Bourdieu's Approach and Analytical Framework', in Robbins, D. (ed.) *Pierre Bourdieu*, vol. 6, London: Sage, 220–43.

Ross, A. (2000) *The Celebration Chronicles*, London: Verso.

Rothschild, J. with Cheng, A. (1999) *Design and Feminism: Re-Visioning Spaces, Places, and Everyday Things*, New Brunswick, NJ: Rutgers University Press.

Ruggieri, G. (1993) 'Comprendere il fenomeno urbano contemporaneo: utilità ed attualità delle analisi di M. Weber, G. Simmel, R E. Park, L. Wirth' (Understanding the Modern Urban Phenomenon: The Usefulness and Relevance of Analyzing M. Weber, G. Simmel, R. E. Park, L. Wirth), *Sociologia* 27, 1–3: 577–88.

Ruggiero, V. (2001) *Movements in the City. Conflict in the European Metropolis*, Harlow: Prentice Hall.

Rusk, D. (1993) *Cities without Suburbs*, Baltimore, MD: Johns Hopkins University Press.

Russell, J.S. (2001) 'Dutch Docklands Renewed: A New New Urbanism is Taking Shape in the Netherlands, Solving Housing Issues and Revamping Formerly Deserted Docklands', *Architectural Record* 189, 4: 94–103.

Russian Federation (2000) *The State of the Cities in the Russian Federation*, Moscow: Gosstroy. Available on-line at http://www.unhabitat.org/Istanbul%2B5/Russia2.doc.

Rutheiser, C. (1996) *Imagineering Atlanta. The Politics of Place in the City of Dreams*, London and New York: Verso.

Rykwert, J. (2000) *The Seduction of Place: The City in the Twenty First Century*, London: Weidenfeld and Nicolson.

Sabel, C. (1994) 'Flexible Specialisation and the Re-Emergence of Regional Economies', in Amin, A. (ed.) *Post-Fordism: A Reader*, Oxford: Blackwell, 101–56.

—— and Piore, M. (1984) *The Second Industrial Divide. Possibilities for Prosperity*, New York: Basic Books.

Safier, M. (1996) 'The Cosmopolitan Challenge in Cities on the Edge of the Millennium. Moving from Conflict to Coexistence', *City* 3–4: 12–29.

Sandercock, L. and Berry, M. (1983) *Urban Political Economy. The Australian Case*, Sydney: George Allen and Unwin Australia.

Sassen, S. (1991) *The Global City. New York, London, Tokyo*, 1st edn, Princeton, NJ: Princeton University Press.

—— (1994) *Cities in a World Economy*, 1st edn, Thousand Oaks, CA: Pine Forge Press.

—— (1999) 'Whose City is It? Globalization and the Formation of New Claims', in Holston, J. (ed.) *Cities and Citizenship*, Durham, NC, and London: Duke University Press, 177–94.

—— (2000) *Cities in a World Economy*, 2nd edn, Thousand Oaks, CA: Pine Forge Press.

—— (2001a) *The Global City: New York, London, Tokyo*, 2nd edn, Princeton, NJ, and Oxford: Princeton University Press.

—— (2001b) 'The Changing Context and Directions of Urban Governance', in Habitat/United Nations Centre for Human Settlements *Cities in a Globalizing World. Global Report on Human Settlements*, London: Earthscan, 58–76.

Saunders, P. (1981) *Social Theory and the Urban Question*, 1st edn, London: Hutchinson.

—— (1995) *Social Theory and the Urban Question*, 2nd edn, London: Routledge.

Savage, M. and Warde, A. (1993) *Urban Sociology, Capitalism and Modernity*, Basingstoke: Macmillan.

—— and —— (2003) *Urban Sociology, Capitalism and Modernity*, 2nd edn, Basingstoke: Palgrave.

Savitch, H.V. (1988) *Postindustrial Cities: Politics and Planning in New York, Paris and London*, Princeton, NJ: Princeton University Press.

Saxenian, A. (1994) 'Lessons from Silicon Valley', *Technology Review* 97, 5 (July): 42–51.

Schama, S. (1987) *The Embarrassment of Riches. An Interpretation of Dutch Culture in the Golden Age*, London: Collins.

Schumpeter, J.A. (1987) *Capitalism, Socialism and Democracy*, 7th edn, London: Unwin Paperbacks.

Scott, A.J. (1981) *Metropolis. From the Division of Labour to Urban Form*, Berkeley, CA: University of California Press.

—— and Soja, E.W. (eds) (1996) *The City. Los Angeles and Urban Theory at the End of the Twentieth Century*, Berkeley, CA: University of California Press.

Scott, M. (1971) *American City Planning since 1890*, Berkeley, CA: University of California Press.

Seabrook, J. (1996) *In the Cities of the South. Scenes from a Developing World*, London: Verso.

Seamon, D. (1979) *A Geography of the Lifeworld: Movement, Rest and Encounter*, London: Croom Helm.

Sennett, R. (1990) *The Conscience of the Eye. The Design and Social Life of Cities*, London: Faber and Faber.

—— (1993) *The Fall of Public Man*, London: Faber and Faber.

—— (1994) *Flesh and Stone. The Body and the City in Western Civilization*, New York: W.W. Norton.

—— (1999) *The Corrosion of Character. The Personal Consequences of Work in the New Capitalism*, New York: W.W. Norton.

—— (2000) *The Art of Making Cities*, London: London School of Economics.

Shaw, C.R. (1930) *The Jack-Roller. A Delinquent Boy's Own Story*, Chicago: University of Chicago Press.

——, McKay, H.D. and McDonald, J.F. (1938) *Brothers in Crime*, Chicago: The University of Chicago Press.

Shaw, J.S. and Utt, R.D. (2001) *A Guide to Smart Growth: Shattering Myths, Providing Solutions*, Washington, DC: Heritage Foundation.

Shefter, M. (1985) *Political Crisis/Fiscal Crisis: The Collapse and Revival of New York City*, New York: Basic Books.

Shepard, B. and Hayduk, R. (eds) (2002) *From ACT UP to the WTO: Urban Protest and Community Building in the Era of Globalization*, London: Verso.

Shepherd, I.D.H. (1999) 'Mapping the Poor in Late-Victorian London: A Multi-scale Approach', in Bradshaw, J. and Sainsbury, R. (eds) *Proceedings of the Conference to Mark the Centenary of Seebohm Rowntree's First Study of Poverty in York*, Cambridge: Polity Press.

Shiel, M. and Fitzmaurice, T. (2001) *Cinema and the City*, Oxford: Blackwell.

Shields, R. (1991) *Places on the Margin: Alternative Geographies of Modernity*, New York: Routledge.

—— (1999) *Lefebvre, Love and Struggle. Spatial Dialectics*, London: Routledge.

Shonfield, K. (2001) *Walls Have Feelings – Architecture, Film and the City*, London: Routledge.

Short, J.R. (1996) *The Urban Order. An Introduction to Cities, Culture and Power*, Oxford: Blackwell.

—— (1999a) *Globalization and the City*, New York: Prentice Hall.

—— (1999b) 'Urban Imagineers: Boosterism and the Representation of Cities', in Jonas, A.E.G. and Wilson, D. (eds) *The Urban Growth Machine. Critical Perspectives Two Decades Later*, Albany, NY: State University of New York Press, 37–54.

Shragge, E. and Church, K. (1998) 'None of Your Business?! Community Economic Development and the Mixed Economy of Welfare', *Canadian Review of Social Policy* 41 (Spring): 33–44.

Simey, T.S. and Simey, M.B. (1960) *Charles Booth, Social Scientist*, London: Oxford University Press.

Simmel, G. (1950) 'The Metropolis and Mental Life', in Wolff, K.H. (ed.) *The Sociology of Georg Simmel*, Glencoe, IL: Free Press, 409–26.

—— (1986) *Schopenhauer and Nietzsche*, Amherst, MA: University of Massachussets Press.

—— (1990) *The Philosophy of Money*, London: Routledge.

—— (1996) 'Roma, un'analisi estetica' (Rome, an Aesthetic Analysis), *La Critica Sociologica* 116 (Jan.–Mar.): 1–7.

Sites, W. (2002) 'The Limits of Urban Regime Theory: New York City Under Koch, Dinkins and Giuliani', in Judd, D.R. and Kantor, P. (eds) *The Politics of Urban America. A Reader*, New York: Longman, 215–32.

Sitte, C. von (1965) *City Planning According to Artistic Principles*, London: Phaidon.

Skocpol, T. (1985) 'Bringing the State Back In: Strategies of Analysis in Current Research', in Evans, P.B., Rueschemeyer, D. and Skocpol, T. (eds) *Bringing the State Back In*, Cambridge: Cambridge University Press, 4–37.

Smith, D. (1988) *The Chicago School. A Liberal Critique of Capitalism*, Basingstoke: Macmillan.

Smith, D.A. and Timberlake, M. (1995) 'Conceptualising and Mapping the Structure of the World Systems City System', *Urban Studies* 32, 2: 287–302.

Smith, G. (ed.) (1991) *On Walter Benjamin. Critical Essays and Recollections*, Cambridge, MA: MIT Press.

Smith, H.L. (1930–1935) *The New Survey of London Life and Labour*, 9 volumes, London: P.S. King and Sons.

Smith, M.P. (1980) *The City and Social Theory*, Oxford: Basil Blackwell.

—— (ed.) (1984) *Cities in Transformation: Class, Capital, and the State*, Beverly Hills, CA: Sage.

—— (2001) *Transnational Urbanism. Locating Globalization*, Oxford and Malden, MA: Blackwell.

—— and Feagin, J.R. (eds) (1987) *The Capitalist City. Global Restructuring and Community Politics*, Oxford and Cambridge, MA: Blackwell.

Smith, N. (1992) 'Blind Man's Bluff, or Hamnett's Philosophical Individualism in Search of Gentrification', *Transactions of the Institute of British Geographers* 17, 1: 110–15.

—— (1996) *The New Urban Frontier. Gentrification and the Revanchist City*, London and New York: Routledge.

—— (1999) 'Which New Urbanism? New York City and the Revanchist 1990s', in Beauregard, R. and Body-Gendrot, S. (eds) *The Urban Moment. Cosmopolitan Essays on the Late-20th Century City, Urban Affairs Annual Reviews*, Thousand Oaks, CA: Sage, 185–208.

—— and Williams, P. (eds) (1986) *Gentrification of the City*, London: Unwin Hyman.

Soja, E.W. (1989) *Postmodern Geographies. The Reassertion of Space in Critical Social Theory*, London and New York: Verso.

—— (1996a) 'The Trialectics of Spatiality', *Österreichische Zeitschrift für Soziologie* 21, 2: 139–64.

—— (1996b) *Thirdspace*, Oxford: Blackwell.

—— (2000a) 'Putting Cities First: Remapping the Origins of Urbanism', in Bridge, G. and Watson, S. (eds) *A Companion to the City*, Oxford: Blackwell, 26–34.

—— (2000b) *Postmetropolis*, Oxford: Blackwell.

Sorkin, M. (ed.) (1992) *Variations on a Theme Park. The New American City and the End of Public Space*, New York: Hill and Wang.

—— (2002) 'The Center Cannot Hold', in Sorkin, M. and Zukin, S. (eds) *After the World Trade Center*, New York: Routledge, 197–207.

—— and Zukin, S. (eds) (2002) *After the World Trade Center*, New York: Routledge.

Spain, D. (2002) 'What Happened to Gender Relations on the Way from Chicago to Los Angeles?', *City and Community* 1, 2: 155–69.

Spak-Lisak, R. (1989) *Pluralism and Progressives: Hull-House and the New Immigrants 1890–1919*, Chicago: University of Chicago Press.

Spivak, G.C. (1987) *In Other Worlds. Essays in Cultural Politics*, London and New York: Methuen.

Stedman-Jones, G. (1971) *Outcast London: A Study in the Relationship between Classes in Victorian Society*, Oxford: Oxford University Press.

—— (1976) *Outcast London. A Study in the Relationship Between Classes in Victorian London*, Harmondsworth: Penguin.

Stein, H. and Nafzier, E.W. (1990) 'Structural Adjustment, Human Needs, and the World Bank Agenda', *Journal of Modern African Studies* 29, 1: 173–89.

Stephens, G.R. and Wikstrom, N. (2000) *Metropolitan Government and Governance. Theoretical Perspectives, Empirical Analysis, and the Future*, New York: Oxford University Press.

Stoker, G. (1995) 'Urban Regime Theory', in Judge, D., Stoker, G. and Wolman, H. *Theories of Urban Politics*, London: Sage, 54–71.

—— and Mossberger K. (1994) 'Urban Regime Theory in Comparative Perspective', *Government and Policy* 12: 195–212.

Stone, C.N. (1989) *Regime Politics. Governing Atlanta 1946–1988*, Lawrence, KA: University Press of Kansas.

—— (1993) 'Urban Regimes and the Capacity to Govern: A Political Economy Approach', *Journal of Urban Affairs* 15, 1: 1–28.

—— (2001) 'Urban Regimes: A Research Perspective', in Judd, D.R. and Kantor, P.B. (eds) *The Politics of Urban America*, New York: Longman, 26–42.

—— and Sanders, H.T. (1987) *The Politics of Urban Development*, Lawrence, KA: University Press of Kansas.

Storper, M. (1997) *The Regional World. Territorial Development in a Global Economy*, London: The Guilford Press.

—— and Walker, R. (1989) *The Capitalist Imperative. Territory, Technology and Industrial Growth*, Oxford: Blackwell.

Strange, S. (1997) *Casino Capitalism*, 2nd edn, Manchester: Manchester University Press.

Sudjic, D. (1995) *The 100 Mile City*, London: Flamingo.

Sumka, H.J. (1979) 'Neighborhood Revitalization and Displacement. A Review of the Evidence', *Journal of the American Planning Association* 45: 480–7.

Susser, I. (2002) *The Castells Reader on Cities and Social Theory*, Oxford: Blackwell.

Suttles, G.D. (1968) *The Social Order of the Slum. Ethnicity and Territory in the Inner-City*, Chicago: University of Chicago Press.

Swanstrom, T. (1988) 'Urban Populism, Uneven Development, and the Space for Reform', in Cummings, S. (ed.) *Business Elites and Urban Development*, Albany, NY: SUNY Press. 121–52.

—— (2001) 'What We Argue About When We Argue About Regionalism', *Journal of Urban Affairs* 23, 5: 479–96.

Swyngedouw, E. (1997) 'Neither Global nor Local: "Glocalization" and the Politics of Scale', in Cox, K. (ed.) *Spaces of Globalization*, London: Arnold, 167–76.

—— and Baeten, G. (2001) 'Scaling the City: The Political Economy of "Glocal" Development – Brussels' Conundrum', *European Planning Studies* 9, 7: 827–49.

Tabb, W.K. and Sawers, L. (1984) *Marxism and the Metropolis*, Oxford: Oxford University Press.

Tafuri, M. (1999) [1976] *Architecture and Utopia: Design and Capitalist Development*, Cambridge, MA: MIT Press.

Talen, E. (1999) 'Sense of Community and Neighbourhood Form: An Assessment of the Social Doctrine of New Urbanism', *Urban Studies* 36, 8: 1361–80.

Tavakoli-Targhi, M. (2001) 'The Homeless Texts of Persianate Modernity', *Cultural Dynamics* 13, 1: 263–92.

Taylor, M. (1992) 'Can You Go Home Again? Black Gentrification and the Dilemma of Difference', *Berkeley Journal of Sociology* 37: 121–38.

Taylor, P.J. and Walker, D.R.F. (2001) 'World Cities: A First Multivariate Analysis of their Service Complexes', *Urban Studies* 38, 1: 23–47.

——, Beaverstock, J., Cook, G., Pain, K., Greenwood, H. and Pandit, N. (2003) *Financial Service Clustering and its Significance for London*, London: Corporation of London.

Testa, M., Astone, N.M., Krogh, M. and Neckerman, K.M. (1993) 'Employment and Marriage among Inner-City Fathers', in Wilson, W.J. (ed.) *The Ghetto Underclass. Social Science Perspectives*, Newbury Park, CA: Sage, 96–108.

Thomas, A.D., Kam, V. and Gibson, M. (1984) *Research On Urban Renewal. A Consultancy Report to the ESRC Environment and Planning Committee*, London: ESRC.

Thomas, W.I. and Znaniecki, F. (1984) *The Polish Peasant in Europe and America*, Urbana, IL: University of Illinois Press.

Thompson, F.M.I. (1982) *The Rise of Suburbia*, Leicester: Leicester University Press.

Thrasher, F.M. (1927) *The Gang. A Study of 1,313 Gangs in Chicago*, Chicago: University of Chicago Press.

Thrift, N. (1996) *Spatial Formations*, London and Thousand Oaks, CA: Sage.

Tiebout, C. (1956) 'A Pure Theory of Local Expenditure', *Journal of Political Economy* 64: 416–24.

Tilly, C. (1990) *Coercion, Capital, and European States, A.D. 990–1990*, Oxford: Blackwell.

Topalov, C. (1993) 'The City as *Terra Incognita*: Charles Booth's Poverty Survey and the People Of London, 1886–1891', *Planning Perspectives* 8, 4: 395–425.

Tschumi, B. (1994) *Event-Cities*, Cambridge, MA: MIT Press.

—— (2000) *Event-Cities 2*, Cambridge, MA: MIT Press.

Turner, B.S. (1996) *For Weber. Essays on the Sociology of Fate*, London: Sage.

Turner, S. (ed.) (2000) *The Cambridge Companion to Weber*, Cambridge: Cambridge University Press.

United Nations (1999) *Human Development Report, 1999*, New York: United Nations.

Unwin, R. (1971) *Town Planning in Practice. An Introduction to the Art of Designing Cities and Suburbs*, New York: B. Blom.

Urban Task Force (1999) *Report of the Urban Task Force*, London: Spon.

Urry, J. (1990) *The Tourist Gaze. Leisure and Travel in Contemporary Societies*, London: Sage.

US Bureau of Census (2002) National Population Projections I. Summary Files. Available on-line at http://www.census.gov/population/www/projections/natsum-T1.html (accessed September 2002).

US National Advisory Commission on Civil Disorders (1988) *The Kerner Report*, New York: Pantheon Books.

Valentine, G. (1997) '(Hetero)Sexing Space: Lesbian Perceptions and Experience of Everyday Spaces', in McDowell, L. and Sharp, J.P. (eds) *Space, Gender, Knowledge. Feminist Readings*, London: Arnold, 284–300.

van Kempen, R. and Priemus, H. (1999) *Revolution in Social Housing in the Netherlands: Changing Social Function and Legal Status of Housing Associations*. Paper for the ENHR-CECODHAS-NETHUR Workshop on Social Housing Policy, Nunspeet, the Netherlands, 18/19 February, 1999.

van Weesep, J. and Musterd, S. (eds) (1991) *Urban Housing for the Better-Off: Gentrification in Europe*, Utrecht: Stedelijke Netwerken.

Venkatesh, S. (2001) 'Chicago's Pragmatic Planners: American Sociology and the Myth of Community', *Social Science History* 25, 2: 275–317.

Venturi, R. (1977) *Complexity and Contradiction in Architecture*, 2nd edn, London: The Architectural Press.

—— , Scott Brown, D. and Izenour, S. (1993) *Learning from Las Vegas. The Forgotten Symbolism of Architectural Form*, revised edn, Cambridge, MA: MIT Press.

Vergara, C.J. (1995) *The New American Ghetto*, New Brunswick, NJ: Rutgers University Press.

Vernant, J.-P. (1972) 'Greek Tragedy: Problems of Interpretation', in Macksey, R. and Donato, E. (eds) *The Structuralist Controversy*, Baltimore, MD, and London: Johns Hopkins University Press, 278–88.

Vidler, A. (1991) 'Agoraphobia: Spatial Estrangement in Georg Simmel and Siegfried Kracauer', *New German Critique* 54, (Fall): 31–45.

Virilio, P. (1986) *Speed and Politics. An Essay on Dromology*, New York: Semiotext(e).

Von Neumann, J. and Morgenstern, O. (1944) *The Theory of Games and Economic Behaviour*, Princeton, NJ: Princeton University Press.

Wacquant, L.J.D. (1994) 'The New Urban Color Line: The State and Fate of the Ghetto in Post-Fordist America', in Calhoun, C. (ed.) *Social Theory and the Politics of Identity*, Oxford: Blackwell, 231–76.

—— (1999a) 'America as Social Dystopia', in Bourdieu, P. et al. *The Weight of the World. Social Suffering in Contemporary Society*, Cambridge: Polity Press, 130–9.

—— (1999b) 'Inside "The Zone". The Social Art of the Hustler in the American Ghetto', in Bourdieu, P. et al. (eds) *The Weight of the World. Social Suffering in Contemporary Society*, Cambridge: Polity, 140–67.

—— and Wilson, W.J. (1993) 'The Cost of Racial and Class Exclusion in the Inner City', in Wilson, W.J. (ed.) *The Ghetto Underclass. Social Science Perspectives*, Newbury Park, CA: Sage, 25–42.

Waley, P. (2000) 'Tokyo: Patterns of Familiarity and Partitions of Difference', in Marcuse, P. and van Kempen, R. (eds) *Globalizing Cities. A New Spatial Order?*, Oxford and Malden, MA: Blackwell, 127–57.

Walton, J. (1993) 'Urban Sociology: The Contribution and Limits of Political Economy', *American Review of Sociology* 19, 1: 301–20.

—— (2000) 'Urban Sociology', in Quah, S.R. and Sales, A. (eds) *The International Handbook of Sociology*, London, Thousand Oaks, CA, New Delhi: Sage, 299–318.

Ward, D. (1982) 'The Ethnic Ghetto in the United States: Past and Present', *Transactions of the Institute of British Geographers* 7, 3: 257–75.

Warde, A. (1991) 'Gentrification as Consumption: Issues of Class and Gender', *Environment and Planning D: Society and Space* 9: 223–32.

Warner, L. (1963) *Yankee City*, New Haven, CT, and London: Yale University Press.

Watson, S. and Gibson, K. (1995) *Postmodern Cities and Spaces*, Oxford: Blackwell.

Wax, M.L. (2000) 'Old Chicago and the New France', *The American Sociologist* 31, 4: 65–82.

WCED (1987) World Commission on Environment and Development. *Our Common Future*, Oxford: Oxford University Press.

Webber, M.A. (1964) 'Urban Place and the Nonplace Urban Realm', in Webber, M.A. *et al. Explorations into Urban Structure*, Philadelphia, PA: Pennsylvania University Press, 79–153.

Weber, A. (1899) *The Growth of Cities in the Nineteenth Century. A Study in Statistics*, New York: Columbia College Studies in History, Economics and Public Law. Reprinted by Cornell University Press, Ithaca, 1963.

Weber, M. (1958) *The City*, Glencoe, IL: Free Press.

—— (1968) *Economy and Society. An Outline of Interpretive Sociology*, Vols 1–3, New York: Bedminster Press.

—— (1985) *The Protestant Ethic and the Spirit of Capitalism*, London: Unwin.

Weinstein, D. and Weinstein, M.A. (1993) *Postmodern(ized) Simmel*, London: Routledge.

Werlen, B. (1993) *Society, Action, and Space: An Alternative Human Geography*, London: Routledge.

Wetherell, M., Taylor, S. and Yates, S. (eds) (2001) *Discourse Theory and Practice: A Reader*, London and Thousand Oaks, CA: Sage.

White, H. (1987) *The Content of the Form: Narrative Discourse and Historical Representation*, Baltimore, MD: Johns Hopkins University Press.

White, R.R. (2001) 'Sustainable Development in Urban Areas: An Overview', in Devuyst, D. (ed.) *How Green is the City. Sustainability Assessment and the Management of the Urban Environments*, New York: Columbia University Press, 47–62.

Whyte, W.F. (1955) *Street Corner Society*, Chicago and London: Chicago University Press.

Whyte, W.H. (1956) *The Organization Man*, New York: Simon and Schuster.

—— (ed.) (1958) *The Exploding Metropolis*, New York: Doubleday.

—— (1980) *The Social Life of Small Urban Spaces*, Washington, DC: Conservation Foundation.

Wickersham, J. (2001) 'Jane Jacobs's Critique of Zoning: From Euclid to Portland and Beyond', *Boston College Environmental Affairs Law Review* 28, 4: 547–63.

Wigley, M. (2002) 'Insecurity by Design', in Sorkin, M. and Zukin, S. (eds) *After the World Trade Center*, New York: Routledge, 69–85.

Williams, R. (1961) *The Long Revolution*, London: Chatto and Windus.

—— (1973) *The Country and the City*, New York: Oxford University Press.

Willmott, P. and Young, M. (1960) *Family and Class in a London Suburb*, London: Routledge and Kegan Paul.

—— and —— (1971) *Family and Class in a London Suburb*, 2nd edn, London: New English Library.

Wilson, E. (1992) *The Sphinx in the City. Urban Life, the Control of Disorder, and Women*, Oxford: University of California Press.

—— (1995) 'The Invisible Flaneur', in Watson, S. and Gibson, K. (eds) *Postmodern Cities and Spaces*, Oxford: Blackwell, 59–79.

Wilson, W.J. (1978) *The Declining Significance of Race*, Chicago: University of Chicago Press.

—— (1987) *The Truly Disadvantaged. The Inner City, the Underclass, and Public Policy*, Chicago: University of Chicago Press.

—— (1996) *When Work Disappears. The World of the New Urban Poor*, New York: Knopf.

Wirth, L. (1928) *The Ghetto*, Chicago: University of Chicago Press.

—— (1938) 'Urbanism as a Way of Life', *American Journal of Sociology* 44, 1: 1–24.

—— (1964) *On Cities and Social Life: Selected Papers*, Chicago: Chicago University Press.

—— (1996) 'Urbanism as a Way of Life', in Le Gates, R.T. and Stout, F. (eds) *The City Reader*, London and New York: Routledge, 189–97.

Wohl, A. (ed.) (1970) *The Bitter Cry of Outcast London, by Andrew Mearns*, Leicester: Leicester University Press.

Wolf, E.R. (1997) *Europe and the People Without History*, Cambridge: Cambridge University Press.

Wolff, J. (1990) 'Feminism and Modernism', in Wolff, J. (ed.) *Feminine Sentences: Essays on Women and Culture*, Berkeley, CA: University of California Press.

World Bank (2000) *Poverty in an Age of Globalization*, Washington, DC: World Bank.

Wright, Frank Lloyd (1996) 'Broadacre City: A New Community Plan', *The Architectural Record*, 1935, reprinted in Le Gates, R.T. and Stout, F., *The City Reader*, 1st edn, London: Routledge, 377–381.

Yeoh, B.S.A. (1999) 'Global/Globalizing Cities', *Progress in Human Geography* 23, 4 (Dec.): 607–16.

Young, K. and Mills, L. (1993) *A Portrait of Change*, Luton: Local Government Management Board/ESRC Local Governance Programme.

Young, M. and Willmott, P. (1986) *Family and Kinship in East London*, London: Penguin.

—— and —— (1992) *Family and Kinship in East London*, Berkeley, CA: University of California Press.

Young, P.V. (1932) *The Pilgrims of Russian-Town. The Community of Spiritual Christian Jumpers in America*, Chicago: University of Chicago Press.

Zevi, B. (1950) *Towards an Organic Architecture*, London: Faber and Faber.

Zorbaugh, H.W. (1929) *Gold Coast and Slum. A Sociological Study of Chicago's Near North Side*, Chicago: University of Chicago Press.

Zotti, E. (1997) 'Eyes on Jane Jacobs', in Allen, M. (ed.) *Ideas that Matter. The Worlds of Jane Jacobs*, Owen Sound, Ontario: The Ginger Press, 61–2.

Zukin, S. (1988) *Loft Living. Culture and Capital in Urban Change*, London: Radius.

—— (1991) *Landscapes of Power. From Detroit to Disney World*, Berkeley, CA: University of California Press.

—— (1995) *The Cultures of Cities*, Oxford: Blackwell.

INDEX